MODERN BOTANY

Dr. NAVEEN MALAVIYA

Co-ordinator, Department of Integrative Biology,
Head, Department of Botany,
Govt. P.G. College,
Narsinghpur (M.P.)

SCIENTIFIC PUBLISHERS (INDIA)
P.O. Box 91
Jodhpur (Raj.)

Published by
Scientific Publishers (India)
5-A, New Pali Road, P.O.Box 91
JODHPUR -342 001
E-mail: info@scientificpub.com
www.scientificpub.com

ISBN: 978-81-7233-503-8 (PB)
978-81-7233-507-6 (HB)

Laser typeset: Rajesh Ojha
Printed in India

FOREWORD

Advancement of learning and researches in Botany has been accountable for the vast expansion of this subject from traditional taxonomy to molecular studies. In the present era, all plant studies are made at the molecular level emphasizing on studies of enzymes involved in any particular activity. It has become possible to explore the details while using mutants as an experimental tool. Application of Biotechnology has yielded results, which are fruitful for plant health and productivity. Introduction of a foreign gene to produce a transgenic plant has now become routine in plants to help the mankind by improving productivity and disease resistance. Bioinformatics has also made molecular studies simpler and one can come out with an appropriate design in a very short time.

The present book *Modern Botany* by Dr. Naveen Malaviya is in accordance with the study of present time as it rests on plant physiology and biotechnology. The chapters included in this book are designed specifically to develop a clear understanding of intrinsic aspects in a subtle manner that shall be of immense help for the students both at undergraduate and post-graduate levels. There are a good number of books available in Botany but the present book is an example of comprehensive account of the subject and would certainly cater to the needs of our students.

The hard work by Dr. Malaviya in writing this book, I'm sure, will prove effective for students in understanding about the frontiers of research in Plant sciences. Besides, the text will also help the teachers and young researchers in several ways.

28/4/07

Dr. S.M. Paul Khurana
Vice Chancellor
Rani Durgavati Vishwavidyalaya,
Jabalpur

PREFACE

While teaching in Post-Graduate classes, the author has felt that students feel difficult to understand and to collect literature in certain topics of plant physiology, and biotechnology. The aim to write this book is to collect the latest literature on such topics. The text attempts to focus on the latest knowledge and give a shape to all the chapters so that the fundamentals of the topic are also clear to the students. The latest information presented in the text is based on the original publication, scientific articles, encyclopedia, and monographs. In certain chapters, definitions and basic concepts are mentioned before going in to the detail. The different chapters included in the three sections of this book are selected as per prescribed syllabi from UGC. Since there are many excellent text books on plant physiology, hence the author has omitted some literature and figures/diagrams which are easily available to the students.

It was totally unjustified if a book on metabolism and developmental physiology does not include a topic on the mechanism to transport across the plasma membrane. So, the first chapter of this book is on Plasma membrane. It is written as a separate chapter. The Section I is on metabolism. In this section physiology of all the biochemical pathways is discussed. Emphasis is given to the enzymes involved in these pathways. The molecular aspect is also discussed wherever necessary. Section II covers developmental physiology. Studies on developmental physiology, is also advancing towards molecular level. So, in this section, there is a separate chapter on Signal Transduction. Beside this, the concept of biosignalling is discussed in each chapter. The Section III, includes some important topics like genetic transformation in plants, transgenic plants, and bioinformatics.

The author shall feel satisfied if the students not only gain knowledge from this book but their fundamental concepts on the topics included in this book are made clear.

I am very grateful to my teacher Dr. (Mrs.) Nomita Sen, a renounded plant physiologist, who encouraged me to write this book and has gone very Critically on each topics. She gave her valuable guidance how to give the final shape of each chapter. I also express my thanks to Dr. A.K. Dasgupta for going through this script and pointing out grammatical errors if any and giving a good shape to the script. Thanks also to Shri Shailendra Gupta (Shagun Computers, New Kanchanpur, Jabalpur), who took all pains to type the script and draw the diagram.

I have tried to simplify the text as much as possible, but still I am of the opinion that, when this book will go in hands of learned Professors of this subject they may make some suggestions. I shall be grateful for any suggestions and comments on this book.

Narsinghpur

10th July 2007 **Naveen Malaviya**

CONTENTS

SECTION - I :

BIOCHEMISTRY AND METABOLISM

Chapter - 1

PLASMA MEMBRANE

Plasma membrane completely surrounds the protoplast. The physiological function of plasma membrane is movement of substances. There are basically two means for such movement passively by diffusion or actively by some type of energy-coupled transport process. In 1950s the technique for preparing and staining tissue had progressed. J.D. Robertson made electron micrographs, of the plasma membrane and suggested a three-layered structure, consisting of two darkly staining outer layers and a lightly staining middle layer. The membranes that were examined closely, whether they were plasma, nuclear, or cytoplasmic membranes or taken from plants, animals or microorganisms, showed this same ultra structure.

Structure of the Plasma membrane

The plasma membrane of plant is composed of lipids, proteins and carbohydrate, present in the ratio of 40:40:20. The core of the membrane consists of a sheet of phospholipids arranged in a bimolecular layer-a lipid bilayer. The lipid bilayer serves primarily as a structural framework for the membrane and as a barrier preventing the indiscriminate movements of water-soluble materials into and out of the cell. The proteins of the membrane, on the other hand, carry out most of the specific functions. The lipid-protein are held together in a thin sheet by noncovalent bonds.

Membrane lipid

Membranes contain several types of lipid, all of which are **amphipathic** : that is, they contain both hydrophobic and hydrophilic regions. Most membrane lipids contain a phosphate group, the primary exceptions being cholesterol and glycolipids. Since most membrane phospholipids have a glycerol backbone, they are called phosphoglycerides. Unlike triglycerides, membrane glycerides are diglycerides?only two of the hydroxyl groups of the glycerol are esterified to fatty acids; the third is esterified to a phosphate group. The most common free sterols of plant plasma membrane are campesterol, sitosterol, and stigmasterol. Sterol esters, sterol glycosides, and acylated sterol glycosides are more abundant in plants than in animals.

Phosphatidic acid is a molecule made up of two fatty acyl chains without any additional substitution beyond the phosphate. It is virtually absent in membranes. Membrane diglycerides have an additional group linked to the phosphate. This additional group is most commonly either choline (forming phophatidylcholine), ethanolamine (forming phosphatidylethanolamine), serine (forming phsophatidlyserine), or inositol (forming phsophatidylinositol). Each of these groups is small and hydrophilic, and together with the charged phosphate to which it is attached, forms a highly water-soluble domain at one end of the molecule, called the head group. In contrast, the fatty acyl chains are long unbranched, hydrophobic hydrocarbons. A membrane fatty acid may be either fully saturated, monounsaturated or polyunsaturated. The monounsaturated and polyunsaturated fatty acids have one double bond, and more than one double bond, respectively. Phosphoglycerides often contain one unsaturated and one saturated fatty acyl chain.

A less abundant class of membrane lipids is sphingolipids. They are derivatives of sphinosine, an amino alcohol that contain a long hydrocarbon chain. In animal cells, cholesterol (a type of lipid) may constitute up to 50 percent of the lipid molecules in the plasma membrane. It is absent from the plasma membranes of plant and all bacterial cells.

The lipid bilayer

Dutch scientist E. Gorter and F. Grendel (1925) first proposed that cellular membranes might contain a lipid bilayer. The lipid bilayer is continuous, unbroken structure. The membranes forms extensive interconnected networks that ramify throughout the cell. Because of the flexibility of the lipid bilayer, membranes are deformable and their overall shape can change. The polar groups of each molecular layer (or leaflet) were directed toward the outside of the bilayer. Membrane lipids are asymmetric in their distribution on the two faces of the bilayer. The lipid layer is stable. However, the individual phospholipid and sterol molecules have great freedom of motion within the plane of the membrane. The inner face of the lipid is also fluid. Individual hydrocarbon chains are in constant motion produced by rotation about the carbon-carbon bonds of the long acyl side chains.

The presence of a bimolecular layer of amphipathic lipid molecules has important consequences on the physiological properties of the membrane.

Membrane Proteins

The membrane proteins are asymmetrical so that the outer portion of the membrane is very different from the properties of the inner portion. Membrane proteins can be grouped into three distinct classes distinguished by the intimacy of their relationship to the lipid bilayer. These are

1. Integral proteins are intrinsic proteins, and penetrate into the lipid bilayer. These proteins very firmly bound to the membrane. In fact, virtually all integral proteins pass entirely through the lipid bilayer. Protein domains that protrude from both the extracellualr and cytoplasmic sides of the membrane are included in integral protein. These proteins interact favourably with the fatty acyl chains of membrane phospholipids.

2. Peripheral proteins are extrinsic proteins, located entirely outside of the lipid bilayer, on either the extracellular or cytoplasmic surface. They are bound more

loosely, or reversibly. They are associated with the surface of the membrane by noncovalent bonds.

3. *Lipid-anchored proteins* These proteins are located outside the lipid layer. They are covelantly linked to a lipid molecule which is found within this layer.

Integral membrane proteins generally have rich hydrophobic amino acid domains. These amino acids may be present in the middle of the polypeptide chain or at the carboxyl or amino terminus. Some proteins span the width of the membrane and carry nutrients into and waste products out of the cell. These proteins are named as *Transporters*. There are proteins with highly specific binding sites for extracellular signaling molecules, such proteins are present on the surface membrane and are known as *Signal Receptors*. A single ligand molecule bound to a signal receptor may cause the flux of thousands ions. Such flux may be through an opened channel, or the synthesis of thousands of molecules of an intracellular messenger molecule by an activated enzyme. This phenomenon is *Signal Transduction*.

Membrane carbohydrates

The plasma membranes of eukaryotic cells possess carbohydrates that are covalently linked to both lipid and protein components. Depending on the species and cell type, the carbohydrate content of the plasma membrane ranges between 2 and 10 percent by weight. 8% of carbohydrates are covalently linked to lipids to form glycolipids, and the remaining 93% are covalently linked to proteins to form glycoproteins. In glycoproteins, the carbohydrate is present in short, branched oligosaccharides, typically having less than about 15 sugars per chain. Sugars commonly found in these carbohydrates chains are D-glucose, D-mannose, D-galactose, L-fructose, N-acetyl-D-glucosamine and N-acetylneuraminic acid (sialic acid). Sialic acid is commonly the terminal sugar of the oligosaccharides, which gives the projecting carbohydrate chains a negative charge. Oligosaccharides may be attached to several different amino acids by two major types of linkages. These carbohydrate projections are thought to play a role in mediating the interactions of a cell with other cell as well as with its nonliving environment.

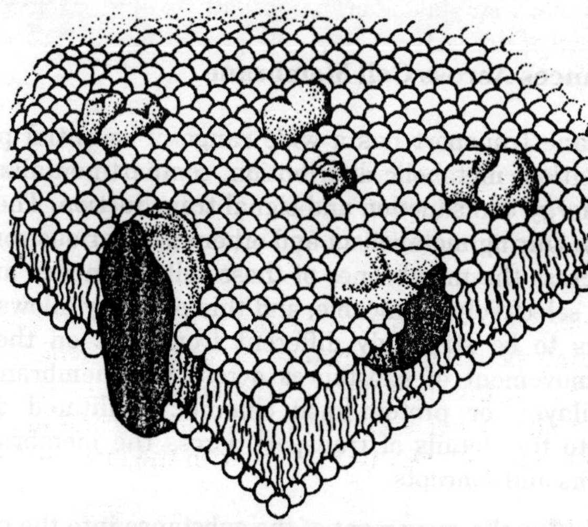

Fig. 1.1. The structure of the plasma membrane.

Membrane Model

Early models of membrane structure, particularly that proposed in 1935 by Hugh Davson and James Danielli, placed the proteins of the membranes on the outer surfaces of the lipid bilayer. S. Jonathan Singer and Garth Nicolson (1972) proposed the fluid-mosaic model, which has served as the "central dogma" of membrane biology for more than two membrane, but attention was focused on the physical state of the lipid. Rather than existing in a frozen, immobile bilayer, the lipid molecules are present in a fluid state capable of rotating and moving laterally within the membrane (Fig. 1.1).

The Fluid-Mosaic Membrane Model

The structure and arrangement of membrane proteins in the fluid-mosaic model are different from those of previous models. The individual lipid and protein subunits in a membrane form a *Fluid Mosaic*; i.e., its pattern is unlike a mosaic of ceramic tile and mortar. The simple evidence for fluid-mosaic model comes from motion of individual protein and lipid molecules within membranes. The electron microscopic studies, chemical composition, and physical studies of permeability also support fluid-mosaic model. Amphipathic phospholipids and sterols form a lipid bilayer, with the nonpolar regions facing each other and the polar head groups facing outward. The globular proteins are embedded at irregular intervals. The hydrophobic domains of protein are held by hydrophobic interactions with the membrane lipid. Some proteins span on the entire width of membrane, while some protrude from one or the other face of the membrane. Most importantly, the fluid-mosaic model presents cellular membranes as dynamic structures in which the components are mobile and capable of coming together to engage in various types of transient or semipermeable interactions. Proteins are oriented in an asymmetric manner. The protein domains exposed on one side of the bilayer are different from those of exposed on the other side. This gives *Sidedness*, and due to this those parts of membrane proteins that interact with other cells or with extra ?cellular ligands (such as hormones or growth factors), face outward. Whereas those parts of membrane proteins that interact with cytoplasmic molecules (such as G proteins or protein kinases), face the cell interior.

The Movement of Substances Across Cell Membrane

In plant cells, membrane transport has a wide range of essential processes, like turgor generation, nutrient acquisition, metabolic distribution, compartmentalization of metabolites, waste product excretion, energy transduction and signal transduction. The cell organelles have specific metabolic functions. During metabolism and otherwise, various compartments requires a transfer of substances across the membranes of these cell organelles and between different cells. Plasma membrane is selectively permeable, and this property allows the concentration of substances inside the cells to be strikingly different from that on the outside. There are different mechanisms of movement of substances across the membrane. These are simple diffusion through lipid bilayer, or protein-lined channel, facilitated diffusion, and active transport. Before going into the details of transport across the membrane, it is essential to understand some basic terms and concepts.

Influx, Outflux and Netflux the movement of the substance into the cell and out of the cell is known as influx and outflux, respectively. These terms relate to net flux.

Exergonic and Endergonic Transport the downhill and uphill movement flow of substance across the membrane is known as exergonic and endergonic, respectively.

Scalar and Vectoral events in membrane, proteins are present within the lipid bilayer and take part in transport of substances. In general such transport can be regarded as conventional enzymes in almost all respect. Such enzyme catalyzed reactions defined entirely by magnitude are Scalar, whereas exceptions of transport system defined by a magnitude and a direction in space are Vectoral.

Membrane Potential it is the difference in electrical potential between two aqueous media separated by a membrane. electrophysiologist encounter it by V_m and bioenergeticists by $\Delta\psi$.

Driving Force and Solute potential the net direction of transport across a membrane is dictated by driving force on the transport system. The quantification of this force is in terms of the free energy relationships inherent in transmembrane.

Carriers, Antiporters and Symporters (Fig. 1.2) transport system that couples downhill flow of ions (H^+ or Na^+) to uphill flow of inorganic ions and solutes are known as Carriers. Transporters involved in a system, where solutes and protons are exchanged are Antiporters. Carriers that catalyze solute flux in the same direction as H^+ or Na^+ flux are called Symporters. Protons moves passively across the membrane and it is symporters which typically energize uptake of solutes in the same direction as movement of H^+ or Na^+.

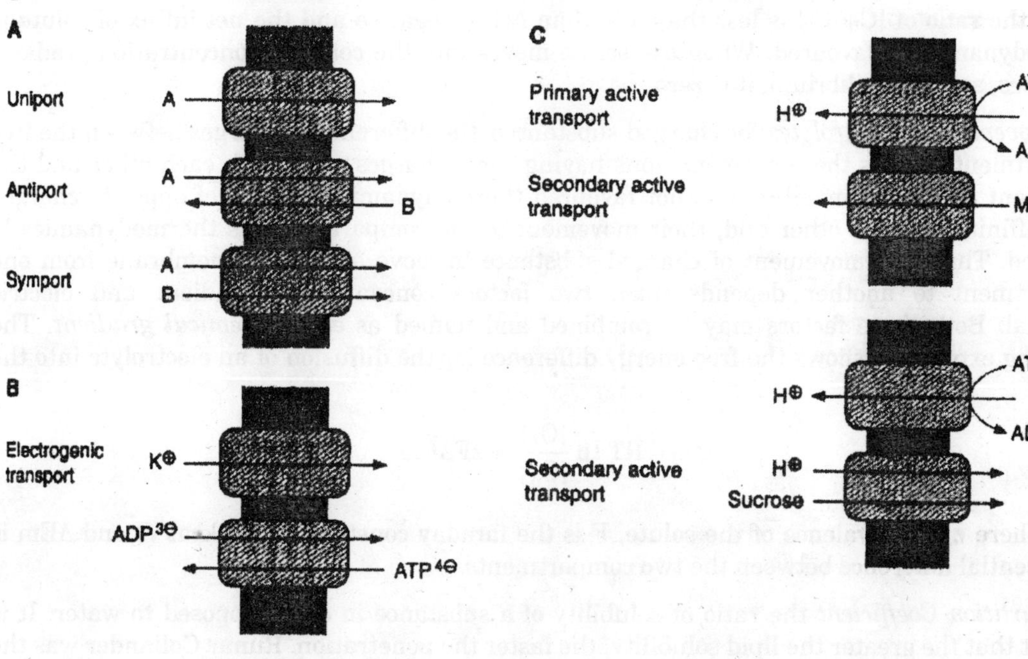

Fig. 1.2. Different types of transport processes through membrane.

Carrier mediated transport the carrier mediated transport process does not involve chemical modification of any of the compounds bound to the carrier. They undergo confromational changes during transport. Carriers transport inorganic ions and simple organic solutes across membrane. Carrier enzymes are relatively specific for their substrates and obey Michaelis-Menton kinetics. The carrier substrate is translocated against its own electrochemical potential. Most carriers are energized by coupling to the pmf i.e., there is H^+-coupled transport system. Carriers are made up of one subunit, with molecular mass ranging from 40- to 50-kDa.

Diffusion is movement of substances from a region of high concentration to a region of low concentration. It depends on the random movement of solutes and is an exergonic process driven by the increase in entropy represented by the random distribution of molecules. When a solute diffuses across the membrane, the free-energy change during this depends on the extent of the concentration gradient.

Electrogenic transport transport of substances in which charges also moves simultaneously is electrogenic transport. The movement may be via uniport, antiport, or symport.

Movement of Non-Electrolytes uncharged substance (non-electrolyte) move inside the cell showing following relationship:

$$\Delta G = RT \ln \frac{[C_i]}{[C_o]} - \text{or } \Delta G = 2.303\, RT \log_{10} \frac{[C_i]}{[C_o]}$$

where ΔG is the free energy difference, R is the gas constant, T is the absolute temperature, $[C_i]/[C_o]$ is the ratio of the concentration of the solute on the inside and outside at the surface of the membrane. For uncharged substances, it is expressed in molar concentration.

If the ratio of $[C_i]/[C_o]$ is less than 1.0, than ΔG is negative and the net influx of solute is thermodynamically favoured. When the solute moves into the cell, the concentration gradient decreases, and at equilibrium, it is zero.

Movement of Electrolytes for charged substances, the difference of charges between the two compartments effects the movement. Ions having same charges repel with each other and the movement across the membrane is not favoured thermodynamically. Ions of opposite charges show affinity for each other and, their movement to the compartments is thermodynamically favoured. Thus, the movement of charged substance to move across the membrane from one compartment to another depends upon two factors concentration gradient and electric potential. Both these factors may be combined and named as *electrochemical gradient*. The following expression shows the free-energy difference for the diffusion of an electrolyte into the cell:

$$\Delta G = RT \ln \frac{[C_i]}{[C_o]} + zF\Delta Em$$

Where z is the valence of the solute, F is the faraday constant (23.06 kcal/V) and ΔEm is the potential difference between the two compartments.

Partition Coefficient the ratio of solubility of a substance in oil as opposed to water. It is evident that the greater the lipid solubility, the faster the penetration. Runar Collander was the first to establish that a molecule which can move across the membrane correlates directly with their oil:water partition coefficient.

Passive Transport in passive transport, the substance moves across the membrane down its concentration gradient. There is no accumulation of such substance. The rate of flow is a direct function of the magnitude of the gradient. Passive transport of molecules occurs in the region of the form where it can move freely.

Simple Diffusion : when the compartments of two solutions of different concentration are separated by a permeable divider, the solute moves from the region of higher concentration to the region of lower concentration till the two compartments have equal solute concentration. Such movements are defined as simple diffusion.

Facilitated Diffusion Plasma membrane is rather impermeable for some substances. These substances bind to a membrane-spanning protein the facilitative transporter. The substance moves down a concentration gradient from a higher to lower concentration. Such binding is not coupled with release of energy but brings conformational change in the protein. Facilitated diffusion is particularly important in mediating the entry and exit of polar solutes that are not able to penetrate the lipid bilayer. This type of diffusion differs from passive diffusion by its enzymatic nature and substrate specificity. The Michaelis-Menton parameters are obeyed. Since no metabolic energy is used in facilitated diffusion, the Michaelis-Menton parameters for entry and exit of a substrate are equal (Fig. 1.3).

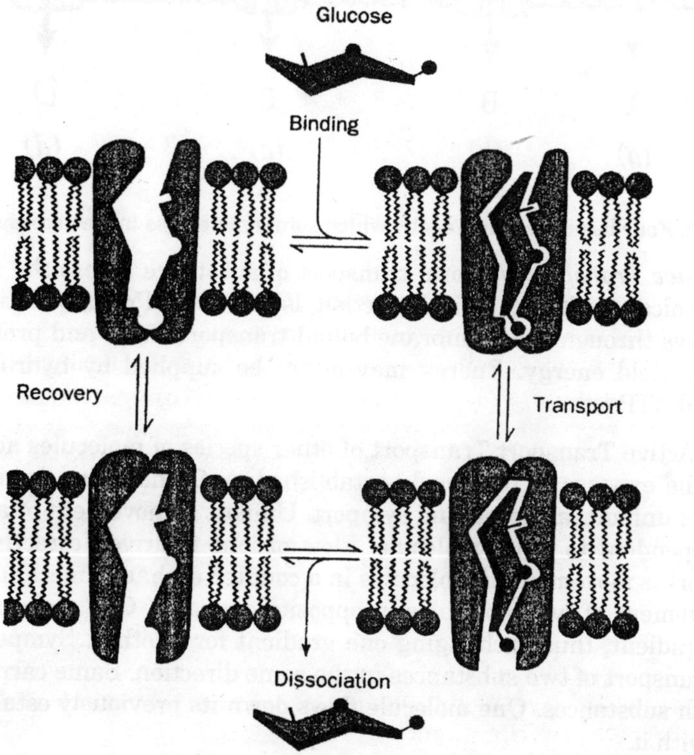

Fig. 1.3. Facilitated diffusion

Active Transport Movement of substance across the membrane against a thermodynamically unfavourable gradient with the coupled input of energy is known as active transport. This type of transport results in accumulation of a solute on one side of a membrane. Active transport is endergonic. In general for such transport energy required i.e., it occurs only when it is coupled with exergonic process. The energy for such transport is generally supplied by hydrolysis of ATP. The active transport may be primary or secondary. When solute accumulation is coupled directly to an exergonic reaction, like hydrolysis of ATP, it is known as Primary Active Transport. In Secondary Active Transport, one solute moves uphill (endergonic) across the membrane is coupled to the downhill (exergonic) flow of a different solute. This different solute was originally pumped uphill by primary active transport (Fig. 1.4).

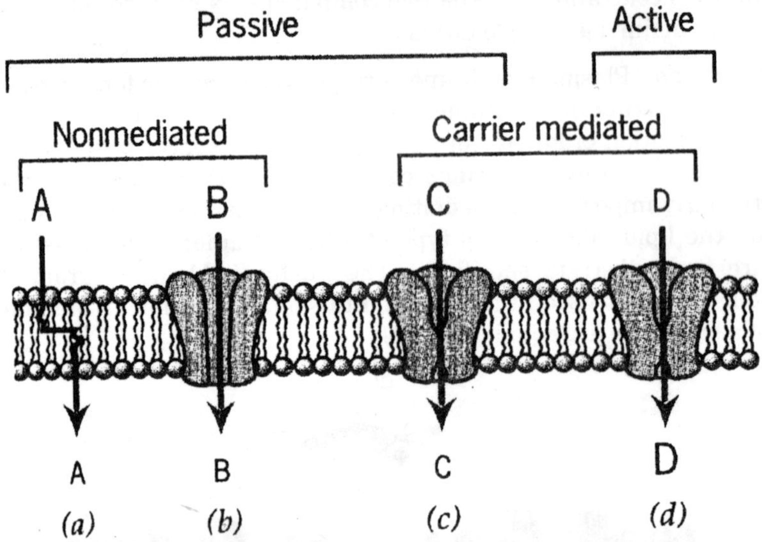

Fig. 1.4. Four basic mechanisms by which solute molecules move across membranes

Primary Active Transport Vectoral transport of substance across the membrane, which is coupled to a chemical or photochemical reaction, is known as Primary Active Transport. While the electron moves through the membrane bound transport chain and proton is extruded, pmf is established to yield energy. Energy may either be supplied by hydrolysis of ATP by the membrane bound ATPase.

Secondary Active Transport Transport of other species of molecules across the membrane, takes place at the expense of previously established pmf. There are three types of secondary active transport : uniport, antiport, and symport. Uniport is movement of a molecule across the membrane independent to other molecule. Flow of ions is driven directly by an electrostatic gradient. Antiport is movement of molecules in a counter-exchange mechanism. Same carrier is involved in movement of two molecules in opposite direction. One molecule flows down in its concentration gradient, thus exchanging one gradient for another. Symport is the mandatory simultaneous transport of two substances in the same direction. Same carrier is involved in the transport of both substances. One molecule flows down its previously established gradient and the other flow with it.

Group Translocation certain substances as they enter the cell get chemically modified. The membrane is impermeable to such modified chemicals. As a result of this modification, high concentration of the modified form is generated inside the cell and low concentration of unmodified substance outside the cell. Thus a metabolic function equivalent to active transport is accomplished by a trapping reaction. Such mechanism of movement of substance is known as Group Translocation. A major fact is that due to modification of substance inside the membrane establishment of the concentration gradient of same species of molecule is not possible. This fact is considered as a difference of group translocation from true active transport, as in the latter concentration gradient of same species of molecule is established on the either side of the membrane. In group translocation, the entering molecule is modified for metabolic function, thus metabolic energy is involved in such transport.

Some sugars enter inside the cell and get phosphorylated. The phosphate moiety is donated by phosphoenol pyruvate. Movement of such sugars is known as phosphotransferase system (PTS). In this system four distinct phosphate-carrying proteins, viz. HPr, Enzyme I (E I), Enzyme III (E III) and Enzyme II (E II) take part. E II a carrier protein and is located in membrane. It is responsible for transfer of sugars. Transfer of phosphate moiety to sugar is catalyzed by E III. E II and E III HPr are specific for sugars and phosphates, while E I are non specific. The later two take part in all the PTSs of a particular cell. Phosphoribosyl transferases (a group of enzymes) and acetyl-CoA-synthetase also act in group translocation.

Translocators

Substances move across the membrane through a specific pore catalyzed by proteins. These proteins are known as Translocators. Translocaters are specific for substrate. Translocators are integral membrane proteins, traversing the lipid bilayer in the form of ?-helix (Fig. 1.5). The side chains contain hydrophobic amino acids like alanine, valine, leucine, isoleucine, or phenylalanine. Like enzyme-substrate action mechanism, translocators form complex with substrate (ST) which is transported through membrane and converts into TS and ultimately dissociates into T and S. The substrate is bound to a specific binding site of the translocator protein (T), transported through the membrane, and then released from the translocator. Translocators also exhibit V_{max} and K_m values. When all the binding site of the translocator are saturated with substrate, this state corresponds to maximal velocity (V_{max}), and K_m value for a translocator corresponds to the substrate concentration at which transport occurs at half maximal velocity.

The triose phosphate-phosphate translocator of chloroplasts enables the export of photo-assimilates from the chloroplasts. In this system translocator catalyzes a counter-exchange of phosphate with glyceraldehydes-3-phosphate or dihydroxyacetone phosphate. This translocator is an antiport. Thus for each molecule transported (e.g., phosphate), another molecule (like dihydroxyacetone phosphate) must be transported out of the chloroplasts. This translocator transports orthophosphate, dihydroxyacetone phosphate, glyceraldehyde-3-phosphate, and 3-phosphoglycerate, but not 2-phosphoglycerate. These substrates compete with each other (competitive inhibitor) and hence, if one substrate is bound to translocator, it will inhibit binding of other substrate.

The mechanism of transport with transporter is shown in Fig. 1.5. The two subunit of transporter forms a gated pore. These subunits form a common substrate binding site, which is accessible either from the inside or from the outside, depending on the conformation of the translocator protein. First the substrate binds with translocator's binding site accessible from the outside. This results in a conformational change in the translocator and finally the substrate is released to the inner side. As the translocators are in antiport, another substrate will bind to the free binding site and it will be transported outside. This mode of antiport is called ping-pong mechanism. Triose phosphate-phosphate translocator of chloroplasts shows ping-pong mechanism. Mitochondrial translocator is operated by simultaneous mechanism. In this process two translocators are linked to each other in such a way that a change of conformation can occur only if both binding sites are either occupied or unoccupied.

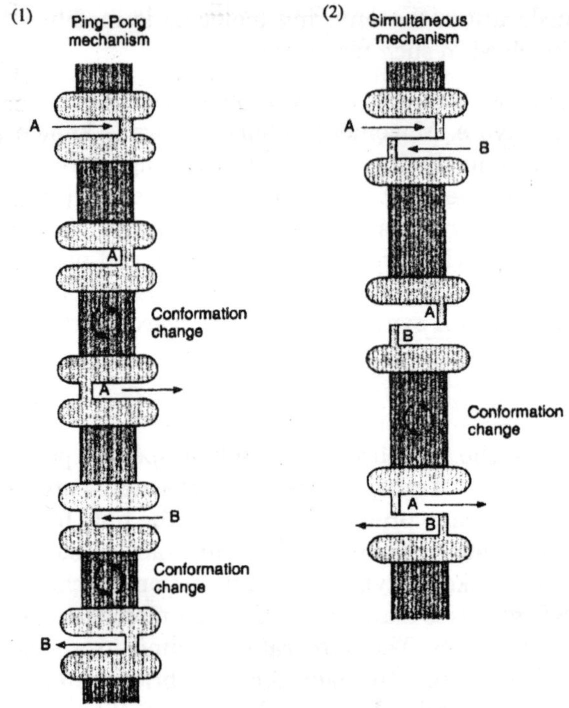

Fig. 1.5. Different types of the counter-exchange of two substrate molecules: (1) A translocator molecule catalyzes the transport of A and B sequentially. (2) A and B are transported simultaneously by two translocator molecules tightly coupled to each other.

Coupling Active Transport to ATP Hydrolysis

ATPase cycle is involved in transport of substrate across the membrane coupled with hydrolysis of ATP. There are three general types of transport of ATPase, viz. P-type ATPase, V-type ATPase and F-type ATPase. Transporters which are reversibly phosphorylated as part of the transport cycle are known as *P-type ATPase*. Each transporter is an integral membrane protein. They share amino acid homology. This homology is especially observed near the Asp residue. This residue is the site of phosphorylation. One of the P-type ATPase common in higher plants is P-type H^+ATPase. This transporter pumps proton out of the cell, and for the transport of each proton one ATP is consumed. *V-type ATPase-* proton pump is found within the vacuole, hence the name V-type (V=vacuole). This type of ATPase is responsible for acidifying intracellular compartments. The V-type enzymes can be separated into a soluble V1 sector and a membrane bound V_o sector. These are analogous to F_1 and F_o. *F-type ATPase* takes part in energy conserving reactions in mitochondria and chloroplast. It performs the same function in bacteria also. The name F-type is due to identification as energy-coupling <F128M>å<F255D> factors. These trasportors catalyze the reversible transmembrane passage of protons, driven by ATP hydrolysis.

Jens Skou (1957) discovered an ATP-hydrolyzing enzyme in the nerve cells of a crab and called it as Na^+-K^+ATPase (Fig. 1.6 A-B), or the sodium-potassium pump. Though such transport is observed in animal cell, however, it is discussed here as it is a prototype for a class of company transporter. Na^+-K^+ATPase is a tetramer consisting of two types of subunits α and β. α is larger subunit and carries out the transport activity, while β is the smaller subunit and functions primarily in the maturation and assembly of the pump within the membrane. α subunit contains 8 to 10 likely transmembrane segments, while the β subunit contains only one. Na^+-K^+ATPase is a P-type ion pump (P means phosphorylation). During pumping cycle, ATP is hydrolyzed, and transfers a phosphate group to one of the amino acids of the transport protein. This results in an essential conformational change within the protein. Conformational changes are necessary to change the affinity of the protein for the two cations that are transported. During Na^+-K^+ATPase pumping cycle, the protein binds three Na+ ions on the inside of the cell, it becomes phosphorylated and changes conformation. In doing so, the binding site becomes exposed to the extracellular compartment, and the protein loses its affinity for Na+ ions, which is then released outside the cell. After the release of sodium ions, the protein picks up two potassium ions. At this stage the protein is dephosphorylated and comes to its original conformation state. In this state, the binding site is open to the inner surface of the membrane, and its affinity for K^+ ions is decreased, leading to the release of these ions into the cell.

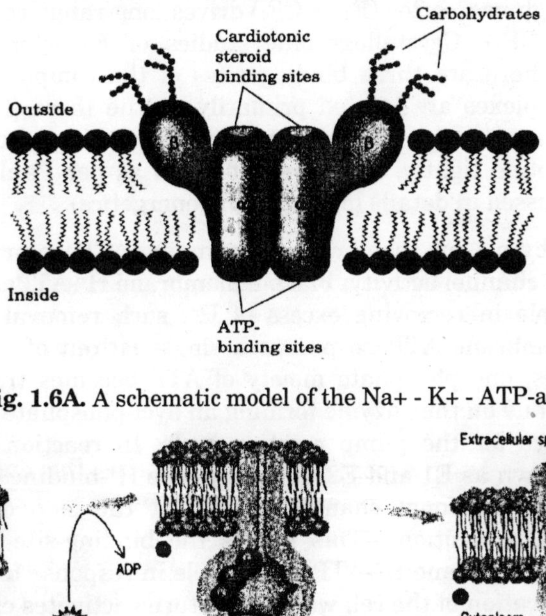

Fig. 1.6A. A schematic model of the Na+ - K+ - ATP-ase

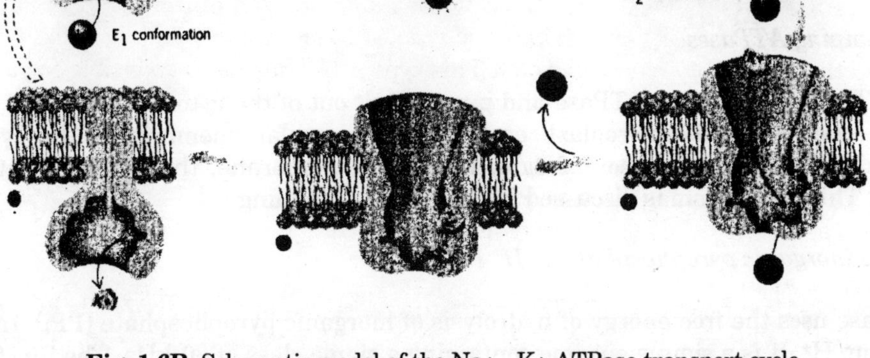

Fig. 1.6B. Schematic model of the Na+ - K+ ATPase transport cycle

Pumps

Some proteins of long range pump the substrates into the cell against a concentration gradient, while some protein pump out certain substances so that the concentration of these substances is lower than those in the surrounding medium. The turn over rate of pumps is correspondingly slow-on the order of 10^2 transported per second. They generate the pmf for an array of symporters and antiportors.

H^+-pumping ATPase

Plasma membrane H^+-ATPase is a single polypeptide of about 100 kDa. It binds ATP and catalyzes H^+ transport. The inner membranes of mitochondria and chloroplast thylakoid synthesize ATP by proton-pumping electron transport system (ETS). The pmf established during ETS serves to drive H^+ flow back through the F-type ATPase. The two sectors of F-type ATPase are F_o and F_1 in mitochondria, while those of chloroplast are known as CF_o and CF_1. The flow of proton through one sector (F_o or CF_o) drives long-range conformational transitions in other sector (F_1 or CF_1). Crystallographic studies of F1sector of mitochondria $\alpha_3\beta_3\gamma$, complexes reveals that there are three binding sites of the complex showing three distinct conformation. These complexes are located primarily on the three β subunits. At any given time, one binding site is in an open conformation, second is binding a nucleotide loosely, and the third is binding a nucleotide tightly. ATP is synthesized as per Paul Boyer's conformational model (this model is discussed in details in chapter Bioenergetics).

The pmf generated by plasma membrane H^+-pumping ATPase provides energy to various carriers. It influences ion channel activity. Plasma membrane H^+-ATPase plays other important roles. It plays critical role in removing excess of H^+, such removal is essential to prevent acidification. Plasma membrane ATPase pumps a single H^+ out of the cell for each MgATP hydrolysis. On hydrolysis, one phosphate moiety of ATP becomes transiently but covalently bound to an aspartyl residue on the enzyme forming an aycl-phosphate bond. This phenomenon provides the driving force for the pump reaction cycle. In reaction cycle, enzyme has two conformational states known as E1 and E2. It exposes the H^+-binding site to alternate sides of the membrane. The conformational change exposed at the side of apoplast is linked to additional conformational transitions. This lowers the binding-site affinity, permitting H+ dissociation. The plasma membrane H^+-ATPase has role in response to auxin. Exposure to this hormone results in acidification of the cell wall and in turns activates expansions. The later is a protein and loosen H-bonds within the wall and allow turgor-generated growth.

Calcium-pumping ATPases

Ca^{2+}-ATPases is a P-type ATPase and pumps Ca^{2+} out of the cytosol. It is found in plasma membrane, the ER, the chloroplast envelop and vacuolar membranes. In cytosol, the concentration of free Ca^{2+} is about 0.2μM. In plasma membrane, this pump is activated by calmodulin. This mechanism is discussed in chapter Biosignalling.

H^+-pumping inorganic pyrophosphatase (H^+-PPase)

H^+-PPase uses the free energy of hydrolysis of inorganic pyrophosphate (PPi), rather than ATP, to pump H^+. It is a simple enzyme containing a single class of 80-kDa. The functional unit of the pump is probably a homodimer. The substrate of this enzyme is dimagnesium PPi.

Ion Channels

Lipid bilayer is impermeable to electrolytes, and for the transport of such substance, there is a passage in membranes which is permeable to specific ions. A.Hodgkin and A.Huxley observed that there are elements in the plasma membrane that responded to electrical stimuli by opening and facilitating selective fluxes of ions during action potentials. They named it as channels. Ion channels are ubiquitous in biological membrane, and show selectivity for their ionic substrate. The ion channels are highly specific allowing only one particular type of ion to pass through the pore. These ion channels are formed by integral membrane proteins that surround an aqueous pore. The diffusion of ions through a channel is always "downhill". Ion flow through the channels is passive. Channel-mediated currents flowing towards the cytosol is negative. Outward current carried by an influx of cations efflux or anion influx is positive. The opening of many ion channels is regulated by the electric membrane potential Ion channels can transport various ions at the rate of 10^6 to10^8 ions per second.

Ion channels do not undergo conformational changes during translocation reaction. However, they show shift between permeable and nonpermeable states. These alterations are named as O and C, respectively. This shift is known as gating. Following equation expresses the contribution of an individual type of channel to whole-cell or whole-vacuole currents

$$I = N \rightarrow i \rightarrow P_0$$

Where, I is the ionic current across the whole cell or vacuole, N is the number of channels, I is the current flow through the channel when open and Po is the probability that the channel is in its open state.

The Potassium Channel

Plant cells contain very specific ion channels which are highly selective for cations (H^+, K^+, Ca^{++}), for anion (Cl^-) and for dicarboxylates (malate). Most classes of ion channels in plant cells usually favour flow of cations or anions. Prominent cations channels are K^+ (monovalent) and Ca^{++} (divalents). It is typical that the amino acid sequence of K^+ channel protein is similar in bacteria, plants, and animals. The channel protein is made up of two identical subunits, each has two transmembrane helices connected by a loop of about 30 amino acids sequence. The loop is responsible for the ion selectivity of the channel. Roderick MacKinnon, a recipient of Nobel Prize (2003) performed X-ray structure analysis of the three dimensional structure of K^+ channel for the bacterium (*Streptomyces livians*) and recognized the molecular function of an ion channel. Structure analysis showed that a K^+ channel is built of four of these subunits. One helix of the subunit present towards inner side lines the channel and the other is directed toward the lipid membrane. The pore's interior consists of a channel filled with water, which is separated from the outside by a selection filter. Loops of the four subunits (as described earlier) form the filters. The pores are very small, and K^+ ions can only pass through it when they are dehydrated i.e., their hydrate coat should be removed. It requires energy for such type of dehydration. To compensate this energy, the pore is lined with a circular array of oxygen atoms that act as a `water substitute' and form a complex with the K^+ ions. The pore is also negatively charged to bind the cations. The bound K^+ ions are pushed by the entering K^+ ions through the other side of the filter. Na^+ ions are too small and the process of dehydration and movement across the membrane is not possible. Thus the channel is selective only for K^+ ions.

The Calcium Channel

Voltage gated Ca^{++} permeable channels are present in the plasma membrane of different types of cells. In cytosol free Ca^{++} increases by the activation of Ca^{++} permeable channels. In plants, four different classes of Ca^{++} permeable channels are present in intracellular membranes. The vacuolar membrane has two types of Ca^{++} channels, which are activated by ligands-inositol 1,4,5-triphosphate (IP_3) and cyclic ADP-ribose (cADPR). The latter compound releases Ca^{++} through ryanodine receptors. Ca^{++} induced Ca^{++} release might be generated through the activity of one of the two voltage-activated Ca^{++} permeable channels at the vacuolar membrane. One of the channel is activated by membrane hyperpolarization and is not sensitive to cytosolic Ca^{++}. The second channel exhibits slow time-dependent activation.

The Chloride Channel

A major function of anion channel is controlling salt release during turgor adjustment. Studies of three dimensional structure of chloride channels of *Salmonella* and *E.coli* reveals that these channels are made up of transmembrane <F128M>a<F255D>-helices. These helices of interior loops are responsible for the selectivity filters by electrostatic interaction with and coordinating binding of the chloride anion.

In plants cellular loss of Cl^- is affected by Ca^{++}-activated anion channel. In guard cells there are two distinct classes viz. rapidly activating R-type and the slowly activating S-type. They are gated by voltage. In cytosol the concentration of Cl^- is higher as compared to the extracellular concentration. When the channel is open, Cl^- flows out and the membrane gets depolarized. The depolarization depolization probably activates outward rectifying K^+ channels during salt loss.

Ion Channels of Guard Cells

The guard cell ion channels were determined to play key roles in mediating the solute fluxes that accompany stomatal opening and closing. In guard cells the plasma membrane undergoes hyperpolarization and depolarization. The former opens channel that allows potassium ions to flow into the cell while the latter opens another channel by which potassium ions flow outside the cell. Thus, these channels are named as K^+ inward channel and K^+ outward channel. The inward and outward flow of K^+ has two compartments. The current is time-dependent. The channel carrying these currents are said to rectify and the channels are known as rectifiers. In guard cells other factors also regulate these rectifiers. In these cells outward rectifiers are activated by the small increase in cytosolic pH that can be elicited by abcisic acid. Inward rectifiers are modulated by G-proteins and are inhibited by increased free cytosolic Ca ions and by dephosphorylation by protein phosphatase 2B. These regulatory factors are all components of guard cell signal transduction network. The coordinated action of these factors tightly controls K^+ uptake and loss.

The structural features of inward rectifiers resemble the shaker family (as present in mutant of *Drosophila*). The fourth transmembrane helix (S4 domain) exhibits a regular pattern of positively charged residues lysine or arginine. The outward and inward rectifiers are regulated by membrane voltage. As a hyperpolarization voltage is imposed across the membrane, the S4 helix is thought to screw out of the membrane slightly. This brings conformational transition opening the `gate' that controls the ion flow through a separate part of the protein. Inward rectifying K^+ channels are tetramers. The Shaker subunit of this tetramer contributes to the pore through which ions flow. The pore domain (P domain) of each subunit is present between the fifth and sixth transmembrane spans. These domain loop into the membrane and come out again. The P- domain endows the channel with its ionic selectivity.

Ion Channels of Vacuolar Membrane

In general the vacuoles accumulate and store solutes. However, under certain physiological conditions (e.g., osmotic adjustment), there is loss of solutes from these organelles. During hypoosmotic stress ions move from vacuoles, so that the turgidity of cells is restored. Even at the time of closure of guard cells, ions come out from the vacuoles of the guard cells. Loss of ions occurs from ion channels, which are activated instantaneously in response to an imposed voltage. The fast vacuolar channels show little selectivity for monovalent cations. They are inhibited when cytosolic Ca^{++} concentration exceeds 1 μM and is activated when cytosolic pH increases. Vacuolar K^+ channel is highly selective for K^+ is activated by cytosolic Ca^{++} and inhibited by increasing cytosolic pH.

Aquaporins

Aquaporins are highly selective for water transport. Johns Hopkins a recipient of 2003 Nobel Prize in chemistry, detected membrane channels for water in kidney and blood cells protein. Such structures were also observed in plasma membrane and membranes of vacuoles. X-ray analysis by electron cryomicroscopy shows the aquaporin is made up of subunits present in the form of tetramer. Each subunit has six transmembrane helices. The monomers form a water channel. These channels consist of a very narrow pore. Although aquaproins exist as tetramer, a pore is believed to be formed within each subunit, the *NAP* boxes. The *NPA* are with Asn-Pro-Ala residues present in both the N- and C-terminal halves of the protein. In aquaporins, the NPA boxes being folded into the membrane between transmembrane spans 2 to 3 and 5 to 6 to provide the constriction that confers water selectivity. This pore is primarily hydrophobic and acts as a selection filter. The pores can bind four water molecules. The channel can transport water molecules at the rate of 10^9 to 10^{11} per second.

A plant contains many aquaporin isoform. Aquaporins are present on plasma membrane as well as on vacuolar membrane. The intrinsic enzymes of aquaporins of this family are known as plasma membrane intrinsic proteins and tonoplast intrinsic proteins respectively.

When the solute concentrations of two sides of a membrane are very different, there is a concentration gradient of solvent (water) molecules. This osmotic imbalance results in transmembrane flux of water, and it continues until the osmotic strength equalizes on both sides of the membrane. The direction of flow of water is determined by two forces, the hydrostatic pressure difference (ΔP) across the plasma membrane or cell wall and the osmotic pressure difference (Δp) between the inside and the outside of the cell. The flux of water is proportional to driving force (J_v); it is expressed by following equation:

$$J_v = L_p (\Delta P - \sigma\Delta\pi)$$

In the above equation L_p is a constant of proportionality, σ is known as the reflection coefficient.

There are two approaches for defining membrane permeability to water. They are an osmotic coefficient (P_f) and diffusional coefficient (P_d). Osmotic coefficient involves net flow of water. The flow of water is in one direction. The diffusional coefficient is related to flow of water inside the pore or even back to solution. The diffusional coefficient accounts for the flow of water through lipid bilayer. Thus, the water can move across the membrane through lipid bilayer as well as through water channel proteins.

SUMMARY

The physiological function of plasma membrane is movement of substance. J.D. Robertson suggested a three-layered structure, consisting of two darkly staining outer layers and a lightly staining middle layer. Membranes contain several types of lipid, all of which are amphipathic. The membrane proteins are asymmetrically so that the outer portion of Interagal proteins, Peripheral, Integral membrane proteins, Transporters. The plasma membranes of eukaryotic cells possess carbohydrates that are covalently linked to both lipid and protein component. The movements of the substance into the cell and out of the cell are known as influx and outflux, respectively. ATPase cycle is involved in transport of substrate across the membrane coupled with hydrolysis of ATP. Some proteins of long range pump the substrates into the cell against a concentration gradient, while some protein pump out certain substances so that the concentration of these substances is lower than those in the surrounding medium. There are ion channels for transport of specific ions through lipid layer, which is impermeable in nature. Aquaporins are highly selective for water transport. A plant contains many aquaporin isoform. Water can move across the membrane through lipid bilayer as well as through water channel proteins.

SECTION- I :

BIOCHEMISTRY AND METABOLISM

Chapter - 2

METABOLISM AND BIOENERGETICS

Living organisms can be divided into two large groups: autotrophs and heterotrophs. This classification is based on the chemical form carbon, which they obtain from the environment. Autotrophs (such as photosynthetic bacteria and higher plants) can use carbon dioxide from the atmosphere as their sole source of carbon. Heterotrophs cannot use atmospheric carbon di oxide. They obtain carbon in the form of relatively complex organic molecules, such as glucose. For these which depends on other organic substances. Some autotrophic organisms, such as cyanobacteria, can also use atmospheric nitrogen to generate all their nitrogenous components.

Metabolism (categorized in anabolism and catabolism), is the sum of all the chemical transformations that occur in a cell or organism. Metabolism is a well directed cell activity in which biochemical pathways are highly coordinated. It involves many multienzyme systems, which cooperate to serve four functions : (1) to obtain chemical energy: by virtue of trapping energy and converting it into chemical energy or by degrading energy-rich nutrients from the environment, (2) to convert nutrient molecules into the cell's macro-molecular precursors, i.e., nutrients are converted into simple compounds, which acts as precursors for biosynthesis of macromolecules (3) to polymerize monomeric precursors into biomolecules like proteins, nucleic acids, lipids, polysaccharides, as well as other cell components, and (4) to synthesize and degrade biomolecules required in specialized cellular functions.

In a biochemical pathway, a precursor is converted into a product through a series of metabolic intermediates. These intermediates are simply termed as metabolites. During a biochemical pathway the consecutive steps brings about a small, specific chemical change, in which there is removal, transfer, or addition of a specific atom, functional group, or molecule. The term intermediatry metabolism is often applied to the combined activities of all of the metabolic pathways.

Anabolism, is solely biosynthetic phase of metabolism, in which small, simple precursors synthesizes larger and more complex molecules, including lipids, polysaccharides, proteins, and nucleic acids. Anabolic reactions require the input of energy, generally in the forms of the free energy of hydrolysis of ATP and the reducing power of NADH and NADPH.

Catabolism is the degradative phase of metabolism, in which organic nutrient molecules (carbohydrates, fats, and proteins) are convert into smaller, simpler end products (e.g., lactic acid, CO_2, NH_3). Catabolic pathways release free energy, some of which is conserved in the formation of ATP and reduced electron carriers (NADH and NADPH).

The electromagnetic radiations of the sun are the ultimate source of energy to the earth. This energy is utilized by autotrophs in carbon di oxide assimilation. During photosynthesis, there is transfer of reducing equivalents. The reducing equivalents come from H_2O or H_2S, which are accepted by NAD^+ or $NADP^+$ and ultimately goes to CO_2. The transport of electron during photosynthesis is known as photosynthetic electron transfer. Such transfer converts solar energy into biologically universal form of energy, ATP and this type of phosphorylation is called photophosphorylation. In mitochondria, phosphorylation takes place by energy released from electrons removed during substrate oxidation; the process is called oxidative phophorylation.

BIOENERGETICS

The Energy State of Cell

Energy is defined as the capacity to do work, that is the capacity to change or move something. Energy exists in two alternate states: potential and kinetic. Bioenergetics is the quantitative study of the energy transductions that occur in living cells and of the nature and function of the chemical processes underlying these transductions.

The Laws of Thermodynamics Thermodynamics is the study of the change in energy that accompanies events in the universe. *The First Law of Thermodynamics* is the law of conservation of energy. It states that energy can neither be created nor destroyed. Energy can, however, be converted (transduced) from one form to another.

To discuss energy transformation involving matter, we need to divide the universe into two parts: the *system* (part of universe under study) and the *surrounding* remainder of the universe, excluding the system. In most thermodynamic systems, if there is no exchange of *matter* with its environment; it is a *closed system*. In contrast, the open system it may exchange *energy* with its surroundings. The changes in a system's energy occurs during an event. These changes are manifested in two ways - as a change in the heat content of a system (may gain or lose energy), this is termed as *internal energy* (E), and the change in the internal energy during a transformation is ΔE. Thus, the first law of thermodynamics may be expressed as:

$$\Delta E = Q - W$$

Where, Q is the heat energy and W is the work energy. If heat is absorbed into the system during the event from the surroundings, Q is negative. If the system does work on the surroundings (by increasing in volume and compressing the surroundings), then W is positive, when work is done on the system, W is negative.

In a process, the internal energy of the system at the end of the work can be greater than, equal to or less than its internal energy which was at the start. This depends on its relationship to its surroundings. In other words, ΔE can be positive, zero, or negative. Under conditions of constant pressure and volume, reactions that lose heat are termed exothermic, and ones that gain heat are *endothermic*. *The Second Law of Thermodynamics* expresses the concept that events in the universe have direction; they always proceed "downhill" from a state of higher energy to a state of lower energy.

Thus, in any energy transformation, there is a decreasing availability of energy for doing additional work.

FREE ENERGY

The energy that can do some work is known as Free energy. J. Willard Gibbs (1878) put together the concepts inherent in the first two laws of thermodynamics into the expression $\Delta H = \Delta G + T\Delta S$, where ΔG is the change in free energy, that is, the change during a process in the energy available to do work. ΔH is the change in enthalpy or total energy content of the system (equivalent to ?E for our purpose): T is the absolute temperature (°K = °C + 273); and ΔS is the change in the entropy of the system. The equation states that the total energy change is equal to the sum of the changes in useful energy (ΔG) and energy that is unavailable to do further work ($T\Delta S$).

This is rearranged and written as $\Delta G = \Delta H - T\Delta S$, expresses a measure of the spontaneity of a particular process. It allows one to predict the direction in which a process will proceed and the extent to which that process will occur. All spontaneous energy transformations must have a negative ΔG; that is, the process must proceed from a state of higher free energy to a state of lower free energy. The magnitude of ΔG indicates the maximum amount of energy that can occur spontaneously. A reaction may be exergonic or endergonic. The biochemical reactions that are thermodynamically favoured are described as exergonic. Such processes have a $-\Delta G$. In contrast, those processes, which are thermodynamically unfavourable and are described as endergonic. The ΔG for a given process is positive. It cannot occur spontaneously.

Free-Energy Changes in Chemical reactions

All chemical reactions within the cell are reversible and, therefore, we must consider two reactions occurring simultaneously, one in the forward direction and the other in reverse. According to the law of Mass Action, the rate of reaction is proportional to the concentration of the reactants. In a following hypothetical reaction:

$$A + B \;=\; C + D$$

The rate of the forward reaction is directly proportional to the product of the molar concentrations of A and B. This can be expressed as the rate of forward reaction equals $k1$[A][B], where k1 is a rate constant for the forward reaction. The rate of the backward reaction equals $k2$[C][D]. All chemical reactions proceed, however slowly, toward a state of equilibrium; that is, toward a point at which the rates of the forward and backward reactions are equal. At equilibrium, the same number of A and B molecules per unit time as will be formed from them. At equilibrium, therefore,

$$k1[A][B] = k2[C][D]$$

this can be rearranged to

$$\frac{k1}{k2} = \frac{[C]\,[D]}{[A]\,[B]}$$

thus the equilibrium depends on a predictable ratio of the concentration of products to the concentration of reactants. This ratio, which is equal to *k1*/k2, is termed the equilibrium constant, Keq. The equilibrium constant allows one to predict the direction (forward or reverse) in which the reaction is favoured under a given set of conditions. If the Keq is greater than one, the reaction will proceed at a greater rate toward the formation of the products C and D. If, on the other hand, the Keq is less than one, the reverse reaction will proceed at a greater rate than the forward reaction, so that the concentration of A and B will rise at the expense of C and D. Thus, the Keq gives a picture about the direction in which the reaction is proceeding at any moment. This dependents on the relative concentration of all molecules present at that moment. In terms of energetics, the ratio of reactants and of products present at equilibrium is determined by the relative free-energy levels of the substances on either side of the equation. As long as the total free energy of the reactants is greater than the total free energy of the products, the ΔG will have a negative value and the reaction will proceed in the direction of formation of products. The greater the ΔG, the further the reaction is from equilibrium. As the reaction proceeds, the difference in free-energy content between the reactants and products decreases (ΔG becomes less negative), and reduces until, at equilibrium the difference becomes zero ($\Delta G = 0$).

The standard free-energy difference ($\Delta G^{o\prime}$) describes the difference in free energy when one mole of each reactant is converted to one mole of each product under standard conditions. Standard conditions means when all the reactants and products are present at 1.0 M concentration, except for water, which is present at 55.6 M and H^+ at 10^{-7} M (pH 7.0) the reaction takes place at 25°C (298°K) and 1 atm of pressure. As the standard conditions do not prevail in cell, and therefore, one must be cautious in the use of values for standard free-energy differences in calculations of cellular energetics.

The relationship between the equilibrium constant and the standard free-energy difference is given by the equation

$$\Delta G^{o\prime} = -RT \, ln \, K'eq$$

When the natural log (*ln*) in converted to log 10, the equation becomes

$$\Delta G^{o\prime} = -2.303 \, RT \log K'eq$$

Where R is the gas constant (1.987 cal/mol.°K), and T is the absolute temperature (298°K)[3]. Recall that the log of 1.0 is zero. Consequently, it follows from the above equation that reactions having equilibrium constants greater than one will have negative $\Delta G^{o\prime}$ values, which indicates can occur spontaneously under standard conditions. In other words, given the reaction written as follows: A + B = C + D, if the $\Delta G^{o\prime}$ is negative, the reaction will go to the right direction when reactants and products are all present at 1.0 M concentration at pH 7. The greater the negative value, the farther to the right the reaction will proceed before equilibrium is reached. Under the same conditions, if the $\Delta G^{o\prime}$ is positive, the reaction will proceed to the left direction that is, the reverse reaction is favoured.

COMPONENTS OF ELECTRON TRANSPORT SYSTEM

Pyridine linked dehydrogenase these enzymes contain nicotinamide rings resembling pyridine as coenzyme (Fig. 2.1). One of the soluble vitamin niacin provides the nicotinamide moiety for the synthesis of the pyridine nucleotides. Nicotinamide is the site of oxidation and reduction. In a cell there are two pyridine nucleotide, namely NAD (nicotinamide adenine dinucleotide originally called coenzyme I, cozymase or DPN^+) and NADP (nicotinamide adenine dinucleotide phosphate originally coenzyme II, or TPN). They are composed of two nucleotides joined through their phosphate groups by a phosphoric acid anhydride bond. The intracellular pool of NAD is maintained largely in the oxidized state, and the pool of NADP is maintained largely in the reduced state. It is interesting to note that in catabolic reactions oxidized pyridine nucleotides (NAD) are the usual reactant, and in the biosynthetic reactions the reduced form (NADP) is the usual reactant. These nucleotides can readily undergo reversible oxidation and reduction. When NAD^+ or $NADP^+$ comes in contact with reduced substrate the reducing equivalent is transferred to it and it is converted into $NADH+H^+$ or $NADPH+H^+$. It can be seen that the oxidized form of the pyridine nucleotides carries one hydrogen atom less than the reduced form. In addition, it has a positive charge on the nitrogen atom, which enables it to accept a second electron upon reduction. The reducing equivalent is accepted as hydride ion. It attaches at 4 end of nicotinamide ring. The standard redox potential of the NAD^+/ $NADH+H^+$ and $NADP^+$/ $NADPH+H^+$ is – 320mV and – 324mV, respectively.

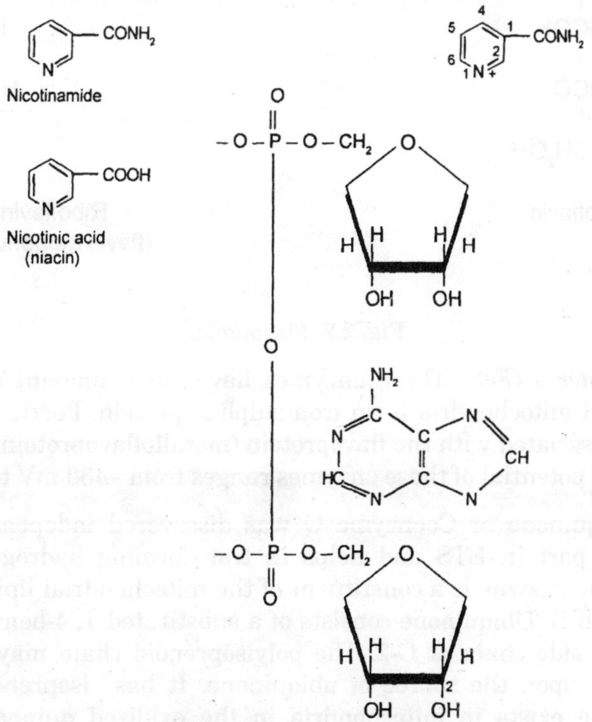

Fig. 2.1. Nicotinamide, Nicotinic acid

Flavoprotein These enzymes have FMN (flavin mononucleotide) and FAD (flavin adenine dinucleo-5-phosphate) coenzyme (Fig. 2.2). These groups are tightly bound to the protein and hence act as prosthetic group. FAD and FMN are formed in the body from the vitamin riboflavin. The coenzyme is tightly bound by covalent bond to protein and never leaves it during the reaction. Isoalloxazine ring of the riboflavin moiety accepts reducing equivalents. After reduction, the coenzyme is converted into $FADH_2$ and $FMNH_2$, respectively. Flavoprotein enzymes can receive either one or two reducing equivalents. FAD or FMN when receives only one reducing equivalent the semiquinone form of isoalloxzine ring is produced. The standard redox potential of the $FMNH_2/FMN$ and $FADH_2/FAD$ couples are – 190 and – 210mV, respectively. But these values are often quite different when the flavin is bound to the apoprotion.

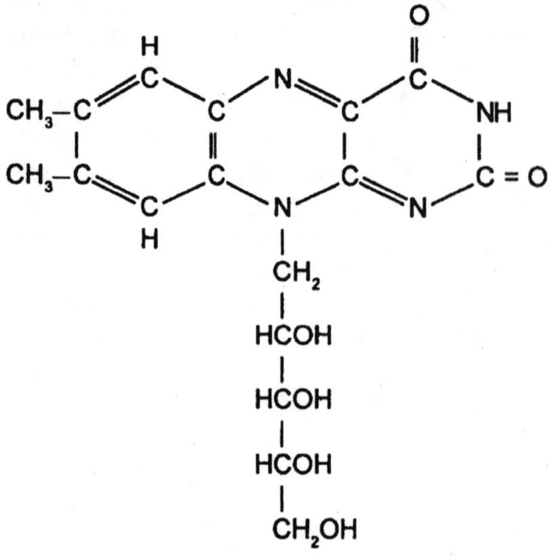

Riboflavin

2.2A

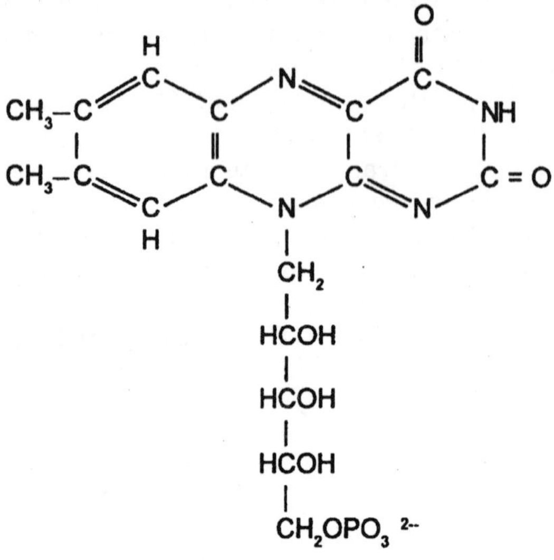

Riboflavin phosphate
(flavin mononucleotide; FMN)

2.2B

Fig. 2.2. Flavoprotein

Iron sulphur proteins (FeS) These enzymes have equal amount of iron and acid liable sulphur. Ferredoxin of mitochondria is an iron sulphur protein. Ferric ion takes part in redox reactions. They are associated with the flavoprotein (metalloflavoproteins) and with cytochrome b. The standard redox potential of these enzymes ranges from –430 mV to + 350mV.

Ubiquinone Ubiquinone or Coenzyme Q was discovered independently by Mortan and Crane. It also takes part in ETS and helps in transforming hydrogen of the substrate to molecular oxygen. This enzyme is a constituent of the mitochondrial lipids. It is very similar to vitamin K and vitamin E. Ubiquinone consists of a substituted 1, 4-benzoquinone nucleus plus a long polyisoprenoid side chain at C-2. The polyisoprenoid chain may contain 6-12 isoprene units, which depends upon the source of ubiquinone. It has isoprenoid side chain in their coenzyme. Ubiquinone exists in mitochondria in the oxidized quinone form under aerobic conditions and in the reduced quinol form under anaerobic conditions. Redox potential is + 810 mV. Ubiquinone reduced by the NADH dehydrogenase complex or succinate dehydrogenase is oxidized by the cyt-b/c1 complex. In mitochondria this complex consists of 11 subunits, only one of which is encoded in the mitochondria. Plastoquinone, quite related to ubiquinone takes part in photosynthesis.

Cytochromes in 1884 MacMunn discovered cytochrome (Fig. 2.3). Yakushiji and Okunuki (1940) first distinguished cytc1 and separated it from cytc. The prosthetic group is heme. They are made up of tetrapyrrole ring, the porphyrin. They have Fe nucleus. It carries a single electron and oscillates between the ferrous (Fe^{2+}) and ferric (Fe^{3+}) states. Five types of cytochromes are present in mitochondria. They are named as Cyt b, Cytc, Cytc1, Cyt a and Cyta3. The individual cytochrome differ principally in the nature of their substituent groups on carbon atoms 2, 4, 5 and 8 of the heme moieties. The redox potential varies from +260 mV to +290 mV. The b- type cytochromes are firmly bound to the respiratory membrane. Besides its redox functioning, it also contributes to the structure. ATP is required for the complete reduction of cyt b565, while cyt b562 acts predominantly as a single redox carrier. Cyt c is loosely bound to the respiratory membrane and is readily soluble in water. The surface of the molecule is highly charged and exhibits distinct areas of acidity and basicity. This separation of surface change is unusual in a globular protein. Cyt a3 Cytaa3 (cytochrome oxidase) is firmly bound to the respiratory membrane. This cytochrome contains two molecules of heme, two molecules of copper and up to 100 moles of phospholipid per mole of protein. The protein is composed of at least six polypeptide chains. The copper atoms act as a one electron redox carrier. The standard redox potential of cyt a is +230mV and of a3 is +375mV. These two cytochromes are generally regarded as being separate cytochromes although they have never been physically separated from one another. Reduced cyt a3 donates electron to molecular oxygen.

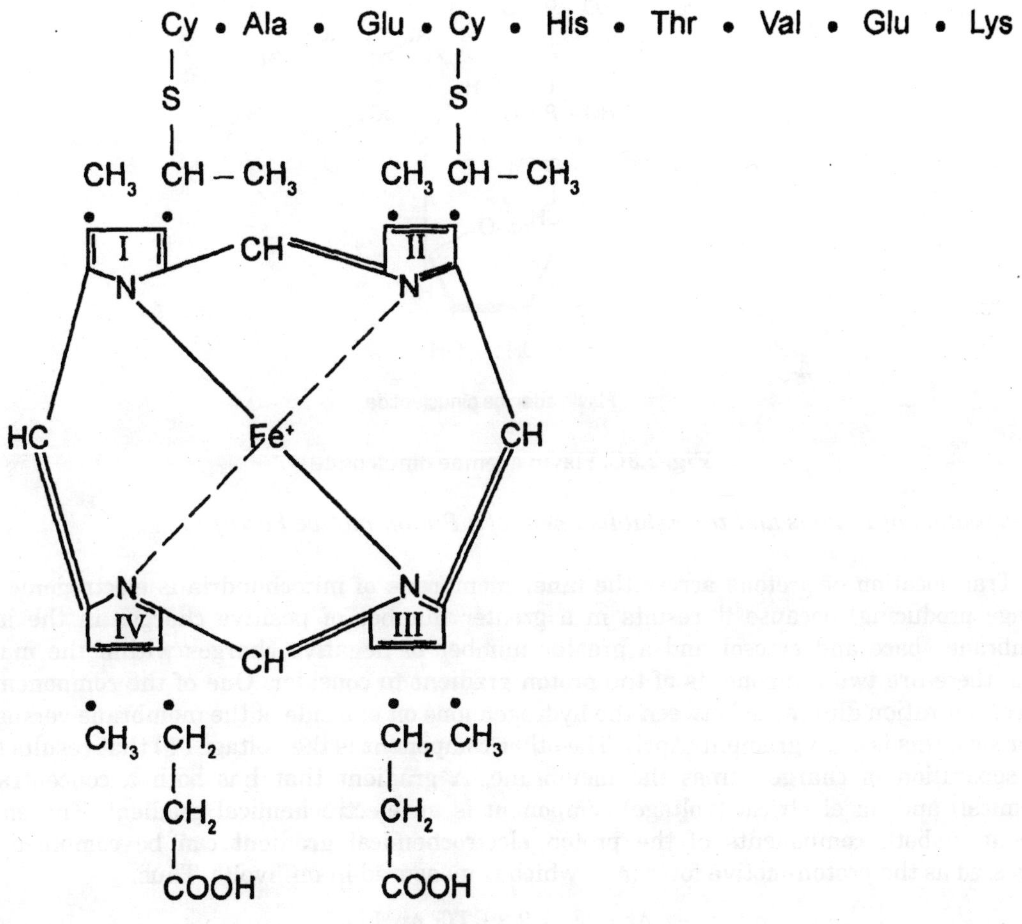

Fig. 2.3. Cytochrome C-Peptide

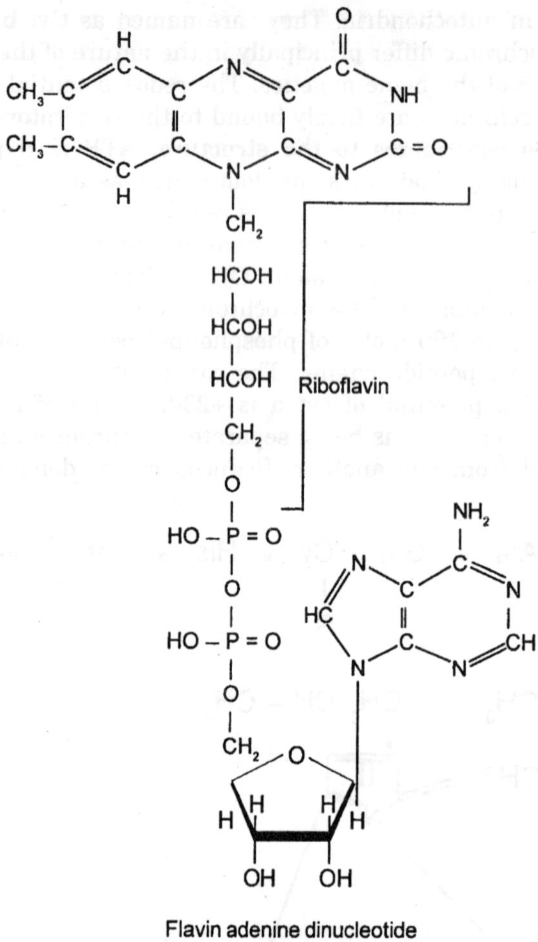

Fig. 2.3C. Flavin adenine dinuleotide

Translocation of Protons and the Establishment of a Proton-motive Force

Translocation of protons across the inner membrane of mitochondria is electrogenic (i.e., voltage producing) because it results in a greater number of positive charges in the inter-membrane space and cytosol and a greater number of negative charges within the matrix. Thus, there are two components of the proton gradient to consider. One of the components is the concentration difference between the hydrogen ions on one side of the membrane versus the other side; this is a pH gradient (ΔpH). The other component is the voltage (Ψ) that results from the separation of charge across the membrane. A gradient that has both a concentration (chemical) and an electrical (voltage) component is an electrochemical gradient. The energy present in both components of the proton electrochemical gradient can be combined and expressed as the proton-motive force (Δp), which is measured in millivolts. Thus,

$$\Delta p = \Psi - 2.3\ RT/F\ \Delta pH$$

Since 2.3 RT/F is equal to 59 mV at 25°C, the equation can be rewritten as

$$\Delta p = \Psi - 59\ \Delta pH$$

Where the ΔpH is expressed as a negative value as long as the [H^+] is higher in the inter-membrane space than in the matrix.

The contribution to the proton-motive force made is by the permeability properties of the inner membrane.

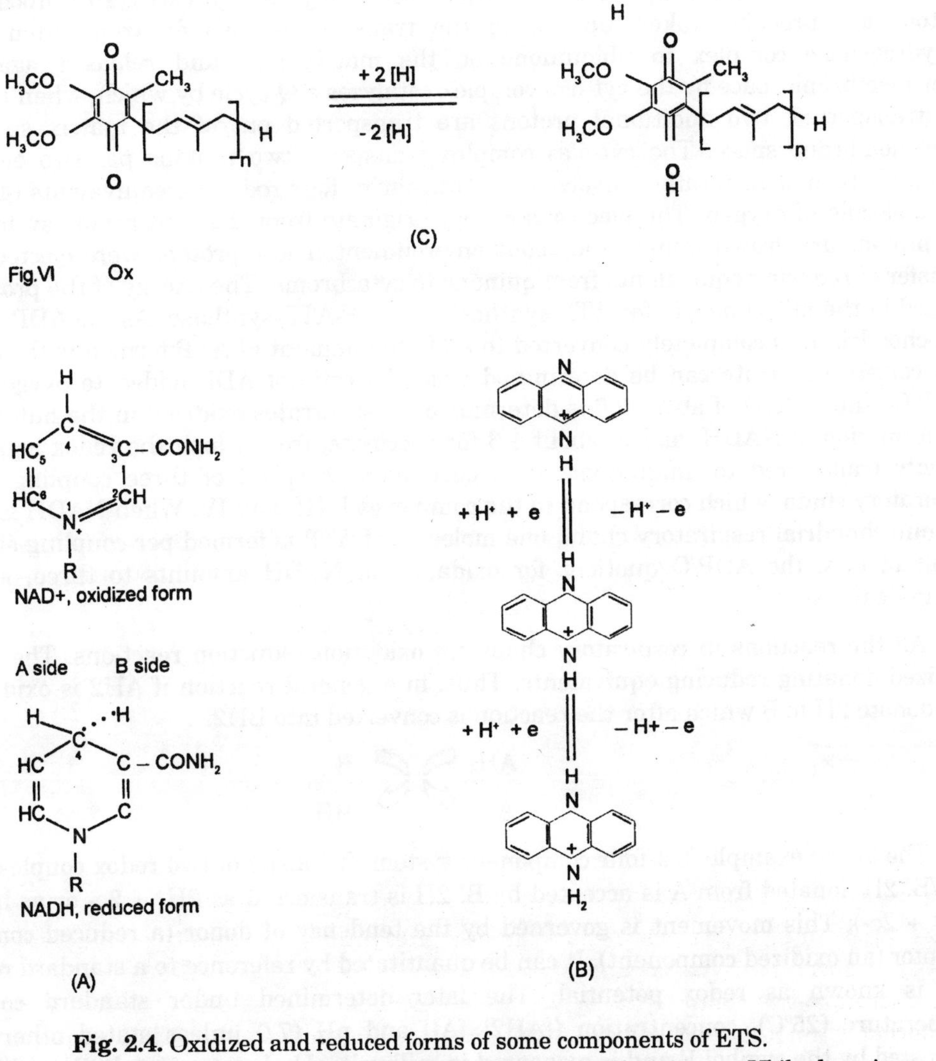

Fig. 2.4. **Oxidized and reduced forms of some components of ETS.**
(A) NAD+ ----NADH (B) Oxidized and reduced from riboflavin (C) Ubiquinone - Ubihydroquinone

ELECTRON TRANSPORT CHAIN

Electron transport chain is transfer of electron(s) by the carriers (Fig. 2.4). The presence of metal (Fe, Cu) and non-metal (NAD, FAD etc.) are significant in Electron Transport System (ETS). The metals are associated with electron and oxygen transfer and never the hydrogen transfer. Iron is a transitional metal and it can exhibit several oxidation states. It can operate over a wide range of redox potential. The nonmetals (organic) components take part in hydrogen transfer. The electron transport and associated reaction lead to synthesis of ATP. The flow of electrons through the transport chain allows a stepwise release of this energy, some of which is conserved in the form of ATP. At these specific steps the difference in Eo values is great enough to permit sufficient energy to be liberated for phosphorylation to occur. When a pair of electrons or hydrogen atoms (which contains electrons) from an oxidizable substrate is coupled with the reduction, there is a large free energy change (ΔG^o). In mitochondria, four protons are probably taken up during the transport of two electrons from the NADH dehydrogenase complex to ubiquinone on the matrix side and released again into the inter-membrane space by the cyt-b/c_1 complex catalyzes a Q-cycle by which, when two electrons are transported, two additional protons are transported out of the matrix space into the inter-membrane space. The cyt-a/a_3 complex transports two protons per two electrons. The terminal stage of respiration involves the transfer of four reducing equivalents ($4H^+ + 4e^-$) to one molecule of oxygen. The electrons clearly originate from the cytochrome system, whereas the protons are drawn from the aqueous environment. These protons were ejected during the transfer of reducing equivalents from quinone to cytochrome. The energy of the proton gradient is used in the mitochondria for ATP synthesis by an F-ATP-synthase. As the ADP added to the mitochondria it is completely converted to ATP, the amount of ATP formed with the oxidation of a certain substrate can be determined from the ratio of ADP added to oxygen consumed (ADP/O) and ADP/O of about 2.5 is determined for substrates oxidized in the mitochondria via the formation of NADH, and of about 1.6 for succinate, from which the redox equivalents are directly transferred to uniquinone. It is customary to speak of three coupling sites of the respiratory chain, which correspond to the complexes I, III, and IV. When NADH is oxidized by the mitochondrial respiratory chain, one molecule of ATP is formed per coupling site, and as a result of this, the ADP/O quotient for oxidation of NADH amounts to three, and that for succinate to two.

All the reactions in respiratory chain are oxidation reduction reactions. The substrate is oxidized donating reducing equivalents. Thus, in a general reaction if AH2 is oxidized to A, it will donate 2H to B which after the reaction is converted into BH2.

The above example is a four component system. It consist of two redox couples AH_2/A and BH_2/B. 2H donated from A is accepted by B. 2H is transferred as $2H^+$ + 2e- or as hydride ions ($2H^+$ + 2e-); This movement is governed by the tendency of donor (a reduced component) to acceptor (an oxidized component). It can be quantitated by reference to a standard redox couple and is known as redox potential. The later determined under standard conditions of temperature (25°C), concentration ([AH2]=[A]) and pH (7.0 unless stated otherwise). It is expressed by the symbol E and is measured in millivolts. Under non standard conditions, redox potential is expressed by the symbol Eh. $NADH^+$ H^+/NAD^+ is the first redox component of the respiratory chain. The reducing equivalents are ultimately accepted by molecular oxygen. The redox potential of NADH+H^+/NAD is -320 mV, thus the reducing equivalent travels to increasing positivity and reaches the ultimate acceptor whose redox potential is +820 mV. The standard free energy change ($\Delta G^{o'}$) of this reaction is related to the difference in the standard redox potential of the oxidizing and reducing couples (Eo). It is expressed by a simple equation

$$\Delta G'_o = -n \times F \times \Delta E'o$$

$$= -n \times 96.5 \times \Delta E'o \text{ (KJ/mol)}$$

Where n is the number of electrons transferred and F is the Faraday constant. In the respiratory chain there is a yield of -219KJ mol-1 amount of free energy.

Role of ATP

ATP is a shared intermediate between the catabolic and anabolic pathways catalyzed by enzyme. In 1929, Fiske and Subbarow isolated ATP from extracts of mammalian muscles. It plays an important role in energy cycle. ATP is the chemical form of free energy stored in cells. It is transported to the site wherever energy is required. ATP is a derivative of AMP to which two phosphate moiety are attached by an anhydride linkage. These bonds are indicated by ~ (read ~ as squiggle). The high energy bond should not be confused with the term bond energy. The latter is defined as the energy required to break a bond between two atoms. The terminal phosphate group of ATP is transferred to a variety of acceptors, which are thereby activated for further chemical transformation.

At pH 7, three of the four hydroxyl groups in the triphosphate side chain are fully dissociated. Under physiological conditions, the molecule is strongly negatively charged (ATP^{4-}). It readily binds with divalent cations. The energy released during ATP hydrolysis ($\Delta G^{o'}$ for ATP at 298 K) is -31 KJ mol^{-1}. The synthesis of ATP requires a similar amount of energy.

Phosphorylation

Synthesis of ATP from ADP and inorganic phosphate is known as phosphorylation. In metabolism, phosphorylation takes place by two mechanisms : substrate level phosphorylation and electron transport.

In *substrate level phosphorylation,* an intermediated transfers a high energy phosphate moiety to ADP. The following biochemical reaction of glycolysis is an example of substrate level phosphorylation

$$\text{2-phosphoglyceric acid} \xrightarrow{-H_2O} \text{phosphoenol pyruvic acid} \xrightarrow{+ADP} \text{Pyruvic acid + ATP.}$$

Removal of water results in changing low-energy ester linkage of phosphate in 2-phosphoglyceric acid to high-energy linked phosphate. This high-energy linked phosphate can then be transferred to ADP and ATP is synthesized.

Respiration and Photosynthesis are basic metabolic process in which electron transfer takes place. In course of electron transport chain, phosphorylation takes place. In photosynthesis phosphorylation takes place in presence of light, hence it is known as photophosphorylation, while in respiration, ATP is generated by the electron transport system where presence of oxygen is essential, it is known as oxidative phosphorylation.

When ATP formation is driven by energy that is released from electrons removed during substrate oxidation, the process is called oxidative phosphorylation. The importance of proton movements in the formation of ATP was first proposed in 1961 by Peter Mitchell. It has been observed in intact mitochondria that when substrates are oxidized via an NAD-linked dehydrogenase, approximately 3 molecules of inorganic phosphate are incorporated into 3 molecules of ADP to form 3 molecules of ATP per + molecule of O_2 consumed. The P:O ratio = 3 and if the substrate is oxidized via a flavoprotein-linked dehydrogenase, only two molecules of ATP are formed and P:O=2.

Britton Chance and coworkers worked on the specific sequence of carriers that constitute the electron-transport chain, using a variety of inhibitors.

The hypothesis are proposed to explain the mechanism of phosphorylation. They are chemical hypothesis, chemiosmotic and conformational hypothesis. Chemiosmotic hypothesis is mostly accepted.

Chemical hypotheses Slates (1953) postulated direct chemical coupling at all stages of the process. The hypothesis is based on the principle of substrate level phosphorylation, in which oxidation-reduction leads to ATP synthesis via the sequential formation of nonphosphorylated intermediates. The chemical hypothesis may be expressed as following :–

$$AH_2 + B + I.OH \longrightarrow A{\sim}I + BH_2 + OH^-$$

$$A{\sim}I + X.H \longrightarrow X{\sim}I + A + H^+$$

$$X{\sim}I + P.OH \longrightarrow X{\sim}P + I.OH$$

$$X{\sim}P + ADP\ \ddot{A} \longrightarrow ATP + X.H$$

$$AH_2 + B + ADP + P.OH \longrightarrow A + BH_2 + ATP + H_2O$$

AH2 and B are the respiratory carriers. X and I are hypothetical acceptors of the products of electron flow. I and X are coupling factor ligands of unknown identity but which must be capable of forming anhydride bond with high free energy of hydrolysis. X~I is envisaged as the energized intermediate which is common to all coupling sites. Uncoupling results from a nucleophilic attack by the uncoupler anion (ϕ-) on X~I, causing hydrolysis of the latter with the liberation of X.H, I.OH and heat. This theory may be verified only if A~I, X~I, and X~P are isolated and identified.

Chemiosmotic theory Mitchel proposed this theory in 1961. According to this theory, in respiratory chain oxidation of components is coupled to the translocation of hydrogen ions (proton's, H^+) from the inside to the outside of the inner mitochondrial membrane. This electron transfer produces an electrochemical potential difference of protons across the coupling membrane. This potential was subsequently used to drive the dehydration of ADP and inorganic phosphate. Later on Mitchel and Jennifer Moyle expanded this concept and explained it in four main postulates:-

1. An anisotropic (direction oriented), proton translocating redox systems.
2. An ion impermeable coupling membrane.
3. Specific exchange-diffusion systems and
4. An anisotropic, proton-translocating, reversible ATPase.

Respiration is accompanied by the outward translocation of protons across the coupling membrane. The proton ejection results from the arrangement of the respiratory system into a series of oxidation-reduction loops. There may be up to four proton-translocating respiratory loops. It catalyses the outward translocation of two protons per pair of reducing equivalents transferred. The proton-translocating develops a transmembrane pmf. The magnitude of this pmf should be sufficient to drive the synthesis of ATP via the proton-translocating ATPase. The coupling membranes (of mitochondria, chloroplast or bacterial system) have low conductance nature. It is because of presence of central non-aqueous osmotic barrier. This was named as M-phase. To generate transmembrane pmf (210mV, required for ATP synthesis) in the absence of a significant membrane potential the pH differential across the coupling membrane would be approximately 3.5 units. To maintain such a differential, pH of the matrix may be highly alkaline and unsuitable for enzyme activity. Thus, a part of the pmf is composed of a membrane potential which is produced through the operation or specific exchange-diffusion system. These exchange-diffusion systems may be like electroneutral cation-proton antiports, or anion-proton symports. According to chemiosmotic hypothesis the rate of respiration is controlled by the effective proton conductance of the coupling membrane. F1 particles are present on the inner membrane facing towards the matrix of mitochondria. These are proteinaceous phosphorylating bodies and contain ATP synthase. Here synthase is used as these enzymes catalyze condensation reactions in which no nucleoside triphosphate is required as an energy source. Protons pass through the electrochemical potential difference is used to drive a membrane-located ATP synthase. It forms ATP with ADP and Pi. In the controlled state the inward flow of protons via the ATP synthase is restricted by the lack of ADP. Since the permeiability of the membrane to protons is low, a very high proton-motive-force (pmf) is generated. Movement of ions across the membrance is same as that of chemical hypothesis, except that the initial expulsion of protons results directly from the activity of the proton-translocating ATPase.

In photosynthetic systems, electron transfer is accomplished by the inward translocation of protons via at least one loop. F_1 is located on the outside surface of these membranes. The mechanism is discussed in chapter Photosynthesis.

To generate transmembrane pmf during photosynthesis induced pH differential (acid inside) appears to be the more important component in photophosphorylation. In chloroplasts, any membrane potential is dispersed by the electrogenic entry of chloride ion. An ammonium ions uncouples photophosphorylation in chloroplasts. However, it has no effect on oxidative phosphorylation. This fact reflects on the different polarities of these coupling membranes. Thus ammonium ions only uncouple when added to the F1 side of the membrane (i.e., alkaline side).

Conformational hypothesis Boyer in 1964, proposed this mechanochemical hypothesis. It is a variation of chemical hypothesis. According to this hypothesis, energy released by respiration is conserved initially bringing about conformational changes in certain protein components of the coupling membrane and subsequently as a non-phosphorylated covalent compound. The non-phosphorylated covalent compounds are possibly a thioester. Now this theory has made some changes dropping the idea of covalent and phosphorylated intermediates. When the prosthetic group of conjugated protein is altered under oxidation-reduction state, the apoprotein undergoes conformational changes. Conformational hypothesis suggests the redox-induced, such conformational changes occur within respiratory chain complexes I, III and IV. Thus, conformational changes in complexes I, III and IV constitute the primary energy conserving event. The complex, multi-chain nature of F1 explains the mechanism by which electron transfer-induced conformational changes are transmitted to ATPase resulting in phosphorylation. F_1 undergoes conformational changes when the coupling membrane is energized.

Conformational hypothesis also explains mechanism of ATP synthesis in chloroplasts. The CF_1 particles undergoes quite dramatic conformational changes after energization of the thyalkoid membrane.

SUMMARY

Metabolism is the sum of all the chemical transformations that occur in a cell or organism. It involves many multienzyme systems. Energy is defined as the capacity to do work, that is the capacity to change or move something. Thermodynamics is the study of the change in energy that accompanies events in the universe. The First Law of Thermodynamics is the law of conservation of energy. The Second Law of Thermodynamics expresses the concept that events in the universe have direction. The energy that can do some work is known as Free energy. The standard free-energy difference ($\Delta G^{o\prime}$) describes the difference in free energy when one mole of each reactant is converted to one mole of each product under standard conditions. Electron transport system (ETS) is biologically significant process. The components of ETS are pyridine linked dehydrogenase, Iron sulphur proteins, Ubiquinones, and cytochrome. Translocation of protons across the inner membrane of mitochondria is electrogenic. The energy of the proton gradient is used in the mitochondria for ATP synthesis by an F-ATP-synthase. All the reactions in respiratory chain are oxidation reduction reactions. ATP is the chemical form of free energy stored in cells. Synthesis of ATP from ADP and inorganic phosphate is known as phosphorylation. ATP is transported to the site where ever required.

Chapter - 3

CELL BIOCHEMISTRY

The study of biochemistry includes description of an organism (cell) in terms of chemical structures, mechanisms, and interaction processes amongst the biomolecules. A cell contains a number of molecules. These molecules conform to all the familiar laws of chemistry. The macromoleculs are composed of a few simple compounds. The monosaccharides, amino acids, and mononucleotides are the monomeric subunits of polysaccharides, proteins, and nucleic acid. The simple molecules are generally composed of carbon atoms covalently linked with other carbon atoms and with hydrogen, oxygen or nitrogen.

COVALENT BOND

The atoms that make up a molecule are joined together by covalent bonds in which pairs of electrons are shared between the atoms. The formation of a covalent bond between two atoms is governed by the fundamental principle that an atom is most stable when its outermost electron orbit is filled. Consequently, the number of bonds an atom can form depends on the number of electrons needed to fill its outer orbit. The formation of a covalent bond is accompanied by the release of energy, which must be reabsorbed at some later time if the bond is to be broken. In many cases, two atoms can joined by bonds in which more than one pair of electrons is shared. The covalent bonds may be double or triple, however, quadruple bonds do not occur. The type of bond between atoms has important consequences in determining the shapes of molecules. When the same types of atoms unite with one another, as in H_2, the electron pairs of the outer orbit are equally shared between the two partner atoms. However, when two unlike atoms are covalently bonded, it is inevitable that the positively charged nucleus of one atom exerts a greater attractive force on the outer electrons than does its bonded atom. Consequently, the shared electrons tend to be located more closely to the atom with the greater attractive force, that is the more electronegative atom. The electronegativity of an atom is dependent on two factors : (1) the number of positive charges in its nucleus (the more protons, the more electronegative) and (2) the distance of the outer electrons from the nucleus (the greater the distance, the less electronegative). Among the atoms most commonly present in biological molecules. Nitrogen and oxygen are strongly electronegative.

POLAR AND NONPOLAR MOLECULES

Molecules, such as water, that have an asymmetric distribution of charge are said to be polar molecules. Polar molecules of biological importance contain one or more electronegative atoms, usually O, N, S, or P. Molecules that lack polarized bonds, such as those that consist almost entirely of carbon and hydrogen atoms, are said to be nonpolar. The presence of polarized bonds is of the utmost importance in determining the reactivity of molecules. Molecules that lack electronegative atoms, such as waxes and fats, tend to be relatively inert.

NONCOVALENT BONDS

Interactions between molecules (or between different parts of a large biological molecule) are governed by a variety of weaker linkages called noncovalent bonds. Noncovalent bonds do not depend on shared electrons but rather on attractive forces between positively and negatively charged regions within the same molecule or between two nearby molecules.

IONIC BONDS

A crystal of table salt is held together by an electrostatic attraction between positively charged Na^+ and negatively charged Cl^- ions. This type of attraction between fully charged components is called an ionic bond.

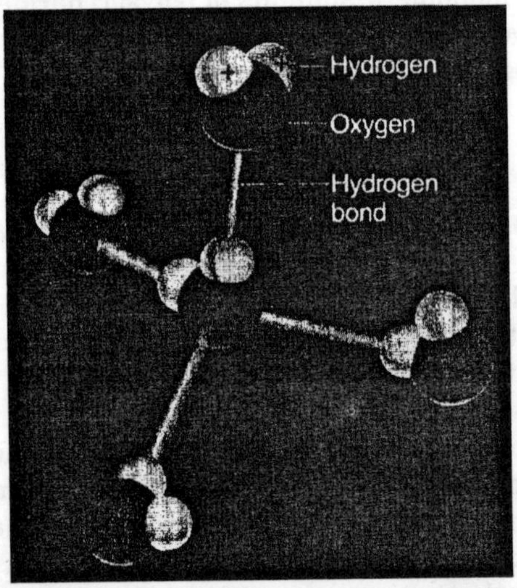

Fig. 3.1. Hydrogen bound formation between neighboring water molecules

HYDROGEN BOND

When a hydrogen atom is covalently bonded to an electronegative atom, particularly an oxygen or nitrogen atom, the single pair of shared electrons is greatly displaced toward the nucleus of the electronegative atom, leaving the hydrogen atom with a partial positive charge (Fig. 3.1). As a result, the bare, positively charged nucleus of the hydrogen atom is able to approach closely enough to form an attractive interaction with an unshared pair of outer electrons of a second electronegative atom. This weak attractive interaction is called a hydrogen bond. Hydrogen bond occurs between most polar molecules (Fig. 3.2). Hydrogen bonds also form between polar groups present in large biological molecules, as occurs between the two strands of a DNA molecule.

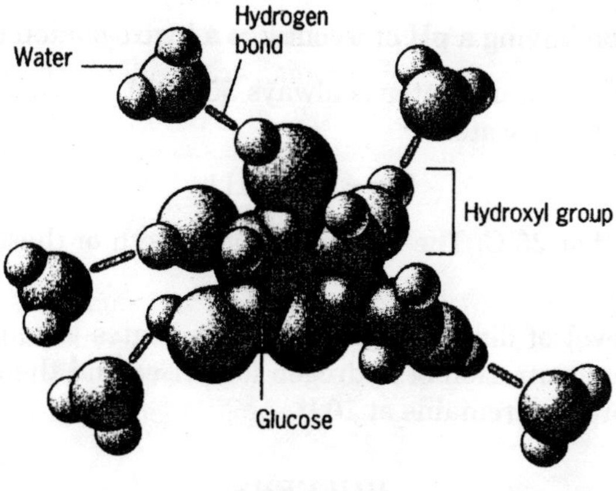

Fig. 3.2. Schematic view of the types of hydrogen bounds

HYDROPHOBIC INTGERACTIONS AND van der Waals FORCES

Because of their ability to interact with water, polar molecules, such as sugars and amino acids to be described shortly, are said to be hydrophilic, or "water loving". Non polar molecules, such as steroids or fats, are essentially insoluble in water because they lack the charged regions that would attract them to the poles of water molecules. When nonpolar compounds are mixed with water, the nonpolar, hydrophobic substances are forced into an aggregate, which minimizes their exposure to the polar surroundings. This association of nonpolar molecules is called hydrophobic interaction.

If two molecules with transitory dipoles are very close to one another and oriented in the appropriate manner, they experience a weak attractive force, called van der Waals force, which can serve to bond them together.

THE LIFE SUPPORTING PROPERTIES OF WATER

Water is a highly asymmetric molecule with the O atom on one side and the two H atoms on the opposite side.

Each of the two covalent bonds in the molecule is highly polarized.

All three atoms in a water molecule are adept at forming hydrogen bonds.

The life-supporting properties of water stem from these properties.

ACIDS, BASES AND BUFFERS

A molecule that is capable of releasing (donating) a hydrogen ion is termed as an acid, any molecule that is capable of accepting a hydrogen ion is defined as a base.

The acidity of a solution is measured by the concentration of hydrogen ion and is expressed in terms of pH.

$$pH = -\log [H^+]$$

For example, a solution having a pH of 5 contains a hydrogen ion concentration of 10^{-5}M.

Since the concentration of pure water is always 55.51 M, we can generate a new constant, Kw, the *ion-product constant* for water.

$$Kw = [H^+][OH^-]$$

Which is equal to 10^{-14} at 25°C. The concentration of both of these species in pure water is approximately 10^{-7} M.

The extremely low level of dissociation of water indicates that it is a very weak acid. In presence of an acid, the concentration of hydrogen ions rises and the concentration of hydroxyl ions drop so that the ion product remains at 10^{-14}.

BUFFERS

Compounds that react with free hydrogen or hydroxyl ions, thereby resisting changes in pH. Buffer solutions usually contain a weak acid together with its conjugate base.

FUNCTIONAL GROUPS

Functional groups are particular grouping of atoms that often behave as a unit and give organic molecules their physical properties, chemical reactivity, and solubility in aqueous solution. Two ester bonds, which form between carboxylic acids and alcohols, and amide bonds which form between carboxylic acids and amines (Fig. 3.3).

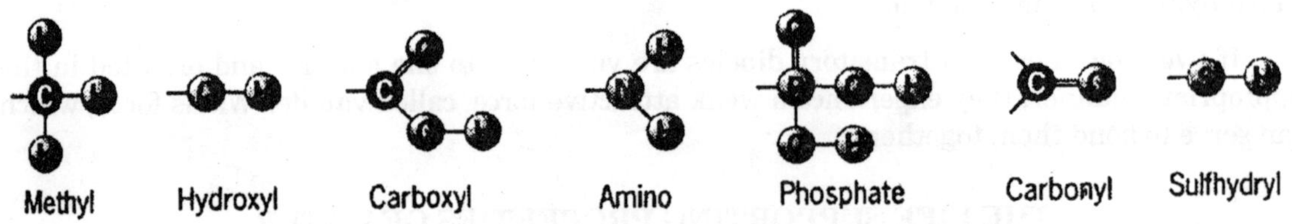

Fig. 3.3. Different functional group

The molecules that form the structure and carry out the activities of cells are huge, highly organized molecules called macromolecules, which contain anywhere from dozens to millions of carbon atoms. Macromolecules can be divided into four major categories: proteins, nucleic acids, polysaccharides, and lipids. The first three types are polymers composed of a large number of low-molecular-weight building blocks, or monomers.

BIOCHEMISTRY OF CARBOHYDRATE

Carbohydrates are defined as polyhydric alcohols of carbon possessing active aldehydic or ketone groups. Carbohydrates functions primarily as stores of chemical energy and as durable building materials for biological construction. Most sugars have the general formula $(CH_2O)n$. The values of n in sugars of important in cellular metabolism range from 3 to 7. Carbohydrates are classified into three categories Monosaccharide, Oligosaccharide and Polysaccharide (the word "saccharide" is derived from the Greek *sakkharon* meaning "sugar"). The monosaccharides cannot be hydrolysed into simple sugars. Oligosaccharides (Oligo-few) on hydrolysis yield three to six monosaccharides. Polysaccharides are sugars with many monosaccharide monomers. The subunit monomers are joined by glycosidic linkage in a long linear or branched chain.

Each sugar molecule consists of a backbone of carbon atoms linked together in a linear array by single bonds, (the hydroxyl group) except for one that bears a carbonyl (C=O) group. If the carbonyl group is located at an internal position (to form a ketone group) the sugar is a ketose, such as fructose. If the carbonyl is located at one end of the sugar, it forms an aldehyde group and the molecule is known as an aldose. The ring forms of sugars are usually depicted as flat (planar) structures lying perpendicular to the plane of the paper with the thickened line situated closest to the reader. The H and OH groups lie in the plane of the paper projecting either above or below the ring of the sugar. In actual fact, the sugar ring is not a planar structure but usually exists in a three-dimensional conformation resembling a chair. On paper, the three-dimensional structure of sugars is represented by Fischer-projection formulas.

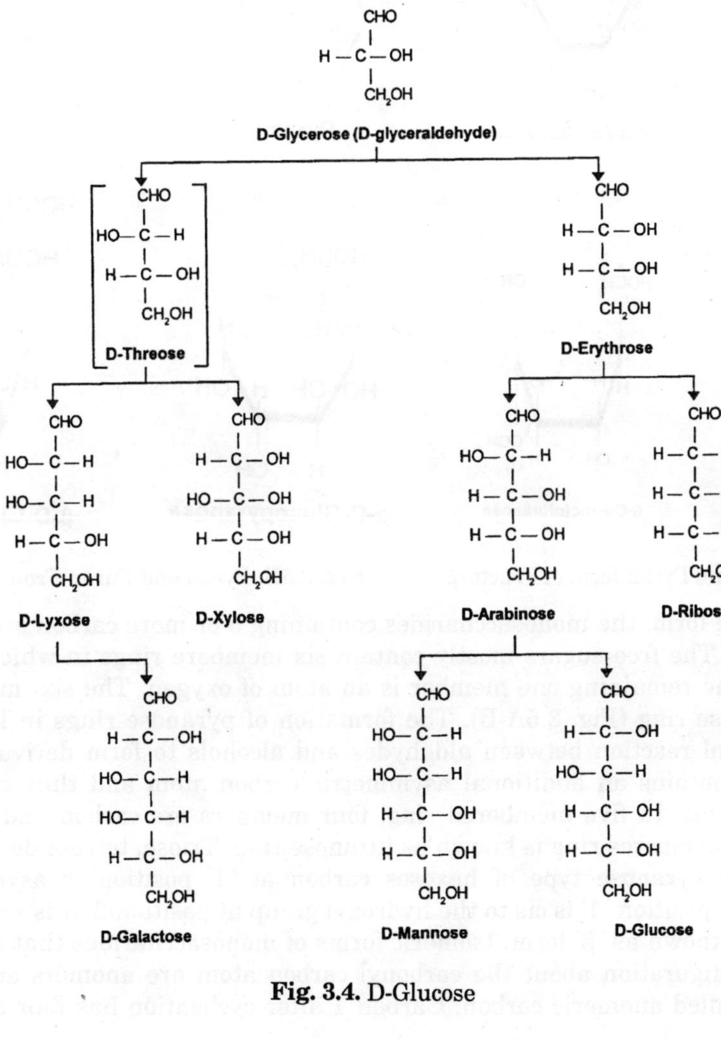

Fig. 3.4. D-Glucose

MONOSACCHARIDE

The monosaccharides are also called as simple sugars and are expressed by empirical formula $(CH_2O)n$. Monosaccharides with four, five, six, and seven carbon atoms in their backbones are called triose, tetrose, pentose, hexose and heptoses, respectively (Fig. 3.4). They contain one asymmetric carbon atoms and thus are Chiral molecules. Suffix "ose" denotes that the particular compound belongs to aldo group of carbohydrate while suffix "ulose" denotes keto sugars.

They can be subdivided into depending upon the number of carbon atom they possess, as trioses, tetroses, pentoses, hexoses etc. and depending upon whether aldehyde (-CHO) or ketone (-CO) groups are present as aldoses or ketoses. e.g., Erythrose ($C_4H_8O_2$ tetroses) Ribose ($C_5H_{10}O_5$ - Pentoses) Glucose ($C_6H_{12}O_6$ - Hexoses).

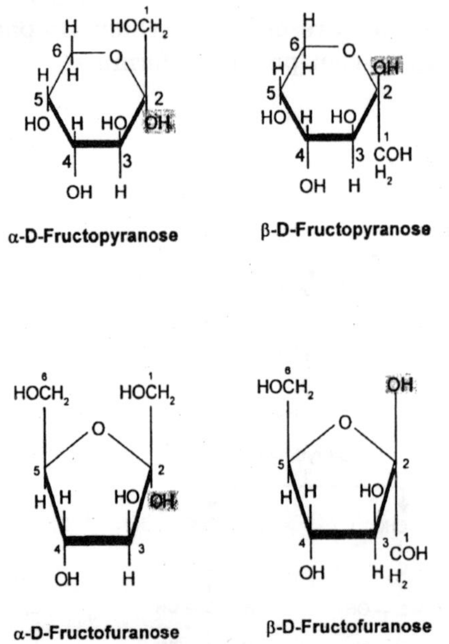

Fig. 3.5A. Furan and Pyran form of Fructose

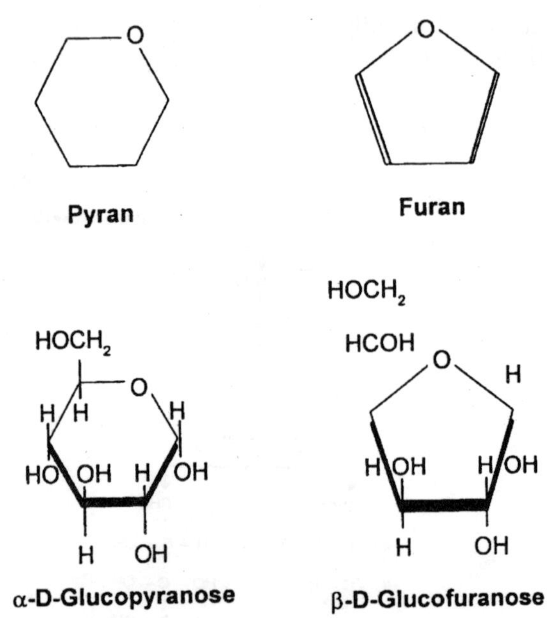

Fig. 3.5B. Pyran and Furan Froms of Glucose

In crystalline form, the monosaccharides containing 5 or more carbon atoms exist in the tautomeric rings. The free sugars mostly contain six membere rings in which five members are carbon and the remaining one member is an atom of oxygen. The six- memberes ring is known as pyranose ring (Fig. 3.5A-B). The formation of pyranose rings in D-glucose is the result of a general reaction between aldehydes and alcohols to form derivatives known as hemiacetals. It contains an additional asymmetric carbon atom and thus can exist in two stereoisomeric forms. In five membered ring, four members are carbon and one member is oxygen. The five memberes ring is known as furanose ring. Triose, tetrose do not possess ring structure. In the pyranose type of hexoses carbon at `1' position is asymmetric. If the hydroxyl group at position `1' is cis to the hydroxyl group at position 2, it is known as `α' form and if trans it is known as `β' form. Isomeric forms of monosaccharides that differ from each other only in configuration about the carbonyl carbon atom are anomers and the carbonyl carbon atom is called anomeric carbon. Carbon 1 after cyclization has four different groups attached to it and thus it becomes now "asymmetric". The two cyclic compound have different optical rotations, but they will not be same because the compounds as a whole are not mirror images of each other. Compounds related in this way are called "anomers" and carbon 1 after cyclization becomes asymmetric is called now "anomeric carbon atom".

The five member ring is known as furanose ring. In five member ring, four members are carbon and one member is oxygen. Ketohexose also occur in α and β anomeric forms. In furanose ring, the hydroxyl group on C-5 (or C-6) reacts with the keto group at C-2. A *hemiketal* is an analogous product formed by reaction of a ketone with an alcohol.

Monosaccharides can be oxidized by relatively mild oxidizing agents such as ferric (Fe^{3+}) or cupric (Cu^{2+}) ions. Sugars with such property are known as reducing sugars. The carbonyl carbon is oxidized to a carboxylic acid. D-glucose in solution mostly exists in the pyranose form and it remains in equilibrium with a small amount of open chain form. The open chain form is capable of reducing various reagents such as Fehling's solution, Benedicts reagent etc.

When an aldohexoses is first dissolved in water and the solution is put in optical path so that plane polarized light is passed the initial optical rotation shown by the sugar gradualy changes until a constant fixed rotation characteristic of the sugar is reached. This phenomenon of change of rotation is called as *"mutarotation"*. If D-glycopyranose is dissolved in water, it first shows rotation of +112 but this value gradually falls to +52. This change of rotation is known as mutarotation and this is due to establishment of equilibrium between α and β forms of the ring structures. Thus, a solution of α-D-glucose and a solution of β-D-glucose finally form identical equilibrium mixtures having identical optical properties. Two sugars differing only in the configuration around one specific carbon atom are called *epimers* of each other.

Haworth Formula is commonly used to show the ring forms of monosaccharides. The pyranose ring exists in two conformations, the chair form and the boat conformations. The chair form is much more stable and relatively rigid. The edges of the ring nearest to the reader are represented in bold lines. The Haworth projections suggest that the five and six member furanose and pyranose ring are planar, it is somewhat misleading and is not a fact. In aqueous solutions of hexose chair form is found predominating. The substituent group in the chair form is not geometrically and chemically equivalent.

The presence of asymmetric carbon atoms in a compound gives rise to the formation of isomers of that compound. Such compounds which are identical in composition and differ only in spatial configuration are called *stereo-isomers*, and the phenomenon is known as *Stereoisomerism. Vant Hoff's Rule `n'* According to the Vant Hoff's rules of `n'; 2 equals the possible isomers of that compound where n represents the number of asymmetric carbon atom in a compound. The number of possible isomers of any given compound depends upon the number of asymmetric carbon atoms the molecule possesses. In general, a molecule with n chiral centers can have 2n stereoisomers. Glyceraldehyde has 2^1. Hence, the number of stereoisomers will be 2. The aldohexoses, with four chiral centers, have 2^4, so the stereoisomers will be 16. *D-series and L-series* the orientation of the H and OH groups around the carbon atom first adjacent to the terminal primary alcohol carbon, e.g. C-atom 5 in glucose determine the series (Fig. 3.6). When the -OH groups on this carbon is on the right, it belongs to D-series, when the OH group on this carbon is on the left it is a member of the L series. The sugars with the same configuration at the reference carbon are D-isomers or L isomers.

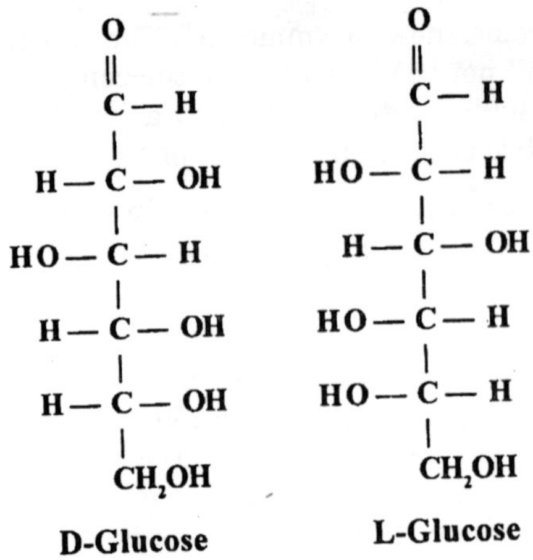

Fig. 3.6. D and L Glucose

Presence of asymmetric carbon atom also confers *optical activity* on the compound when a beam of plane polarized light is passed through a solution exhibiting optical activity, it will be rotated to the right or left in accordance with the type of compound, i.e. the "*optical isomers*" or "*enantiomorphs*"; when it is rotated to right, the compound is called "*dextrorotatory*" (d or + sign), when rotated to left, the compound is called "*Laevorotatory*" (l or -sign). A racemic mixture contains equal amounts of dextrorotatory and laevorotatory isomers. Thus, there is no optical activity of each isomer.

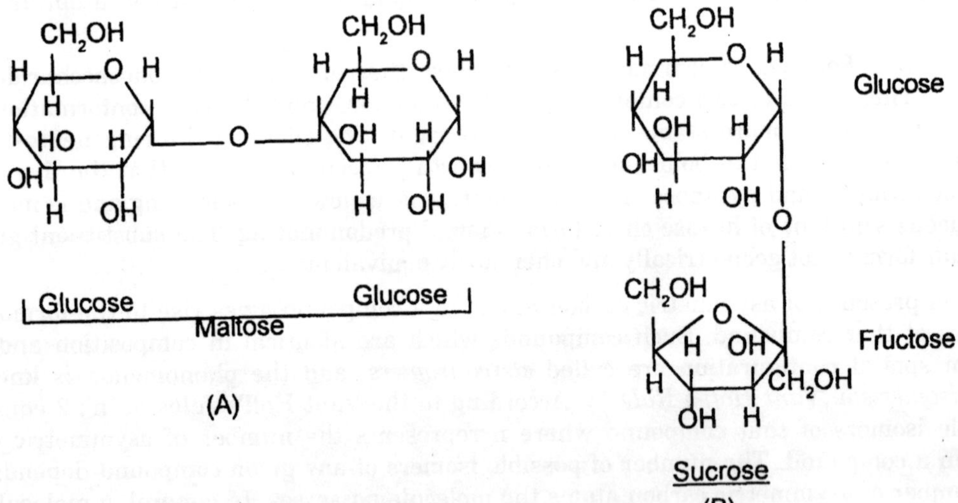

Fig. 3.7. Disaccharides (A) Maltose and (B) Sucrose

DISACCHARIDES

The Disaccharides sugars are composed of two molecules of the same or different molecules of monosaccharides (Fig. 3.7). The disaccharides are expressed by general formula $C_n(H_2O)_{n-1}$. Maltose, sucrose, lactose are disaccharides in which two monosaccharides are joined covalently by an O-glycosidic bond. On hydrolysis Maltose yields two molecules of glucose, sucrose yields one molecule of fructose and one molecule of glucose, while lactose yields one molecule of glucose and one molecule of galactose. This bond is formed when a hydroxyl group on one sugar reacts with the anomeric carbon of the other. The disaccharides are non reducing sugars. A disaccharide can be described with respect to both linkage and anomeric configuration. There are certain rules of nomenclature of disaccharides. First, the compound is written with its nonreducing end to the left. An O precedes the name of the first (left) monosaccharide unit. The configuration at the anomeric carbon (α or β) joining the first (left) monosaccharide unit to the second is mentioned. Terms `furanosyl' or `pyranosyl' is used, respectively to mention five or six membered ring structures. The two carbon atoms joined by the glycosidic bond are then indicated in parentheses, with arrow connecting the two numbers. As per these rules, the name of sucrose is O-α-D- glucopyranosyl-(1$\rightarrow$2)-β-D-fructofuranoside [Glc(α1 $\rightarrow$ 2)Fru].

POLYSACCHARIDES (Glycans)

Polysaccharide yields more than six molecules of monosaccharides on hydrolysis. They have high molecular weight with general formula of polysaccharide as $(C_6H_{10}O_5)_n$. Most of the carbohydrates found in nature occur as polysaccharides. Polysaccharides are further divided into two groups *Homopolysaccharides* (homoglycans):- Polymer of same monosaccharide units. Example - starch, cellulose, dextrins. *Heteropolysaccharides* (heteroglycans):- Polymer of different monosaccharide units or their derivatives. Example - mucopolysaccharides (glycosaminoglycans)

STARCH is a common storage polysaccharide of plant cells. Starch contains two types of glucose polymers, amylose and amylopectin (Fig. 3.8). Amylose consists of long unbranched chains in which all the D-glucose units are bound in (α1 $\rightarrow$4) linkage Amylose is rather insoluble in water and forms hydrated micelles. Amylopectin is highly branched. There are both (α1$\rightarrow$ 4), the backbone glycosidic linkage and ($\alpha \rightarrow$ 6) the branch points linkages.

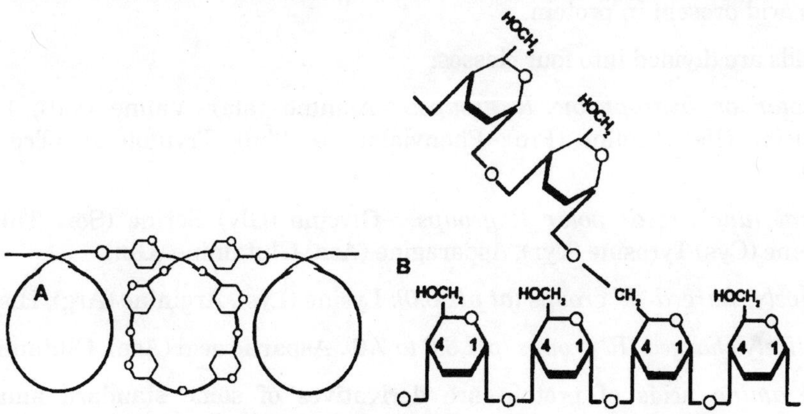

Fig. 3.8. Starch A - Amylose and B - Amylopectin

Cellulose is a fibrous, tough, water-insoluble substance, and a basic structural polysaccharide of cell wall of plant cell. It is a homopolysaccharide with 10,000 to 15,000 D-glucose monomer units. The monomers i.e., D-glucose is linked by (β1$\rightarrow$4) linkage to form the linear polymer. The polysaccharide chains are arranged parallel to each other. The reducing ends of the chains point in the same direction. Partial hydrolysis yields disaccharide cellobiose while complete hydrolysis yields D- glucose. Cellulose is hydrolyzed by enzyme cellulase. Callose differs from cellulose in consisting of (1→3) β-D-glucan chains. These chains form helical duplexes and triplexes.

Plant cell wall is made up of eleven different sugars, 4 different linkage positions, and 2 configurations with respect to the oxygen atom. The cell wall polymer is made up of densely packed cellulose fibrils present in regular parallel arrays, often in cris-cross layers. Cross-linking glycans interlock the cellulosic scaffold. Cross-linking glycans is able to form hydrogen-bond to cellulose microfibrils. Xyloglucans (XyGs) and glucuronoarabinoxylans (GAXs) are major cross-linking glycans of all primary cell walls of flowering plants. These fibrils are cemented together by a matrix of three other polymeric materials, hemicellulose, pectin and extensin. *Hemicellulose* is a polymer of D-xylans, polymers of D-Xylose in β(1→4) linkage with side chains of arabinose and other sugars.

Pectin is a mixture of heterogeneous, branched, and highly hydrated polysaccharides. Pectin is made up of two fundamental constituents viz. homogalacturonan (HGA) and rhamnogalacturonan I (RG I). HGAs are homopolymers of (1→4)α-D-GalA. There are two kinds of structurally modified HGAs, xylogalacturonan and rhamnogalacturonan II. RG I is a rod-like heteropolymer of repeating (1→2) α-L-Rha-(1→4) α-D-GalA disaccharide units.

Extensin is a complex glycoprotein attached covently to the cellulose fibrils. L-rhamnose (6-deoxy-1-mannose) and L-fucose (6-deoxy-L-galactose) are the deoxysugars of cell wall. These deoxysugars are formed by the dehydration of carbon 6 of some hexapyranoses.

BIOCHEMISTRY OF AMINO ACID

Amino acids are monomers of proteins. In 1806, Asparagine was the first amino acid discovered from protein. H. Braconnot (1820) isolated first amino acid Glycine from a protein hydrolyzate of gelatin. In 1935 W.C. Rose isolated Threonine from hydrolyzates of fibrin. It is the 20th amino acid present in protein.

Amino acids are divided into four classes:

1. *Nonpolar or hydrophobic R groups:* Alanine (ala); Valine (val); Leucine (leu); Isoleucine (Ile); Proline (Pro); Phenylalanine (Phe); Tryptophan (Trp); Methionine (Met).
2. *Neutral (uncharged) polar R groups :* Glycine (Gly) Serine (Ser) Threonine (Thr) Cysteine (Cys) Tyrosine (Tyr), Asparagine (Asn) Glutamine (Gln).
3. *Positively charged `R' groups (at pH 6.0):* Lysine (Lys), Arginine (Arg), Histidine (His).
4. *Negatively charged `R' groups (pH 6.0 to 7.0):* Aspartic acid (Asp), Glutamic acid (Glu)

The *rare amino* acids of protein are derivatives of some standard amino acid *e.g.*, 4-hydroxyproline, a derivative of proline found in some abundance in the fibrous protein collagen. Hydroxlysine, the 5-hydroxy derivative of lysine, Sesmosine and isodesmosine occur in fibrous protein elastin. The unusual amino acids N-methyl-lysine, ε-N-trimethylysine and methylhistidine are methyl derivatives of standard amino acids. There are no triplet code words for unusual amino acids. They arise by enzymatic modification after their parent amino acids have already been inserted into the polypeptide chain.

Over 300 amino acids are known to occur biologically in free or combined form but never in proteins. Some non protein amino acids are important precursors or intermediates in metabolism.

The carbon atom with four different substituents is known as asymmetric carbon. The compounds with such asymmetric carbon show *stereoisomerisms* (chirality, handedness). In this phenomenon the compounds have optical activity, and are capable of existing in two forms, that are not super imposable mirror images of each other. They exist as dextorotatory (d) or laevorotatory (l) isomers (Fig. 3.9).

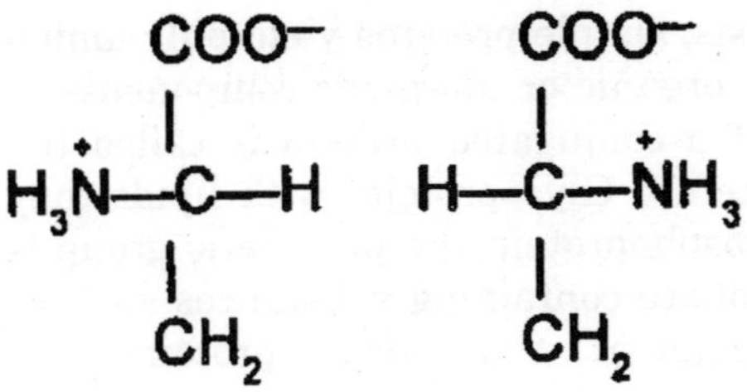

L - Alanine D - Alanine

Fig. 3.9. Stereoisomerism in amino acid (Alanine)

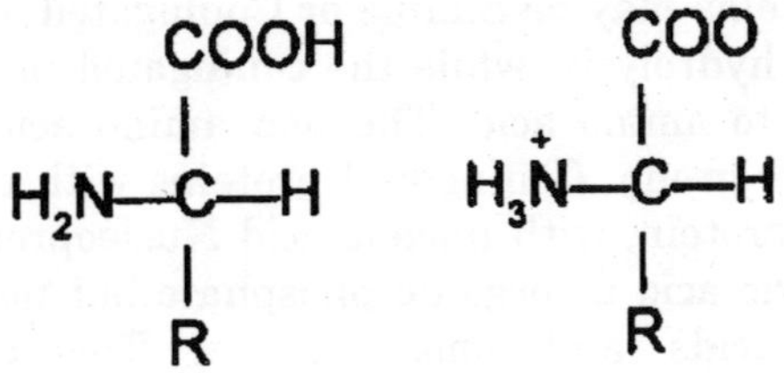

Nonionic form Zwitterionic form

Fig. 3.10. Nonionic and Zwitterionic forms of Amino Acid

Crystalline amino acids have relatively high melting point usually above 200°C. They show affinity towards water and are easily soluble in water than in nonpolar solvents. In crystalline form and in neutral aqueous solutions amino acids occur as dipolar ions or zwitterions (Fig. 3.10), rather than as undissociated molecules. The crystalline zwitter ionic amino acid when dissolved in water shows both acid and base properties. Hence it behaves as amphoteric ions and is called ampholytes. At certain pH amino acids have no net electric charge. This pH is known as isoelectric point. The molecules have no movement towards electric field. Isoelectric pH is symbolized as pHi. The α-carboxyl group of mono amino monocarboxylic acids is a stronger acid than the carboxylic group of comparable aliphatic acids such as acetic acid. While α -amino group of monoamino monocarboxylic acids is a weaker base than the amino group of comparable aliphatic amines.

Amino acids forms esters with alcohol, on decarboxylation they produce amines and amides. These chemical properties are granted to amino acids due to the presence of carboxylic acid residue in them.

Amino acids also have an amino group, due to the presence of this group, they form salts with mineral acid. The acyl derivatives are formed in reaction with acyl anhydride or acyl halides. The oxidation of amino group yields imino acid, which reacts with water to form $-NH_2$ and α-keto acid. Formaldehyde reacts with the -NH2 group to form a methylene compound. Amino acids reacts with Ninhydrin, Millon's test, Sakaguchi test, Hopkin Cole's test amino acids give specific colours.

The presence of both $-NH_2$ and COOH groups in amino acid form chelated, coordination complex with certain heavy metals and other ions. (Cu^{++}, Co^{++}, Mn^{++} and Ca^{++}).

BIOCHEMISTRY OF PROTEINS

Proteins are highly important biomolecules. The proteins contain carbon, hydrogen, nitrogen and oxygen, a few contain sulphur. The acid hydrolysis of proteins yields α amino acid residues which are covalently linked (by peptide bond) to form very long, unbranched chains. There are thousands of different kinds of proteins in a typical cell. They perform different important biological functions. The most significant function is that they are enzymes. Besides, this they are transport proteins, nutrient and storage proteins, contractile or motile proteins, structural proteins, regulatory proteins.

Proteins may be Simple or Conjugated. On hydrolysis, simple proteins yield only amino acids on hydrolysis, while the conjugated proteins yield organic or inorganic components in addition to amino acid. The non amino acid portion of a conjugated protein is called the prosthetic group. Conjugated proteins with carbohydrate are Glycoprotein, with lipids they are Lipoprotein, with nucleic acid Nucleoprotein. In Phosphoprotein the prosthetic group is phosphoric acid as organic phosphate but not the phosphate containing substances such as nucleic acids and phospholipids. The coloured (pigment containing) proteins are Chromoprotein.

Protein products formed from the simple and conjugated proteins are *Derived proteins*. The denatured or coagulated proteins are Primary Derived proteins. They differ in properties like solubility, precipitation and crystallization. The Secondary Derived proteins (roughly called proteoses, peptones and peptides) are formed by the progressive hydrolysis of proteins at their peptide linkages. They represent a great complexity with respect to their size and amino acid composition.

Protein conformation refers to a structural state that can, without breaking any covalent bonds, interconvert with other structural states. Proteins in their functional conformation are called native proteins. In the native state protein molecule has a characteristic three-dimensional shape. Proteins are either fibrous or globular in conformation. Fibrous proteins have chains usually arranged or coiled along a single dimension (axis) to yield long fibers or sheet. The axial ratio of length: width of a protein molecule is more than 10. Fibrous proteins are physically tough and are insoluble in water or dilute salt solutions. Globular proteins have the polypeptide chain tightly folded into compact spherical or globular shapes. The axial ratio of length : width of a protein molecule is less than 10.

PROTEIN STRUCTURE

Proteins have complicated structure having large molecular weight. The simplest protein consists of straight chains of polypeptides but in majority of the proteins there are folds. The connection of protein is due to certain bridges, coiling of chains, folding of such coils and aggregation of such fold entities. The protein structure can be studied at four levels. These levels are referred as primary, secondary, tertiary and quaternary structures.

The *primary structure* refers to the sequence of peptide bonded amino acids and locations of disulfide bonds. In the straight chains, the left hand side end of the chain bears the amino acid which has a free amino group which is known as N- terminal end, whereas the right hand side end bears an amino acid which has a free carboxyl group which is known as C-terminal end. A protein may have branched polypeptide chains. The branching may be due to disulfide bridges (of sulphur containing amino acids) or may be due to amino acid containing two carboxylic group (glutamic acid and aspartic acid) or may be due to amino acid containing two amino group (lysine, arginine). In certain cyclic proteins the C-terminal and N-terminal do not have free ends. Some of the protein contains more than one chain. These chains may be similar or different in amino acid content and their sequence. If more than one chain is present in a protein, the chains are numbered as alpha (α), Beta (β), gamma (γ) and delta (δ).

In 1951 Linus Pauling and Robert Corey predicted the *secondary structure* of proteins. The secondary structures refers to a three dimensional structure by having folding or coiling consisting of a helically-coiled, zigzag, linear or mixed form. It is the secondary structure of proteins, which gives fibrous or globular conformation. X-ray diffraction studies of crystals of amino acids and of simple dipeptides and tripeptides demonstrated that the amide C-N bond in a peptide is shorter than the C-N bond in a simple amine. There is a resonance or partial sharing of two pairs of electrons between the carbonyl oxygen and the amide nitrogen. The electric dipole is established due to partial positive and negative charges in nitrogen and oxygen, respectively. The oxygen atom of the carbonyl group and the hydrogen atom of the amide nitrogen are trans to each other. Rotation takes place about the N-C α, and C α-C bonds. As a result of rotation, bond angles are formed. The bond angles formed by the rotation of N-Cα, and C$_a$-C bond, are designated as ϕ (phi) and Ψ (psi), respectively. The Φ and Ψ may have any value between -180° to +180°F. *The α-Helix* Astbury demonstrated that the hair and wool protein has a regular structure that repeats every 0.54 nm. The amino acid residues in a α-helix have conformations Φ = -60° and Ψ + -42° to -50°. Each helical turn includes 3.6 amino acids. Pauling and Korey proposed the model of twisting and coiling the backbone of the polypeptide chain along one axis. In the α-helix, the backbone is arranged in a helical coil having about 3.6 amino acid residue per turn. The R groups of the amino acids protrude outward from the rather tight helix formed by the backbone.

A single complete turn of the helix (repeat unit) extends about 0.54 nm (5.4) along the long axis. The helix is stabilized by a hydrogen bond. This bond is formed between the hydrogen atom attached to the electronegative nitrogen atom of each peptide linkage and the electro-negative carbonyl oxygen atom of the fourth amino acid on the amino-terminal side of it in the helix (Fig. 3.11). Intrachain hydrogen bonds are found between successive coils of the helix, parallel to the long axis of the helix. The stability of α-helix is affected by five factors, the electrostatic attraction or repulsion between the amino acids of charged R groups, the bulkiness of adjacent R groups, the interaction between amino acid side chains, the occurrence of Pro residues, and the interaction between amino acids at the ends of the helix. All the polypeptides do not have a stable α-helix. It is stabilized or destabilized due to the additional interactions taking place between amino acid side chains. The bulk and certain R groups can also destabilize the α-helix. It may even prevent the formation of α-helix. In general positively and negatively charged amino acids are found at the distance three (and sometimes four) residues. The interaction between such amino acids gives twist to α-helix. Proline amino acid has nitrogen atom in rigid ring and rotation about the N-Cα, bond becomes impossible. In peptide linkage the nitrogen of this amino acid does not provides substituent hydrogen-to-hydrogen bond with other residues. These properties of proline results in a minor constraint on the formation of the α-helix. The identity of amino acids located near the ends of the α-helix segment of a polypeptide. There is a dipole charges at the ends of the helix which is due to the partial positive and negative charges on the peptide amino and carbonyl groups near the amino-terminal and carboxyl-terminal ends of the helix, respectively. The electric dipole in each peptide bond adds across the hydrogen bonds in the helix. The effect of such dipole is that the net dipole increases as helix length increases. The four amino acids at either end of the helix do not participate fully in the helix hydrogen bonding. *The β-Pleated sheath* Puling and Corey found that the polypeptide of fibroin has a zigzag backbone. Such polypeptide chains are arranged side by side to form a structure resembling a series of pleats. This is transition of α-helix into β-helix and is designated as β - conformation. The linkages or bonds involved are hydrogen bonds and disulphide bonds. The disulphide bridges (-S-S-) are formed between two cysteine residues existing in the different chains or in the same chain. Beside disulphide bonds certain non-covalent bonds such as hydrogen bonds are also formed. Side-by-side polypeptide chains in the β-conformation are cross-linked by interchain or intra chain hydrogen bonds. The adjacent polypeptide chains in a β pleated sheet can be either parallel or antiparallel (Fig. 3.12).

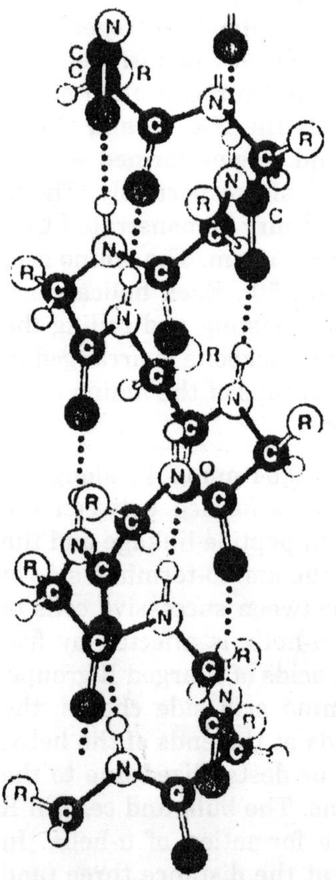

Fig. 3.11. Hydrogen bounds in a polypeptidechain with helica conformation

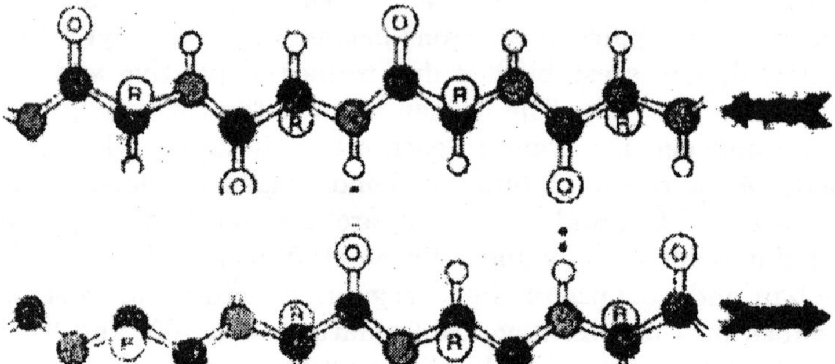

Fig. 3.12. The β-pleated sheet

The *tertiary structure* is three dimensional arrangement of all atoms in a protein. According to Christian Anfinsen amino acid sequence of the polypeptide chain of proteins contains all the information required to fold the chain into its native tertiary structure. Amino acids that are far apart in the polypeptide sequence and reside in different types of secondary structures may interact when the protein is folded. The bonds responsible for interactions between groups of amino acids are hydrophobic interactions, which normally occur between nonpolar side chains of amino acids such as alanine, leucine, methionine, isoleucine and phenylalanine. Hydrogen bonds normally formed by the polar side chains of the amino acids. Ionic or electrostatic interactions are formed between oppositely charged polar side chains of amino acids, such as basic and acidic amino acids. Vander Waal forces occur between non-polar side chains and disulfide bonds. The tertiary level of organization is acquired generally by globular protein. The helices or the twisted structure are further folded to form a globular, ellipsoidal shape.

The evidences from experiments showing that denaturation of some protein are reversible exhibits that the tertiary structure of a globular protein is determined by its amino acid sequence. A classic example is denaturation and renaturation of ribonuclease. The amino acid residues of cytochrome c, from different species explains that at certain positions the amino acid sequence is invariant, whereas at other positions they vary. Many of such invariant amino acid residues of homologous proteins appear to occur at critical points along the polypeptide chain. These amino acids found at or near bonds in the chain, others at cross-linking points between loops in the tertiary structure. The native tertiary conformation of a globular protein is the thermodynamically most stable of form its polypeptide chain. However, this conformation must not be regarded as absolutely rigid. Myoglobin is a small globular protein. It is oxygen-binding protein of muscle cells. In 1950s J.C. Kendrew and his collegues carried out X-ray studies of sperm whale myoglobin. The protein conformation may be described with the backbone of eight relatively straight segment set off by bends. Each straight segment is a length of helix, the longest consisting of 23 aminoacids and the shortest only 7; all are right handed. The positioning of amino acid side chains reflects a structure that derives much of its stability from hydrophobic interactions. The polar R-groups of the amino acid residues are located on the outer surface of the molecule and are hydrated. The nonpolar R groups are in the interior of the molecule. Proline residues occur only at the bends. There are four common tertiary structural patterns. The first type (α/β barrel) is made up of β eight strands arranged in a circle with each strand connected to its neighbour by an α helix. The regions are arranged in the barrel structure.

In the second type there is a four helix bundle. The helices are slightly tilted to form a pocket in the middle. These pockets are meant for binding metal or other cofactors essential for biological function. The α-helix are connected by three peptide loop. Third type of conformation motif is known as saddle conformation forming a stable core, often surrounded by a number of α-helical regions. Such structures are common in many enzymes. The fourth type is sandwich of β-sheets. It is layered so that the strands of the sheets form a quilt like cross-hatching when viewed from above. Some proteins contain two or more separate polypeptide chains or subunits, which may be identical of different in structure. Two or more peptide chains held together by non-covalent interactions or by covalent cross-links, giving quaternary structure. Some of the protein molecules are formed by polymerization of the subunits of the secondary or tertiary structures. The assembly is often called as oligomer and each constituent peptide chain is called as monomer or subunit. The interaction between subunits are stabilized and guided by the same forces that stabilize tertiary structures. The polymerized protein shows a well defined functional activity. These subunits may be the same or different. The aggregation of the subunits is stabilized by means of the non-covalent forces. A number of enzymes have quaternary structure. RNA polymerase of *E.coli*, aspartate transcarbomoylase haemoglobin is multisubunit proteins having quaternary structure. It may also make substrate specificity in the enzyme molecule. Many multisubunit proteins serve regulatory functions.

Some drastic treatment like heating, treatment with acids, alkaline, salts, alcohol, acetone and other organic solvents results in protein denaturation. In protein denaturation due to cleavage of non covalent bonds results in extensive disruption of the secondary, tertiary and whenever applicable quaternary organization of a protein molecule. However, primary structure is not affected. Denaturation does not involve hydrolysis. Generally denaturation is an irreversible phenomenon. At a later stage of denaturation, proteins are precipitated and this stage is known as coagulation. Denaturation results in loss of their natural properties, like decreased solubility, increased viscosity of the solution. Rate of diffusion is decreased.

LIPIDS

Lipids are a diverse group of nonpolar biological molecules. The only common property is their ability to dissolve in organic solvents, such as chloroform or benzene, and their inability to dissolve in water. It is a property that explains many of their varied biological functions. Lipids of importance in cellular function include fats. steroids, and phospholipids.

Fats consists of a glycerol molecule linked by ester bonds to three fatty acids, the composite molecule is termed a triacylglycerol. Fatty acids are long unbranched hydrocarbon chains with a single carboxyl group at one end. Since the two ends of a fatty acid molecule have a very different structure, they also have different properties. The hydrocarbon chain is hydrophobic, whereas the carboxyl (-COOH) group, which bears a negative charge at physiological pH, is hydrophilic. Molecules having both hydrophobic and hydrophilic regions are said to be amphipathic; such molecules have unusual and biologically important properties.

Fatty acids differ from one another in the length of their hydrocarbon chain and the presence or absence of double bonds. Fatty acids present in cells typically in length from 14 to 20 carbons. Fatty acids that lack double bonds, such as stearic acid, are described as saturated, and those possessing double bonds are unsaturated. Whether or not the fatty acid building blocks of a fat are saturated or unsaturated, and the degree of their unsaturation (the number of double bonds), have important consequences. Double bonds (of the *cis*-configuration) produce kinks in a fatty acid chain.

Tristearate, whose fatty acids lack double bonds, is a common component of animal fats and remains in a solid state well above room temperature. In contrast, the profusion of double bonds in vegetable fats accounts for their liquid state. They are being labeled as "polyunsa-turated". Fats that are liquid at room temperature are described as oils.

Solid shortenings, such as margarine, are formed from unsaturated vegetable oils by chemically reducing the double bonds with hydrogen atoms (a process termed hydrogenation). A molecule of fat can contain three identical fatty acids, or it can be a mixed fat, containing more than one fatty acid species. Most natural fats, such as olive oil or butterfat, are mixtures of molecules having different fatty acid species.

Fats are very rich in chemical energy; a gram of fat contains over twice the energy content of a gram of carbohydrate.

STEROIDS are built around a characteristic four-ringed hydrocarbon skeleton. One of the most important steroids is cholesterol.

PHOSPHOLIPIDS The structure of a common phospholipid- a molecule resembles a fat (triglycerol) but has only two fatty acid chains rather than three; it is a diacylglycerol. The third hydroxyl group of the glycerol backbone is covalently bonded to a phosphate group (rather than a third fatty acid), which in turn is covalently bonded to a small polar group, such as choline. Phospholipids contain two end containing the phosphate groups. It has a distinctly hydrophilic character. The other end composed of the two fatty acid tails has a distinctly hydrophobic character.

BIOCHEMISTRY OF NUCLEIC ACID

Nucleic acid is made up of nitrogenous base, pentose sugar and phosphoric acid moiety. Nitogenous base forms nucleoside by forming N-glycosidic bond with pentose sugar (cyclized sugar). By the addition of one or more phosphoryl groups esterified to hydroxyl groups of the sugar, nucleotides are formed. Sugar moiety is present in the furanose ring form. The difference between ribose and deoxyribose sugar is that in the latter, oxygen is absent at 2' position.

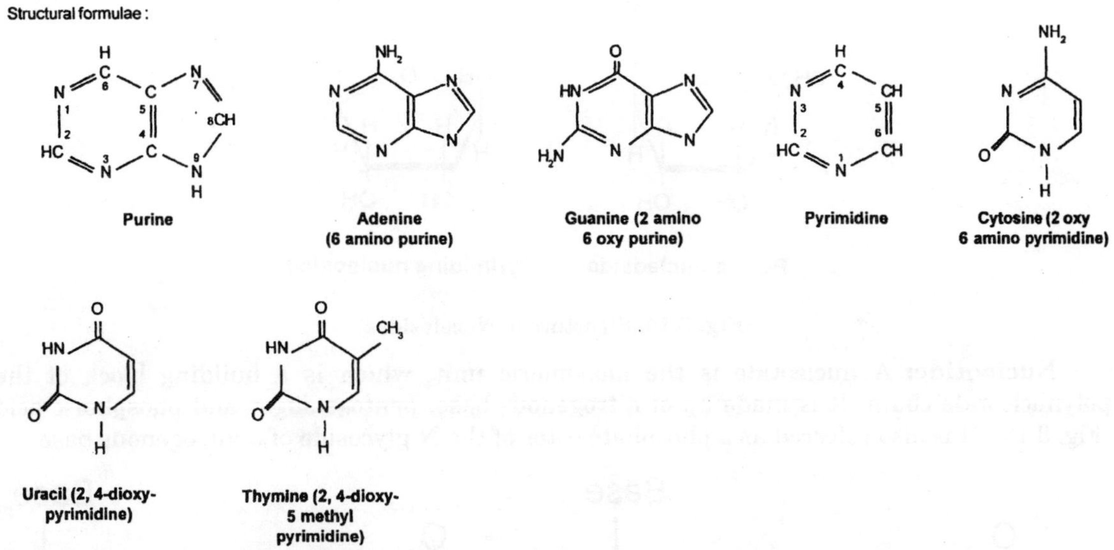

Fig. 3.13. Structures of Nitrogenous bases of Nucleic acid

Nitrogenous base

Purines and pyrimidines are the nitrogenous base of nucleic acid (Fig. 3.13). These bases are heterocyclic compounds having both carbon and noncarbon atoms. These are planer molecules. Pyrimidine (cytosine, thymine and uracil) is six-atom and purine (adenine and guanine) is nine atom compound. The nitrogenous base undergoes keto-enol and amine-imine tautomerism. These functional groups exist as tautomeric mixtures of amino-imino and oxo-hydroxy pairs. Some minor or unusual bases are present both, in proakryotes and eukaryotes. These bases are functionally and physiologically important. 5-Methylcytosine and 5- Hydroxymethylcytosine are uncommon naturally occurring pyrimidines. While, Dimethyla-minoadenine and 7-methylguanine are uncommon naturally occurring purines.

Nucleoside and Nucleotide

A nucleoside is made up of nitrogenous base linked covalently to a pentose sugar by N-C glycosidic bond (Fig. 3.14), which is in the (β)-configuration. The nucleosides of the five nitrogenous bases are named as Adenosine (nucleoside of adenine base), Guanosine (nucleoside of guanine base), Thymidine (nucleoside of thymine base), Cytidine (nucleoside of cytosine base) and Uridine (nucleoside of uracil base). The glycosidic linkage is present between N1 of the pyrimidine ring and N9 of the purine ring and the C'1 of the pentose sugar. One unusual nucleoside is pseudouridine, which differs from uridine in that the glycosidic bond is between C of uracil and the C'1 of the ribose.

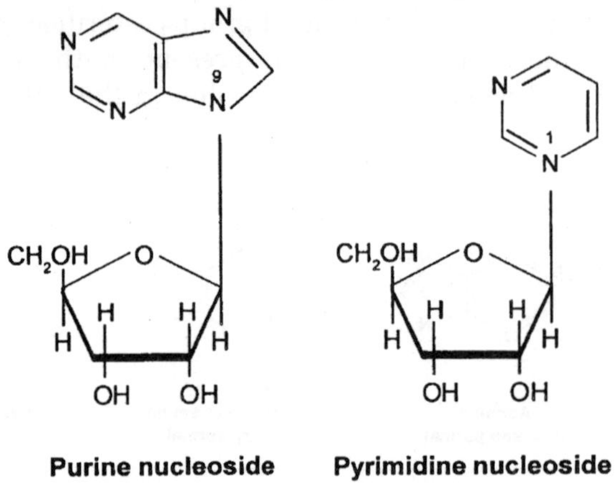

Fig. 3.14. Structure of Nucelosides

Nucleotide: A nucleotide is the monomeric unit, which is a building block of the polynucleotide chain. It is made up of nitrogenous base, pentose sugar and phosphoric acid (Fig. 3.15). It is also referred as a phosphate ester of the N-glycoside of a nitrogenous base.

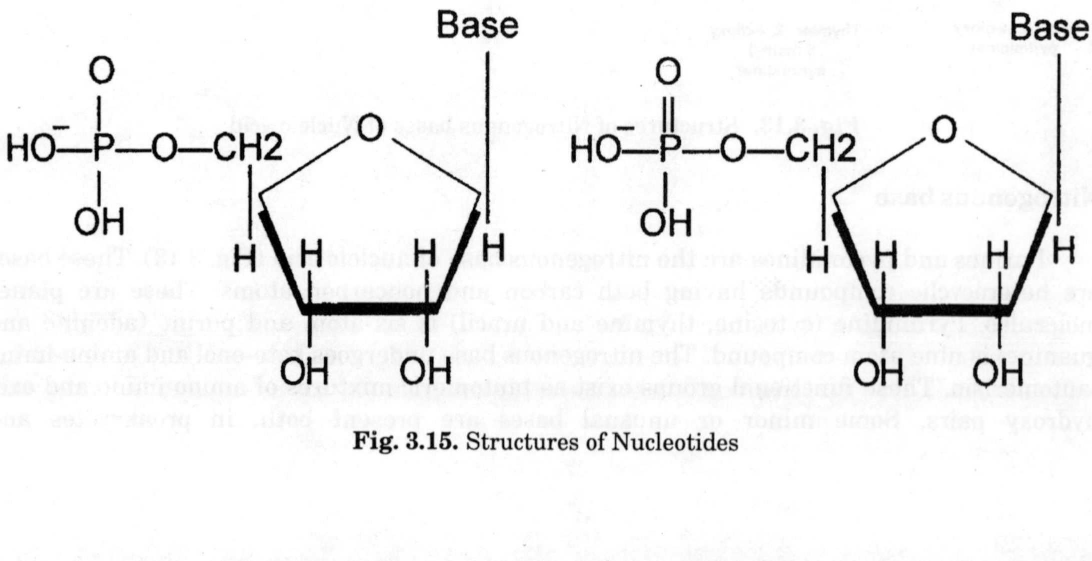

Fig. 3.15. Structures of Nucleotides

Biosynthesis of purine nucleotides

Synthesis of purine nucleotides takes place by three processes (1) synthesis from amphibolic intermediates (synthesis de novo), (2) phosphoribosylation of purines, and (3) phosphorylation of purines.

Adenosine mono phosphate and Guanosine monophosphate are synthesized by a common parent nucleotide, inosine monophosphate (IMP) (Fig. 3.16A). While, the latter is synthesized from the amphibolic intermediate α-D-ribose-5-phosphate (I) in a linear sequence of 11 reactions. Intermediates are 5-phsophoribosyl-1- pyrophosphate (PRPP) (II), 5-phospho-β-D- ribosylamine (III), Glycinamide ribosyl-5-phosphate (IV), formylglycinamide ribosyl- 5-phso-phate (V), formylglycinamidine ribosyl-5-phosphate (VI), aminoimidazole ribosyl-5-phosphate (VII), aminoimidazole carboxylate ribosyl-5-phosphate (VIII), succinyl carboxamide ribosyl-5- phosphate (IX), aminoimidazole carboxamide ribosyl-5-phosphate (X) and formim-idoimidazole carboxamide ribosyl-5- phosphate (XI). EnzymeIMP cyclohydrolase catalyzes the closure of ring (XI) intermediate and forms inosine monophosphate, the first purine nucleotide.

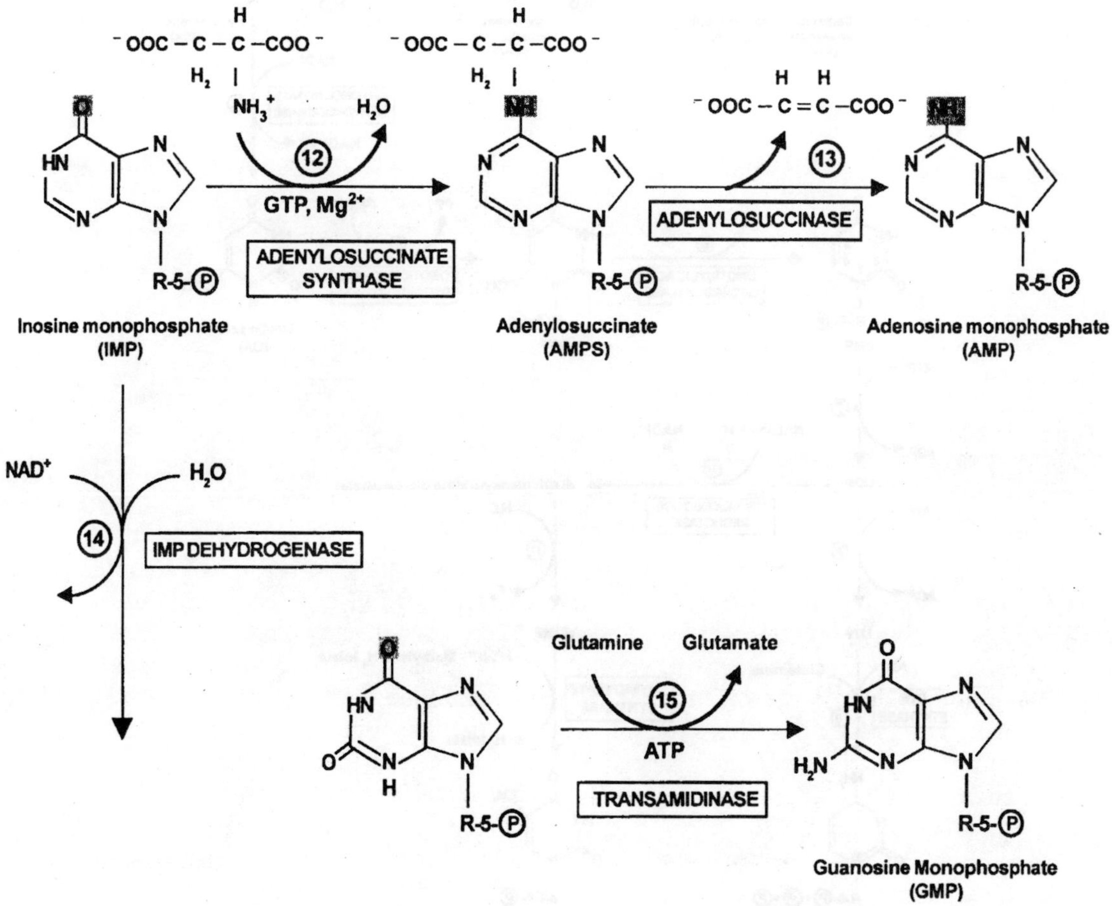

Fig. 3.16. Conversion of IMP to AMP and GMP

Conversion of IMP to AMP and GMP

After the formation of IMP, the pathway branches and two short reaction sequences lead to the formation of AMP and GMP (Fig. 3.16B). Aspartate adds to IMP and forms adenylosuccinate. Enzyme adenylosuccinate synthase catalyzes this reaction. This reaction is superficially similar to reaction 8. GTP is required. Release of fumarate forming adenosine 5'-monopho-sphate (AMP) is catalyzed by adenylosuccinase, the same enzyme that catalyzes reaction 9. IMP is oxidized by NAD^+ with the help of enzyme dehydrogenase and xanthosine monophosphate (XMP) is formed. 6-oxo group of XMP is transamidated by the amide nitrogen of glutamine. Transamidation proceeds by analogy with reaction 5.

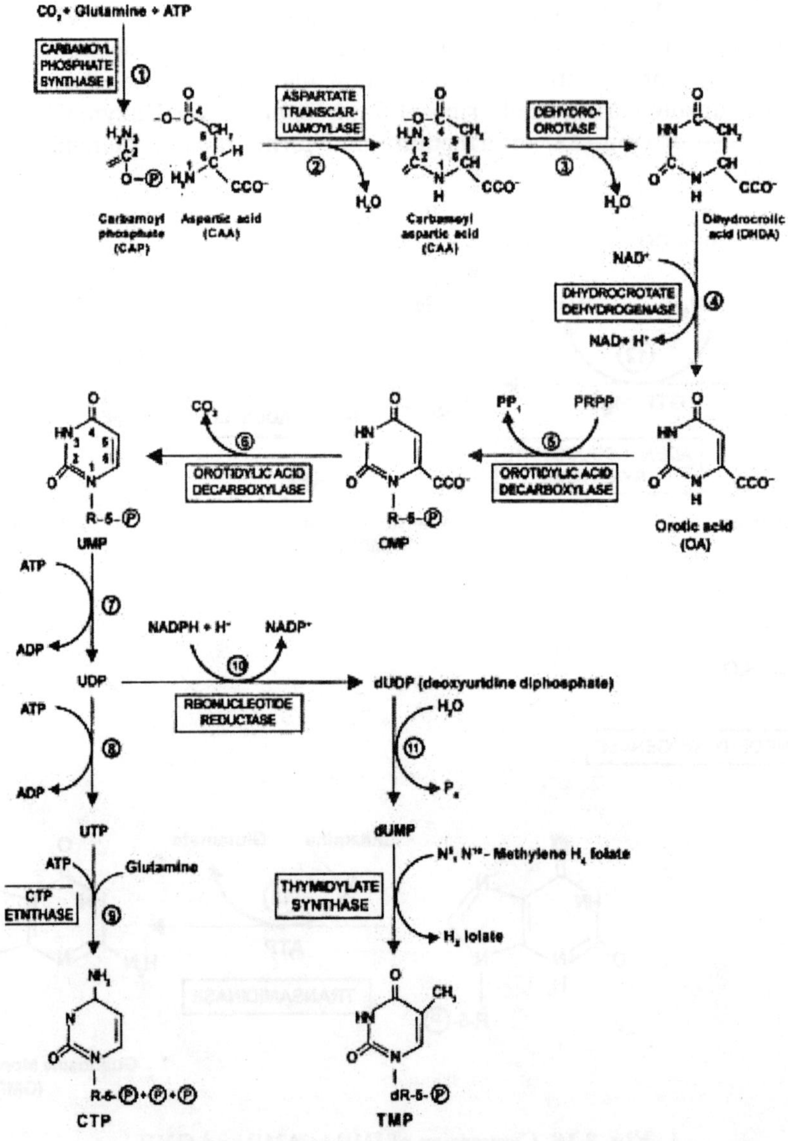

Fig. 3.17. The Biosynthetic pathway for pyrimidine nucleotides

Synthesis of di and triphosphate nucleotide

Enzyme nucleoside monophosphate kinase catalyzes formation of diphosphate nucleoside from their monophosphate nucleotides. The phosphate moiety comes from ATP. GDP is again phosphorylated by ATP with the help of enzyme, nucleoside diphsophate kinase. ADP is phosphorylated to ATP by oxidative phsophorylation and secondarily by reactions of glycolysis and the citric acid cycle.

Biosynthesis of pyrimidine nucleotides

During biosynthesis of pyrimidine nucleotide (Fig. 3.17), first the six membere pyrimidine ring is made and then it is attached to ribose-5-phosphate. Carbamoyl phosphate is needed in this process. It reacts with aspartate to yield N-carbamoylaspartate. This reaction is catalyzed by enzyme aspartate transcarbamoylase. Enzyme dihydroorotase removes water from this intermediate and the pyrimidine ring is closed to form L-dihydroorotate. This compound is oxidized to yield the pyrimidine derivative orotate, a reaction in which NAD^+ is the ultimate electron receptor. PRPP, provides the ribose-5-phosphate side chain and orotidylate is synthesized. The latter is decarobxylated and converted into uridylate. Uridylate is phospho-rylated to give UTP. Enzyme cytidylate synthetase catalyzes the formation of CTP from UTP.

Chapter - 4

ENZYME

Enzymes are catalyst of biochemical reactions showing high degree of specificity for their substrates. They are protein in nature, have molecular weight ranging from 12,000 to over 1 million, and are found as colloids in a living cell.

The concept of biological catalyst was developed in the early 1800s. In 1850s Louis Pasteur opined that fermentation of sugar into alcohol by yeast is catalyzed by "ferments" and in 1897 E. Buchner extracted it from yeast and hence the name enzyme (en=in + zyme= yeast). J.B.S. Haldane wrote a treatise entitled "Enzyme". In 1935, J.J. Berzelius published the general theory of chemical catalysis and included diastase in it.

Enzymes are protein molecules. They require only amino acid residues for activity. However, certain enzymes need some additional groups, other than amino acid residues for their activity. These groups may be organic compound or inorganic metallic ions. Such enzyme molecules with protein and non-protein components are known as holoenzyme and its protein part as apoenzyme. The groups are cofactors or coenzymes. Cofactor may be either one or more inorganic ions such as Fe^{2+}, Mg^{2+}, Mn^{2+} or Zn^{2+} , the coenzymes are complex organic or metalloorganic molecule. Many vitamins are precursors of coenzymes. Some enzyme or metal ion that is covalently bound to the enzyme protein is called a prosthetic group. It is essential for the functioning of enzyme. Many times these terms (coenzyme and prosthetic group) are used as synonym. However, these terms are differentiated on the basis of the binding tendency with the apoenzyme. The tightly bound groups that cannot be removed (e.g., by dialysis) are recognized as "prosthetic groups" and those which dissociate easily are "coenzymes".

NOMENCLATURE AND CLASSIFICATION OF ENZYMES

The simple and convenient way developed for naming an enzyme was to add suffix `ase'. This suffix was used either after the name of the substrate (e.g., enzyme maltase, sucrase, proteinase and lipase acting upon maltose, sucrose protein and lipid, respectively) or after the name of the type of reaction (e.g., dehydrogenase, hydrolyase etc. conducting oxidation reaction and hydrolysis reaction, respectively).

In 1961, International unit of biochemists has adopted a new system of classification and nomenclature of enzymes. Dixon and Webb (1964); Karlson (1965) have outlined the details or rules for systematic and trivial nomenclature of enzymes. The name of enzyme consists of two parts. The first part denotes the name of the substrate and second part denotes the type of reaction catalyzed. Enzymes were divided into following six major classes.

(1) Oxidoreductases : Enzymes catalyzing oxidation reduction are included in this class, e.g., cytochrome oxidase, alcohol dehydrogenases, catalase peroxidase etc.

(2) Transferases : Enzymes responsible for transferring of a group other than hydrogen (e.g., amino, ethyl, formyl, acyl, phosphate, glyoxyl, thiol, ketol aldol etc.) are included in this class.

(3) Hydrolases : Enzymes responsible for hydrolysis of substrate (e.g., ester, peptide, glycosidase, lipases, phosphatases, deaminases nucleases) are included in this class.

(4) Lysases : Enzymes responsible for removal of groups from the substrate by the mechanism other than hydrolysis leaving double bonds in the substrate (e.g., fumarases, aldolase, decarboxylase) are included in this class.

(5) Isomerases : Enzymes responsible for isomerism like substrate is converted into its positional or optical isomer by intramolecular rearrangement of various atoms or groups (e.g., aconitase, phosphohexoisomerase, triose phosphate isomerase) are included in this class.

(6) Ligases (Synthetases) : Enzymes which catalyze linking together of compound utilizing energy made available by simultaneous breakdown of ATP or by similar high energy bond compound (e.g., Succinate thiokinase, Glutamine synthetase, Acetyl CoA carboxylase) are included in this class.

According to this system each enzyme is given a systematic code number which denotes class, subclass, sub-subclass and the name of the enzyme e.g., an enzyme hexokinase is denoted by code number 2.7.1.1, where 2 denotes class transferase, 7denotes subclass transferase of phosphate, 1 denotes sub-subclass alcohol phosphate as acceptor and the last digit 1 denotes the name of the enzyme i.e. Hexokinase.

MECHANISM OF ENZYMATIC REACTION

Enzymes are highly specific in their catalytic activity. The active site of enzyme is involved in enzyme-catalyzed reactions, which is amino acid residue(s). To this active site substrate attaches and forms enzyme substrate complex. The enzyme substrate complex is central to the action of enzymes, and it is the starting point for mathematical treatments defining the kinetic behavior of enzyme-catalyzed reactions.

Two theories have been proposed to explain the catalytic activity of an enzyme. These theories are known as Rigid model and Flexible model.

Rigid Model: In 1894, Emil Fischer proposed a Lock and Key theory which is also known as rigid template model. According to this model enzyme combines with substrate to form enzyme substrate complex. This complex breakdown to give the product and enzyme (Fig. 4.1).

$$E + S \text{ -- } ES \rightarrow E + P.$$

Michealis and Menton (1913) modified this theory and proposed that enzyme has some special points on its surface. These sites were called as active sites and they in turn have special groups that bind the substrate molecule. The substrate binds with this site and the site undergoes chemical changes to form enzyme substrate complex. The number of substrate molecule changed per minute by a molecule of the enzyme is called turnover number.

Fig. 4.1. Formation of Enzyme-Substrate complex (ES) and later ES complex into enzyme and product as proposed by Fisher.

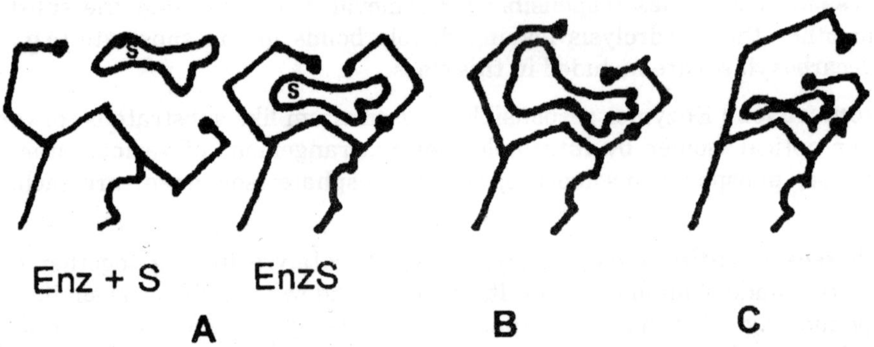

Fig. 4.2. Conformation changes in enzymes

Flexible Model or Induced Fit Models

This model was proposed by Koshland (Fig. 4.2). This model explains that the substrate induces a conformational change in enzyme. This change occurs in the amino acid residue or in the prosthetic group of the holoenzyme. The conformational changes takes place in its correct special orientation for substrate binding, catalysis or both simultaneously other amino acid may become buried in the interior of the molecule. To explain the mechanism Koshland distinguished three types of amino acid residues in an enzyme. These three residues are –

(a) Contact residue : It includes both the catalytic and specific residue and designated to the amino acid residue with one bond distance (0.2 nm) of the substrate or other ligand concerned. The ligands are described as any small molecule bound to the enzyme by non covalent forces (activators, substrate and inhibitor).

(b) Specificity residue : The amino acid residue which takes part for binding the substrate with enzyme. It helps in subsequent catalytic process.

(c) Catalytic residue : An amino acid residue directly involved in covalent bond chain during enzyme action.

Koshland explained his model by enzyme Chymotrypsin (Serine protease). This enzyme has catalytic property on His-57 and Ser-195 present near each other in peptide bond. Beside these active site amino acid residue a covalent intermediate is formed involving OH group of serine 195.

MICHAELIS MENTON EQUATION

In 1913 L. Michaelis and M.L. Menton proposed a general theory of enzyme action and kinetics (Fig. 4.3). According to this theory the enzymatic reaction proceeds with the formation of an intermediate enzyme substrate complex. First the enzyme combines with the substrate to form an intermediate enzyme substrate complex. This forms product and enzyme is regenerated. These reactions are assumed to be reversible. Thus, the rate constant for the forward and reversible direction have positive and negative subscript, respectively. The reaction kinetics may be deduced as follows.

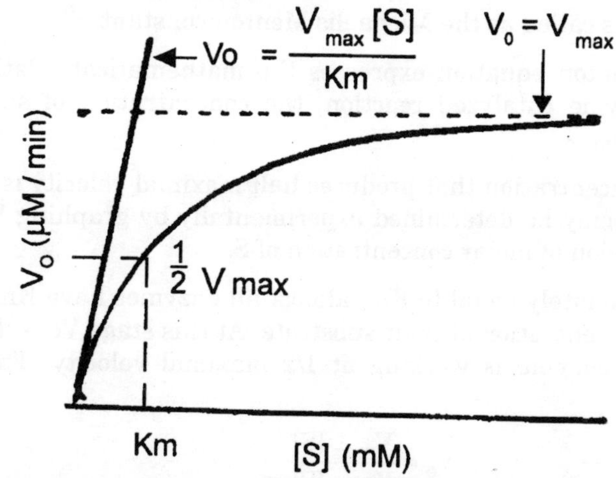

Fig. 4.3. The Kinetic parameters of enzymatic reaction

$$E + S \frac{K1}{K2} ES\ K3\ P + E \qquad \text{eq. (1)}$$

The dissociation constant (Km) of the enzyme substrate complex is equal to the

$$K_M = \frac{[E]\,[S]}{[ES]} \qquad \text{eq. (2)}$$

Where, [E], [S] and [ES] represent the molar concentration of E,S and ES, respectively.

If, the initial total concentration of enzyme is represented as [Eo], the concentration of free enzyme is [E] and the concentration of bound enzyme is [ES]. The equation (2) may be represented as following.

$$K_M = \frac{[E_o] - [ES]\,[S]}{[ES]} \qquad \text{eq. (3)}$$

The rate of forward reaction will be

$= [Eo - E.S] [S] K_1$

The rate of backward reaction will be

$= [E.S] K_2$

The rate of dissociation of ES (i.e., formation of product) is equivalent to

$= [E.S] K_3$

When the reactions are in equilibrium

$$[Eo - ES] [S] K1 = [E.S] K2 + [E.S] K3.$$

$$= [E.S] [K2 + K3]$$

$$\frac{[E_o] - [E.S.][S]}{[S]} = \frac{K2 + K3}{K1} = Km$$

The constant Km is called as the Michaelis Menton constant.

The Michaelis Menton equation expresses the mathematical relationship between the initial rate of an enzyme catalyzed reaction, the concentration of substance and certain characteristic of enzyme.

The substrate concentration that produces half maximal velocity is the Km value or the Michaelis constant. It may be determined experimentally by graphing Vo as the function of [S]. Km has the dimension of molar concentration of S.

When S is approximately equal to Km, almost all enzymes have Km values. It is related to the physiological concentration of their substrate. At this stage Vo is directly related to the change in S and the enzyme is working at 1/2 maximal velocity. The Michealis Menton expression will be

$$V_o = \frac{V_{max} [S]}{Km + [S]}$$

This expression describes the behavior of many enzymes and as the substrate concentration will vary, it will directly affect the velocity of reaction. This effect and the relationship between [S] and Km at any concentration may be illustrated by evaluating Michaelis Menton equation.

ENZYME ACTIVITY : ENERGY CONCEPT

Enzymes are biocatalyst, hence they increase the rate of reaction and do not affect reaction equilibrium. During the reaction substrate is converted into product ($S \rightarrow P$). This reaction may be expressed by reaction coordinate diagram, which is the energetic course of the reaction. In reaction coordinate diagram the free energy of the system is plotted against the progress of the reaction. Every molecule has a free energy at its ground state i.e., when it is in the normal stable form. Thus, the substrate and product also have their free energy. The free-energy change for the reacting system is the standard free energy. The free energy of product is lower than the free energy of the substrate, so $\Delta G^{o'}$ for the reaction is negative. Hence, the equilibrium favours formation of product. This is not affected by any catalyst.

It should not be confused that at favourable equilibrium the conversion of substrate to product is fast, because the rate of reaction depends upon the energetic barrier between substrate and product. The energy barrier represents the energy required for alignment of reacting groups, formation of transient unstable charges, bond rearrangements and other transformations required for the reaction to occur in either direction. The molecules can only react with each other if these barriers are removed, and for this purpose molecules must be raised to a higher energy level. At the top of energetic hill shows a transient state. This point is not related to the chemical state of the substrate or product but it is a point at which decay to the S or P state is equally probable i.e., the S or P may go to downhill either way. It is simply a fleeting molecular moment in which events such as bond breakage, bond formation takes place. The difference between the energy levels of the ground state and the transition state is called the activation energy ($\Delta G^{\ddagger}$). The rate of reaction depends upon the activation energy. A higher activation energy corresponds to a slower reaction. An enzymes enhances the rate of reaction by lowering activation energy. Reaction equilibrium is regulated by free energy and the rate of reaction is due to the activation energy.

The residues of active site of enzyme (amino acid residue, metals ion, and co-enzyme) can interact transiently with a substrate and activate it for reaction. In many cases, these groups lower the activation energy by providing a lower-energy reaction path. The energy required to lower activation energies is generally derived from weak, noncovalent interactions between the substrate and the enzyme. The most important event in enzymatic reaction is the formation of enzyme substrate complex. The interaction between substrate and enzyme formation of a specific ES complex is mediated by the same forces that stabilizes protein structure, including hydrogen bonds and hydrophobic, ionic, and Van der Waals interactions. When such bond(s) is formed between enzyme and substrate there is release of some free energy. The energy derived from enzyme-substrate interaction is called binding energy. Binding energy (ΔG_B) is the major source of free energy used by enzyme to lower the activation, energies of reactions. It provides specificity as well as catalysis.

GENERAL PROPERTIES OF ENZYMES

Enzymes are protein molecule present in colloidal form in a cell. It exhibits all the general properties of a colloid. Some of the important general properties of enzymes are discussed here.

(1) Substrate specificity : Enzymes are highly specific for their substrate they act on their specific substrate only. Specificity is derived from the formation of multiple weak interactions between the enzyme and many or all parts of its specific substrate molecules.

(2) Effect of pH : Every enzyme is optimally active at a specific pH. The enzyme activity can be measured as maximum at their specific pH. There may be a slight variation in the activity of enzyme at pH in vivo and in vitro conditions. In general, for all enzymes the pH ranges between 5.0 to 9.0. However, pepsin shows its activity outside this range.

(3) Effect of temperature : Each enzyme has its own liking of temperature and the temperature co-efficient (Q_{10}) varies from one enzyme to another. As the enzymes are protein molecule they are denatured at higher temperature and become inactive. This temperature may vary from 55 to 60°C. Enzymes of thermophilic bacteria are active up to 85°C.

(4) Effect of substrate concentration : At a constant enzyme concentration an increase in substrate concentration increases the rate of enzymatic reaction. This increase is observed up to a particular concentration of substrate and above this concentration the rate may be inhibited. At the optimal concentration substrate fully occupies the catalytic site of enzyme.

(5) Denaturation : When the protein molecule of enzyme undergoes conformational changes, it becomes denatured. Such denaturation may be brought about by acid, high salt concentration, heavy metal salts, alkaloid reagent and even by uv light. Denatured enzyme shows loss of catalytic activity.

(6) Activators and Inhibitors : A number of molecules or chemicals act as activator and increases enzyme activity that is catalytic activity. Likewise some chemicals inhibit enzyme activity, they are known as inhibitors.

SUMMARY

Enzymes are catalyst of biochemical reactions showing high degree of specificity for their substrates. They are protein molecules and require amino acid residue for catalytic activity. Trivial system nomenclature is adopted to name enzymes. Enzymes are highly specific for catalytic activity. Michelis and Menton proposed a general theory of enzyme action and Kinetics and suggested that the enzymatic reaction proceeds with the formation of enzyme substrate complex as an intermediate. Enzymes are biocatalyst; hence they increase the rate of reaction and do not affect reaction equilibrium. Enzymes exhibit all general properties of colloids.

Chapter - 5

PHOTOSYNTHESIS

Photosynthesis is the most important physiological process taking place in plants, and the only mechanism of energy input into the living world. A simple literal meaning of photosynthesis is `synthesis with the help of light'. It was previously defined as the assimilation of CO_2 in the light to form carbohydrate and oxygen. Photosynthesis is an overall oxidation-reduction reaction taking place between hydrogen donor (water; H_2S in bacteria) and CO_2. The oxidation of water releases O_2 and CO_2 is reduced to carbohydrate. The electrons and protons serve two different functions. The electron reduce $NADP^+$ to NADPH, and the second function is to set up an electrochemical change that provides the energy for pumping protons from the stroma of the chloroplast into the interior of the thylakoid. The protons participate in the reduction of $NADP^+$ to NADPH. Another function of proton is that as they flow back out from the interior of the thylakoid (by facilitated diffusion), passing down their concentration gradient, the energy they give up is harnessed to the conversion of ADP to ATP. Reduction of each CO_2 taken up by the Calvin cycle requires 2 NADPH and 3 ATP molecules.

Historical Background

In the end of eighteenth century scientists started thinking and working on the gases, as before this man knew only solid and liquid form of matter. The simple finding that gases are found in air and their possible importance for living beings, gave birth to pneumochemistry. The concept of photosynthesis came between 1750 and 1775. Priestly, Ingen-Housz and Senebier, are the three names of scientists, who were associated with the discovery of photosynthesis. Joseph Priestley (1771) discovered oxygen (dephlogisticated air) and he also opined that the obnoxious gas released by animals is purified by plants. Jan Ingenhousz, demonstrated that light was necessary for this purification of air. He found that plant also produce the same gas in dark. Jean Senebier (1782) showed that the noxious gas produced by plants (in darkness) and animals stimulated production of purified air by plants in the light. Ultimately, Lavoisier made it apparent that these gases are O_2 and CO_2. It is interesting to note the assumption of Priestly and Ingen-Housz, that the plants, in the process of transforming "impure air" into "purified air" acquire nourishment for themselves.

Liebig (1843) first presented the chemical theory of photosynthesis. He thought that the formation of acids must precede that of sugars. However, Baeyer (1870) was of the opinion that the first stage in photosynthesis is the reduction of CO_2 to formaldehyde, and is polymerization of the latter to sugar. Sachs (1862), observed the formation of starch grains in chloroplast during photosynthesis proves that these bodies are the site of the photosynthetic processes. Stokes (1864) briefly reported isolation of chlorophyll a and b, and two yellow carotenoids. Engelmann (1883) showed the role of accessory pigments in photosynthesis. Warburg (1919) first gave the idea of dark reaction. Ruben (1943), proposed a hypothesis about the role of ATP in photosynthesis.

Photosynthesis has two phases. First phase is photochemical phase and the second is carbon fixing phase. Electron flow and phosphorylation are the events of the first phase, which takes place in the inner membrane system (thylakoids). The products of this phase are NADPH and ATP. The second phase is a biochemical pathway and enzymes of this pathway are found in the matrix of chloroplast. In this metabolic pathway CO_2 is assimilated as carbohydrate.

First Four Quanta theory Van Neil formulated oxidation-reduction theory of photosynthesis and postulated that there is a single photochemical reaction (decomposition of water). The reduction of CO_2 to carbohydrate requires the transfer of four hydrogen atoms. In van Neil's theory, four water molecules and one CO_2 molecules participate in the formation of one (CH_2O) group, and three water molecules are recovered in the end-two by the dismutation of first, oxidation product, and one by the dehydration of an intermediate reduction product.

First Eight Quanta theory after a long controversy the quantum yield of photosynthesis is 1/4 was rejected and it was proved to be 1/8, that is at least eight quanta must be absorbed to reduce one molecule of carbon dioxide. This raises the question as how eight primary photochemical processes can be utilized for the transfer of four hydrogen atoms. It is simple that each hydrogen atom can be achieved twice, thus enhancing its reducing power. Previously it was assumed that there is only one photochemical reaction, but now it is well proved that there are two photochemical reactions. The number of O_2 molecules released per light quanta absorbed is known as quantum yield. In photosynthesis, one molecule of O_2 is evolved utilizing eight quanta energy. Thus the quantum yield is 1/8 or 12%.

The Hill Reaction

C.B. Van Niel (1941) worked on the green sulphur bacteria and proposed following reaction

$$CO_2 + 2H_2S \rightarrow (CHOH) + H_2O + 2S$$

In green plants the analogous reaction is

$$CO_2 + 2H_2O \rightarrow (CHOH) + H_2O + O_2$$

Ruben and his coworkers (1941) worked with Chlorella. They supplied bicarbonate or H_2O labeled with ^{18}O. The evolved oxygen was measured in a mass spectrophotometer. It was observed that isotopic oxygen evolved from labeled water.

R. Hill (1937) observed that when isolated chloroplasts are supplied with ferric salts in the light, the ferric salts are reduced to the ferrous form and O_2 is evolved. Illuminated isolated chloroplasts were able to produce O_2 in an apparent reversal of dark reaction, in the presence of a number of suitable hydrogen acceptors. The hydrogen acceptors may be ferricyanide (one-electron acceptor) and quinone or dyes (two-electron acceptors). Hill experimented with ferricyanide as hydrogen acceptor. The stoichiometry shows that four hydrogen equivalents are transferred for each mole of O_2 produced.

$$4Fe^{2+}(\text{cyanide}) + 2H_2O \xrightarrow[\text{chloroplast}]{\text{light}} 4Fe^{3+}(\text{cyanide}) + 4H^+ + O_2$$

O. Warburg (1948) showed that in isolated chloroplasts, chloride is necessary for evolution of oxygen (in presence of artificial electron acceptors).

Red Drop and Emerson Effect

Emerson and Lewis (1943) performed experiments on quantum yield using monochromatic light, and observed that quantum yield declined in the red zone (at wavelength greater than 680 nm). They named this phenomenon as *red drop*. The reason of such drop in quantum yield remained unanswered. Later Emerson and Chalmers observed that this decline in quantum yield can be enhanced if the light of shorter wavelength is exposed along with the light of long wavelength. Such enhancement is known as *Emerson enhancement effect.* In 1957, Robert Emerson and coworkers observed that the quantum yield in the spectral range above 680 nm increased dramatically when orange light of 650nm was irradiated together with red light. The quantum yield was higher than the sum of the yields obtained when alga was irradiated separately with the light of each wavelength. The Emerson effect (E) can be expressed as the ratio of the rate of oxygen evolution (ΔO_2) in the far-red light in the presence of supplementary beam, and the same rate in the absence of it.

$$E = \frac{\Delta O_2 \text{ (in combined beams)} - \Delta O_2 \text{ (short - wave beam alone)}}{\Delta O_2 \text{ (long - wave beam alone)}}$$

The Emerson effect led the conclusion that two different reaction centers are involved in photosynthesis. The simple explanation of red drop and enhancement effect is PS I has reaction center of long wavelength, and is more efficient in far-red light. PS II has reaction center of short wavelength, and hence it operates more efficiently in red light. If photosynthesizing cell is irradiated with visible light, the two photosystems that absorb optimally in different spectral regions. In monochromatic light, if any one type of photosystem stops operating, the quantum yield will decline.

Robert Hill and Fay Bendall (1960) postulated an electron transport chain in which two reaction centers are arranged in tandem and connected by cytochromes (cyt-6 and cyt f). Light of wavelengths 700nm and 680 nm excited reaction center one and two, respectively. A reaction diagram according to the redox potentials show a zigzag arrangement, hence the scheme was named, Z scheme.

By 1960, the basic principles of photosynthesis changed and the modern concepts of photosynthesis appeared. The ideas poured included, operation of two light reactions in sequence, the new electron carriers plastoquinone, plastocyanin, and ferredoxin, the coupling factor, light-driven proton uptake, post-illumination ATP synthesis, dark phosphorylation by an acid base transition, the electric field across the membrane, vectorial electron flow, and step-wise four-charge accumulation in the oxygen evolution path.

PHOTOBIOLOGY

Light is a form of electromagnetic radiation visible to the human eye. It has properties of both particles and waves. Wavelength (l) is the distance between successive wave crest. Frequency (v) is the number of wave crests that pass an observer in a given time. The relation between wavelength, the frequency and the speed of any wave(c) is expressed in following :

$$c = \lambda v.$$

The speed of light is 3.0 x $10^8 m^{-1}$. Electromagnetic radiation of wave length between 400 and 760 nm (4000 to 7600) is visible to human eye.

To explain the role of light in photosynthesis, we consider that light consist of waves of very small particles, the photons or quanta, and each quanta contains an amount of energy known as quantum. The energy is expressed by Planck's law :

$$E = hv \quad \text{or} \quad E = hc/\lambda$$

where h is Planck's constant (1.58 x 10^{-14} cal/s or 6.63 x 10^{14} J/s) and v is frequency.

Energy is expressed as

$$E = Nk\,\lambda$$

where N = Avogadro's number (6.02 x 10^{23} molecules or photons per mole); k = frequency; λ =wavelength. The molar unit of radiant energy is 1 Einstein, i.e., the energy of one molecule of photons.

The Action spectrum : photosynthetic pigments absorb light of a definite wavelength. The excitation is expressed by following formula (as defined in Bougwer-Lambert Beer law)

$$E = Ig\ Io/I$$

Where Io is the intensity of the incident light, I is the intensity of light which passes through the solution.

The relative absorbance of various wavelengths by a purified absorbance of various wavelengths by a purified pigment can be measured with a spectrophotometer. A graph of the intensity of some phenomenon (like rate of photosynthesis) produced by monochromatic light as function of the wavelength of light is called an "*action spectrum*". Engelmann (1882-84) for the first time recorded action spectrum in *Oedogonium* and *Cladophora*.

Photosynthesis can be measured by the rate of number of molecules of oxygen produced (P), by a known number of incident quanta (I_i) or, better, by a certain number of absorbed quanta ($I\alpha$) of a given wavelength.

$$P/\ I_i = f(\lambda) \text{ or } P/\ I_\alpha = f(\lambda)$$

An "action spectrum" is obtained when we compare the effects of different wavelengths on the rate of photosynthesis, always making sure not to add so much energy of any wavelength that the process becomes saturated. The action spectrum is the dependence of photosynthetic activity on radiation of various wavelength.

Photochemistry

Plant pigment absorb light and convert this electromagnetic radiation in potential chemical energy, this phase of photosynthesis is photochemistry. The obvious meaning is that only light absorbed in a reacting system can have a chemical effect. The significance of electron transfer is to generate the reducing equivalents and photophosphorylation. In simple terms, there are two sets of primary photochemical reactions, one of them may involve photoreduction of an organic substrate by one pigment system, and the second is photooxidation of water by another pigment system. If chl-a is the photocatalyst, it will be oxidized in the first reaction and reduced in the second. According to the law of photochemical equivalence, the photochemical transformation of a molecule is induced by the absorption of a single photon. The energy efficiency of a photochemical process is generally less than the quantum efficiency. Pigments absorb light of wavelength between 400 and 700nm. The photoexcitation of photosystem I or II results in the same photochemical transformation regardless of the energy of the absorbed photon. According to Kamen (1963) during the interval between 10^{-15} seconds and 10^{-9} seconds, the quantum absorption and conversion processes that involve radiation and solid state physics take place. Between 10^{-9} seconds and 10^{-4} seconds, the oxidant and reductant are produced and accumulated. Between 10^{-4} seconds and 10 seconds after the light absorption, CO_2 fixation takes place.

Irradiance is the quantified amount of energy that falls on a flat sensor area per unit time. According to Grotheus and Draper (1872) radiation is only photo chemically active when its photons are absorbed by appropriate photoreceptors or pigments. Electromagnetic radiation between 400 to 700 nm is photosynthetically active radiation (PAR), it is expressed on the basis of energy (Wm^{-2}) or quanta ($molm^{-2}s^{-1}$). When PAR is expressed on a quanta basis, it is given the special term photosynthetic photon flux density (PPED).

Energy transfer

A pigment reaches a number of states after absorbing a quantum. A pigment becomes in excited state after absorbing light. A particular state is attained depends on the wavelength of light absorbed. When a pigment molecule absorbs light, the position of electron within that molecule change their positions. Energy states above the ground state is known as excited state. The energy of an electron depends on the orbital that it occupies, and when a molecule is excited, one of its electron moves to an orbital of higher energy. This makes molecule more chemically reactive. A pigment present in second excited (singlet) state return to first excited state, part of energy is lost as temperature. Pigment can lose excitation energy by a combination of heat loss and fluorescence. According to the laws of quantum mechanics transition to an excited state energy (Eo) is the gap between the ground state and the excited state energies of the pigment molecule. Quantitative requirements for pigment excitation is expressed as following :

$$Eo\text{-}Eg = hc/\lambda$$

In molecules, the ground state and the excited states have many closely spaced substances, the result of molecular vibration and rotation, these are in first and second excited singlet states for a molecule that has two major absorption bands. In an atom, electrons move in different orbitals. Generally two electrons are present in the same orbit and move in opposite spins. A molecule shows two types of excitation state (Fig. 5.1). In *singlet state* electrons orbit move in opposite spins (antiparallel). This state is short lived. While if the electrons start moving in the same spin, it is the *triplet state*. This state is long lived and takes longer time to de-excite. The "π electrons" of molecules consisting mainly of carbon atoms absorb visible light. The "σ orbitals" has high energy level, hence their electrons can be excited by energy-rich ultraviolet radiation only. The "π orbitals" give chemical stability to the molecule. When a quantum is absorbed by a molecule, a π electron is raised from ground state (π) to excited state (π*). This is known as π→π* transition. It is characteristic for molecules containing a system of conjugated double bond compounds of the molecular structure strongly absorb visible radiation. Biological pigments also show another electron transition. It involves those electrons which do not participate in bond formation (n electron), with n→n* transition. The specific orbitals are characteristic of oxygen, nitrogen and sulphur atoms, which may be present in a system of conjugated double bonds or linked to such a system.

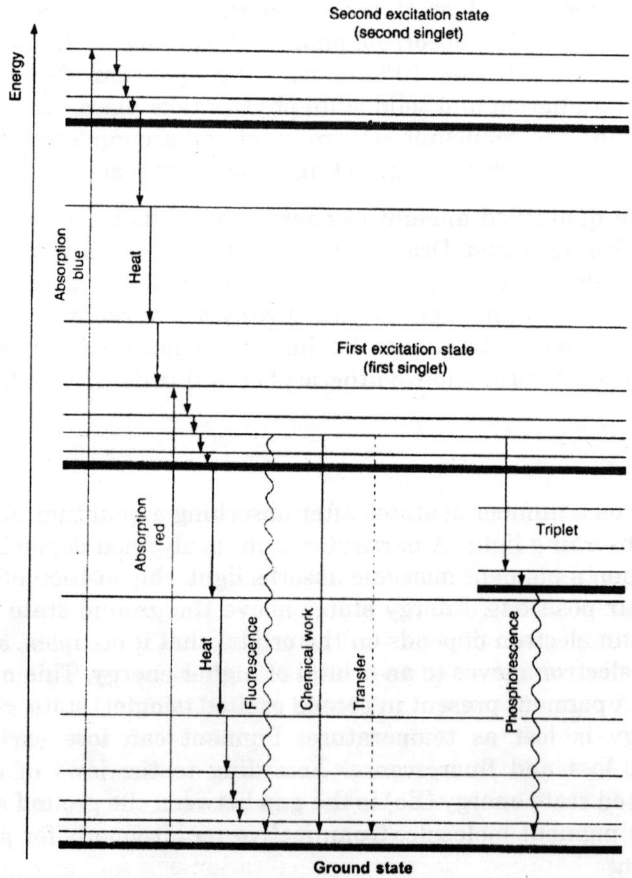

Fig. 5.1. The ground state and different excitation state of chlorophyll a molecule

When the excess energy of a pigment present in triplet state, is transferred to another pigment (acceptor pigment), a photochemical change in the process takes place. This change is known as `*photosenitized*' and the pigment is called as *sensitizer*. When irradiated pigment emits light, it results in the decay of an exited singlet state to the ground state. This phenomenon is known as *fluorescence*. This process is rapid and temperature independent. The average time of a molecule has to spend in the excited state between absorption and fluorescence is known as the *natural life time* of the excited state. If the excited singlet state undergoes a transition to the triplet state without photochemical reaction, decays occur to the ground state by heat dissipation. This process is known as *phosphorescence*. It is temperature dependent.

Photosynthesis makes use only of the metastable triplets which live long enough to allow for all kinds of photochemical energy transfers to take place. This is a peculiar excited electronic state, with a much longer natural life time. The electron loses its energy by fluorescence and reverts to the ground state, or with a further small loss of energy as heat.

The transfer of excitation energy from one pigment molecule to another molecule takes place by *Forster energy transfer mechanism or resonance*. Theoder Forster (1948) analysed the mechanism of resonance transfer of energy between unlike molecules without overlapping absorption bands. In this case, `resonance' does not operate during the absorption process itself. In this mechanism, emission and re-absorption of photon is not required. Resonant energy transfer is directed towards pigments absorbing at longer wavelength. According to Forster, one measure of probability of transfer of energy between unlike pigments is the overlapping of the fluorescence band of the donor and the absorption band of the acceptor. The vibrational energy dissipation in the excited electronic state of the `donor' pigment, is resonance with certain state of the `acceptor' pigment, this causes delayed resonance. This resonance is proportional to r^{-6}. Proximity of the donor and acceptor molecules within the antennae is critical. It is because the efficiency of energy transfer is inversely proportional to the sixth power of the distance separating the two molecules. Interaction between molecules with overlapping absorption bands is a dipole-dipole interaction. The energy of such interaction is proportional to r^{-3}, where r is the distance between the centers of the two dipoles. In very closed aggregated pigments, simple dipole-dipole model of the excitation theory has to be extended by taking overlap of the π-electron orbitals. When pigments strongly interact with each other via their resonant transition dipoles, resonant energy transfer between pigments is more rapid. The degree of interaction depends on the mutual distance and the relative orientation of the chromophores. The extreme cases of energy transfer may be weak or strong interactions. In weaker interaction, when the transfer time is long in comparison with the time for intramolecular vibrational relation, the transfer is unidirectional. The donor molecule releases its excited energy to the vibrational ground state and later this energy is transferred to the acceptor molecule. After the transfer the acceptor molecule releases vibraitonally and drops out of resonance with the donor. If the excited state of the acceptor is thermally uplifted into a higher vibrational level, back transfer of energy from acceptor to donor may take place. This phenomenon is improbable but not impossible. In strong interaction the transfer times are very short in comparison with the times for vibrational relaxation. The transfer is bi-directional. In this case, before vibrational relaxation occurs a quantum may be transferred repeatedly between the resonant vibrational levels of the donor and acceptor pigment molecules. At the extremes the energy is delocalized over a whole aggregate of strongly interacting pigment molecules.

Light Harvesting

Horton *et al.* (1996) developed the concept of light harvesting. It is the process of absorption of light energy and its subsequent transfer to reaction centers of photosystems. The absorption of light energy, and its conservation in the form of excited pigment is the initial reaction of photosynthesis. Chlorophyll and carotenoid molecules bound with the LHCprotein (harvesting complex), are present on the thylakoid, and absorbs light. The absorption of light results in singlet state excitation of pigment. The light energy utilization for photosynthesis involves the sequence of reactions of light harvesting and energy conversion through primary photochemistry at antenna pigments and reaction center. The regulation of the amount of light absorbed depends on the antennae system rather than on the efficiency of transfer process to reaction center. In natural environment, plants have to adjust to changes both in intensity and quality of light. In case, excessive stimulation of one of the two photosystems result in an unequal utilization of light, it may cause damage. The higher plants show a mechanism to mitigate and inappropriate light energy distribution between PS-I and PS-II. This adjustment show a rapid regulatory mechanism, and is called state transition. A specific quantity of light is required by plants at which, the rate of photosynthesis reaches its maximum (P_{max}). The harvesting center continues to absorb light even above this level of saturation, this energy is Excess Excitation Energy (EEE), and causes damage.

PHOTOSYNTHETIC PIGMENT

Pigments contains a chemical group, known as chromophore. The latter is a chemical group capable of absorbing visible light of a particular wavelength. The pigments are organic molecules with conjugated double bonds, consisting of regularly alternating single and double bonds. Conjugated bonds are particularly stable because they are reinforced by `resonance' interaction between several structures with the same arrangement of atoms, but different distribution of electrons. There are three types of pigments involved in photosynthesis, known as photosynthetic pigments. They are chlorophyll, phycobilin and carotenoids.

Chlorophyll

Chlorophyll is the main photosynthetic pigment (Fig. 5.2). It constitutes about 10% of the dry weight of the chloroplast. Richard Willstatter discovered the structural formula of Chlorophyll, for which he was awarded Nobel Prize in Chemistry 1915. Hans fischer (1940) worked out the tetrapyrrole structure. In chemical terms, chlorophylls are methyl phytol esters of the parent dicarboxylic acids, the chlorophyllins, are chemically classified as tetrapyrrole compounds. Chlorophyll molecule has two components: a porphyrin ring and a phytol chain. Porphyrin ring is a tetrapyrrole, containing four convoluted nitrogen containing sub-cycles (pyrrole rings). Mauzerall in his article wrote that the macrocycle of chlorophyll is a chlorin, not a porphyrin. Its ring gives rich optical spectral property to the chlorophyll and richer supply to redox level to the chlorophyll, which are important for the photochemical function. The natural pyrrole is called porphobilinogen and it is itself made up of two molecules of δ-aminolevulinic acid. The two carbon atoms of porphyrin ring adjacent to the nitrogen are designated as "α" and the other two distal carbons are "β", they have free valency. The pyrrole ring has a property to bind metal atoms within their molecular structure. In chlorophyll, magnesium is present in the center of the porphyrin ring as the central atom. It is chelated in the tetrapyrrole ring, Magnesium is covalently bound with two N-atom and coordinately joined with other two N-atoms of porphyrin. There are some basic differences between the tetrapyrrole of chlorophyll and other porphyrin; (1) To the carbon 7 and 8, two additional hydrogen atoms are attached. Thus chlorophyll is a derivative of the dihydroporphyrin; (2) Cyclopentane ring (isocyclic ring) is linked to the III pyrrole ring. The carboxylic group of this ring is esterified with a methyl group. Porphyrin of chlorophyll has certain side chains, attached to the `β' carbon of the pyrrole rings. These are in position 1, 3, 5 and 8 methyl groups, in position 4 an ethyl group, in position 2, a vinyl group. The phytol chain is attached to 7th carbon. Phytol chain ($C_{20}H_{39}OH$) is hydrophobic in nature, and keeps chlorophyll embedded in the photosynthetic membrane. It is a 20 carbon chain made up of repeating unit of isoprenoid units, formed from four isoprene units. A chlorophyll molecule without phytol chain is known as chlorophyllide.

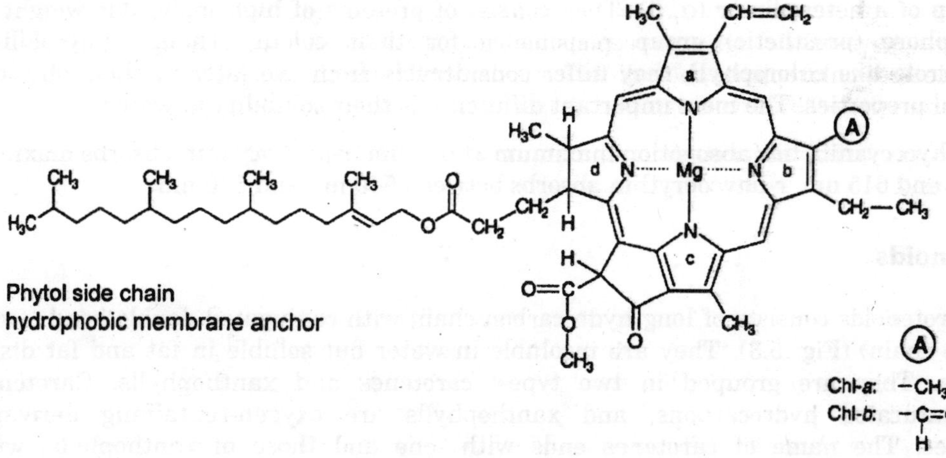

Fig. 5.2. Basic structure of chlorophyll molecule

Chlorophyll is present in different types. Chl `a' and chl `b' are most common in plants. The structural difference between the two types is that in chl `b' at carbon atom 3 is replaced by the aldehyde group (-CHO) of a formyl residue. This simple difference has a large influence on light absorption. Chl `c' is a chlorophyllide. In this type of chlorophyll, 20 carbon phytol chain is replaced by a side chain -CH=CH-COOH. In this type of chlorophyll a double bond is present between the carbon atom 7 and 8. Chl `c' is present in two forms, `c_1' and `$c_2$'. Chl `c_2' has a vinyl group at carbon atom 4.

In alcoholic solutions, chlorophyll exhibits fluorescent property. When chlorophyll in alcoholic solution is permitted to stand in air, it is `allomerized' i.e., oxidized at the C(10) atom. Allomerization of chlorophyll is characterized by a spectral change. A pure alcoholic solution of chl-a is blue green in transmitted light and red in reflected light. In contrast, chl-b is yellow-green in transmitted light and brownish red in reflected light. Chlorophyll a is characterized by two strong absorption bands, located in the blue-violet and in the red region of the spectrum. The first of them (soret band) is blue-violet band and the second is red band. The first band is common to all porphyrin derivative, while the second is peculiar to chlorophyll. Chlorophyll shows different absorption peaks in different solvents. In ether solution chl-a shows two peaks they are 662nm (in the red region) and 435nm (in the violet region). Similarly chl-b also has two absorption peaks. These peaks are at the wavelength of 642 nm and 453 nm. *In vivo*, the absorption peaks for chl-a is 435nm, 670-680nm, and for chl-b, it is 480nm and 650nm. The absorption spectrum of chl-c is also called chlorofucin. It is characterized by a band in the region of 630nm.

Phycobilins

Phycobilins are linear tetrapyrroles, i.e., the tetrapyrrole rings are present in an open chain. The tetrapyrrole rings are linked by methane bridges. Oxygen is attached at 1 and 8 carbons, and propionic acid residues at 4 and 5 positions. Linear tetrapyrrole is present in bile, hence this pigment is named `bilin'. There are two types of phycobilins: phycocyanin (blue pigment) and phycoerythin (red pigment). When these pigments are linked to protein, they are named as phycocyanobilin and phycoerythobilin, respectively. Phycobiliproteins are made up of a heterodimer (α, β). They consist of proteins of high molecular weight and of chromophore (prosthetic) group responsible for their colour. Though phycobilins are tetrapyrrole like chlorophyll, they differ considerably from the later in their physical and chemical properties. The most important difference is their solubility in water.

c-phycocyanin, has absorption maximum at 618 nm. r-phycocyanin absorbs maximally at 552 nm and 615 nm. r-phycoerythin absorbs between 540 nm and 570 nm.

Carotenoids

Carotenoids consists of long hydrocarbon chain with conjugated double bonds (so-called polyene chain) (Fig. 5.3). They are insoluble in water but soluble in fat and fat dissolving solvents. They are grouped in two types: carotenes and xanthophylls. Carotenes are polyunsaturated hydrocarbons, and xanthophylls are oxygen-containing derivative of carotenes. The name of carotenes ends with -ene and those of xanthophylls with -in. Carotenoid molecules are made up of forty carbons. The chemical structure of the building blocks is known as `isoprene'. Latter are unsaturated C_5 hydrocarbon with a methyl group as side chain. The building blocks are arranged to form a chain like central structure with fourteen carbon atoms, seven conjugated double bonds and four methyl groups as side chains. The chain has open or closed ionone rings (hexamethylene ring with one double bond) either at one end or at both ends. The terminal rings may be identical or different. The difference among the carotenoids is due to the terminal rings. Most carotenoids contain 40 carbon atoms, and 11 or more double bonds.

Carotenes are isomeric compounds of the formula $C_{40}H_{56}$. In the α-carotene and β-carotenes, both the ends consists of closed rings, in γ-carotene, however, only one end is cyclic (ionone ring) thus eliminating the terminal double bond. In β- carotene, the double bond of each ring is conjugated with the last one in the straight chain skeleton; in α- carotene this holds true for one ring only; in the other one the double bond has shifted beyond the next carbon atom. By this typical `α-ionone' structure α-carotene differ from β- carotene which has a `β-ionone structure'.

Xanthophylls: In xanthophylls the oxygen is present in a hydroxyl, keto, epoxy, carbonyl, or methoxy groups. Different types of xanthophylls are present in plants. *Lutein* ($C_{40}H_{56}O_2$; 3,3'-dihydroxy-α-carotene) is a derivative of α-carotene with each terminal ring carrying an additional hydroxyl group. *Violaxanthin* ($C_{40}H_{56}O_4$) has closed ionone rings, and epoxy group at C5-6 and 5'-6'; and a hydroxy group at 3 and 3' positions. *Zeaxanthin* (3,3' dihydroxy -β-carotene)has double bonds in the ionone ring as well as two hydroxyl group (3 and 3'). *Cryptoxanthin* ($C_{40}H_{56}O$) is similar to zeaxanthin but for the lack of the hydroxyl group in position 3'. Neoxanthin ($C_{40}H_{56}O_4$) is a derivative of α-carotene. It has additional hydroxy groups at 3 and 3', another one in position 6' and an epoxy group (5-6).

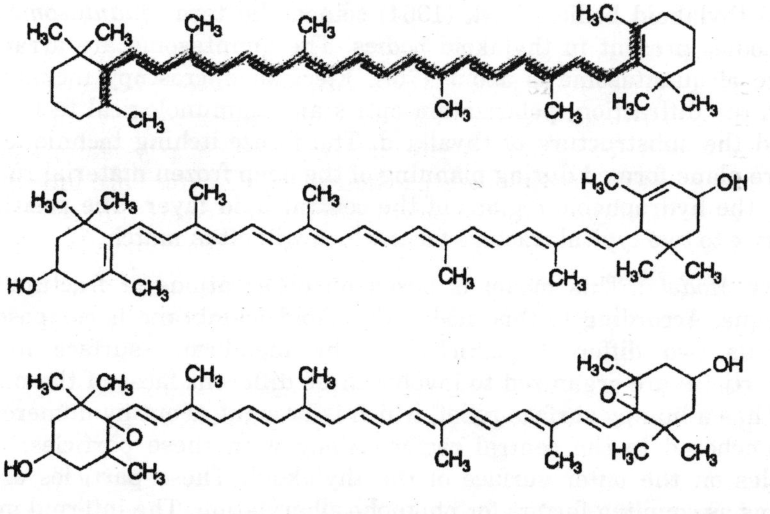

Fig. 5.3. Structural formula of carotene and xanthophyll

The absorption peaks of carotenoids was examined in organic solvents. The absorption peaks of carotenes was observed in hexane. It is for α-carotene 420,440 and 470nm), for β-carotene 425, 450 and 480nm), for γ-carotene 440,460 and 495nm). Absorption peaks of lutein was observed in ethanol, it is 425, 445 and 475nm.

Another diversified class of carotene is carotenols. Some examples of carotenols are luteol (3, 3'-dihydroxy-α-carotene), zeaxanthol (dihydroxy-β-carotene), fucoxanthol.

The major carotenoids of higher plants are β-carotene, Lutein, violaxanthin and neoxanthin.

Photosynthetic Units and Reaction Centers

In 1932, Robert Emerson and William Arnold carried out experiments on *Chlorella*, exposing it to flashing light of very short duration (e.g. 10, μ sec) and saturating intensity. They concluded that not all the chlorophyll molecules in chlorophyll were actively engaged in the conversion of light energy into chemical energy. Based on the number of chlorophyll molecules present in the preparation, they concluded that one molecule of oxygen was being released during a brief flash of light for every 2500 molecules of chlorophyll present. Later, it was established that 300 times as many chlorophyll would appear to be required for photosynthesis. All 300 or so chlorophyll molecules act together as one photosynthetic unit. The aggregation of smallest number of pigments molecules necessary to affect a photochemical act is called as photosynthetic unit. The photochemical act means the absorption and migration of a light quantum to a trapping center where it promotes the release of an electron.

Thylakoid and Location of Pigments

In chromatophores pigments are distributed all over the thylakoid membrane. While in chloroplast of higher plants, pigments are located on the grana thylakoids. Studies on structure of chloroplast has revealed that granum lamellae form sac-like structure. Menke (1962) called these structure as thylakoids. Park and Beggins (1964, 67) observed some small projections known as thylakoid bodies. Park (1964) coined the term *quantasome* to name somewhat rounded bodies present in thylakoid bodies. The quantasome are arranged in a single layer. The size of quantasome is about 175Å. Electron microscopy including freeze itching techniques, X-ray diffraction, polarization optics and immunological techniques has helped to understand the substructure of thyalkoid. The freeze-itching technique findings show that the fracture plane formed during planning of the deep frozen material runs within the membrane along the hydrophobic regions of the central lipid layer. The position of the fracture plane gives rise to two typical fracture faces which will often match.

Braton and Park Model : This model is based on observations of fracture plane of freeze-etching technique. According to this model, thylakoid membrane is composed of two subunits appearing as two different particles in the membrane surface in electron micrograph. These particles are organized to involve three different faces of the membrane. The outer surface `I' has a bumpy surface relief, which is brought about by numerous larger particles obviously anchored to the central region. Along with these particles, there are present some particles on the outer surface of the thylakoid. These particles are loosely attached and functions as coupling factors for photophosphorylation. The internal membrane has two major faces : I, carrying numerous smaller particles (110-120Å). This represents a layer adjacent to the outer membrane surfaces. The undersurface of I is face II with larger particles of about 175Å. The two types of particles probably holds only for thylakoid membranes within the grana. The membranes of grana stacks are considered to contain only the smaller particles of about 120 Å diameter. The smooth surface of face II is composed of glycolipid and chlorophyll is absent in this layer. According to Park (1966) and Braton (1968) particles of 175 Å diameter might be identical to quantosome each of which is likely to be composed of four sub-units of 80-90 Å size.

Kirk presented a model based on the electron microscopic observations and X-ray diffraction. According to this model, both the particles (i.e., 120 Å and 175 Å in size) are embedded within a continuous lipid matrix of the bilayered membrane. The thylakoid membrane is 75-90 Å thick. Layers are not strictly separated but interpenetrate each other. The smaller particles appear to occupy positions nearer the outer membrane surface, the layer particles positions nearer the inner surface penetrate through intrathylakoid space. The rounded structure of 160 x 180 Å in size, had assumed to represent the structural equivalent of the `photosynthetic unit'.

Weier and Benson presented a bit complicated and atypical model. According to this model, the membrane is made up of single sheet of lipoprotein subunit, in which a protein core is coated on the outside with a lipid coat. The areas where two membrane are appressed was named as `partitions'. There are three electron transparent hydrophilic spaces : (1) loculi, space enclosed by the granule membranes; (2) fret channel, the space enclosed by stroma membrane; (3) the space that contains the stroma and thylakoid in the chloroplast. The arrangement of chlorophyll molecule is irregular. The porphyrin head is present in the areas between the subunits, while the phytol tail is buried in the interior of the subunits.

The electrochemical events take place in functional units, located on the inner vesicles of chloroplasts. These are the thylakoid disks, interconnected by intergrana lamellae. Observations of mature spinach shows that disk contains at least 105 chlorophyll molecules. The membrane of this disk is covered by at least 200 electron transport chains. One electron transport chain has two distinct photochemical reaction centers. The electron transport chain has two distinct photochemical reaction centers, which are supplied with quanta by an ensemble of about 300 antenna chlorophylls each.

PHOTOSYSTEM

The light is absorbed by photosynthetic pigments. The latter forms an aggregation in the form of photosystems. Light is absorbed by antenna pigments and transferred into photochemical reaction centers. There are two photosystem with reaction center P700 and P680. Various techniques ranging from magnetic resonance to X-ray scattering are applied to study molecular and function of the photosynthetic apparatus. Boardman (1968) partially separated two pigment systems by differential centrifugation of digitonin-treated chloroplasts. The heavier particles showed Hill's reaction, while the lighter particles could reduce $NADP^+$. Two photosystems can be removed from isolated chloroplasts by proper detergent treatment. Vernon and his co-workers (1967, 1969) used anionic detergent Triton X-100 for the separation of two photosystems. The removed photosystem can be separated by polyacrylamide gel electrophoresis. Junge (1982), opined that each granum contains about 200 units each of PS I and PS II, expressing that both PS are present in equal amount. According to Melis and Brown (1980) the ratio of PS I and PS II varies species to species and growth conditions. The ratio of PS II to PS I is known as photosystem stoichiometry. The latter varies between 0.4 and 1.7 among green and red algae, cyanobacteria, and plants. This may also be regulated by light quality. Both photosystems have a reaction center and antenna pigments. The reaction center is the place of electron transfer and the antenna pigments are light harvesting pigments. In most of the plants, about 250 chlorophyll molecules are associated with each reaction center. There are two types of photosystems: PS II and PS I. PS II is very strong oxidant, and can use light up to the wavelength of 680 nm. PS I is a very strong reductant, and can utilize light of 700 nm. Oxidation of water takes place in PS II, while PS I takes the responsibility of reducing NADP. PS I contains Chl `a', small amount of chl `b', and some β-carotene attached by noncovalent bonds to several proteins. A special type of chl `a' so that it can absorb light of 700 nm as well as at shorter wavelength, this molecule (reaction center) is special in terms to chemical environment. Antenna pigments are present surrounding this reaction center. Beside pigments, there are present at least two iron-containing proteins similar to ferredoxin. These are called Fe-S proteins. Four iron atoms of each protein is bound to two sulphur atoms. The four iron atoms of these proteins first receive electrons released from reaction center. The structure of PS I in a cyanobacterium has been worked out and it closely resembles with that of plants as well. It is a heterotrimer with each subunit in the trimer containing 12 different protein molecules bound to 96 molecules of chlorophyll a, two molecules of the reaction center chl P700, 4 accessory molecules closely associated with them, 90 molecules that serve as antenna pigments, 22 carotenoid molecules, 4 lipid molecules, 3 clusters of Fe4S4 and 2 phylloquinones. PS II is a complex of 20 different protein molecules bound to 50 or more chlorophyll `a' molecules, 2 molecules of the reaction center P 680, 2 accessory molecules close to them, 2 molecules of pheophytin, some molecules of chlorophyll and carotenoid serve as antenna pigments and with all this there are two molecules of plastoquinone.

Two major green bands can be separated from chloroplast by electrophoresis. These bands are in addition to photosystems. Each band contains both chl `a' and `b', xanthophylls and carotenoids and very little amount of α-carotene. These pigments are bound to proteins. The function of these bands is to absorb and harvest light and ultimately transfer it to proper photosystem.

Photolysis of water takes place in PSII

Light induced oxidation of water is known as photolysis of water. It takes place in reaction center at the lumel site of PSII. The oxidation of water releases O_2. This process needs four electrons, and takes place at high electropositive midpoint potential (+820mV) at pH 7. The excited reaction center ($P680^+$) is strong oxidant. The electropositive potential of P680/680+ is much more than required for splitting of water. Four manganese atoms are associated with the oxygen-evolving complex of PSII. Manganese is the main charge accumulator. The S-state model postulates a light-driven charge accumulation mechanism. It postulates that there are five successive states on increasing oxidation in PS II. These states are expressed by S_0 through S_4. After each charge separation state, the reaction center increases charge with +1 (P_{680+}). The fifth state (S_4) is a strong oxidant capable of oxidizing water and oxygen is evolved at this state.

Physiology of PSII

PS II is an integral complex. It contains chl `a' (P680 reaction center), and little chl `b' and β-carotene. This chl `a' molecule of reaction center is colourless, and lacks Mg^{2+}. Klimov, Krasnovski (1981) and Cogdell (1983) named this molecule as pheophytin (Pheo). Horst Witt and Wolfgang Saenger studied the three dimensional structure of PS II by X-ray crystallography. They studied PSII of cyanobacteria *Synechococcus elongates*. The studies of PS II and later of PSI has revealed that these pigment systems have construction on the basic principle same as the reaction center of purple bacteria. PS II has a pair of chl-a molecule at the reaction center, present on two arms, connected by one pheophytin molecule. The sequence is just like purple bacteria. Closely associated with Pheo, P680 and protein that binds P680 is a quinone, called as Q. It is of two types Q_A and Q_B. Quinone has the property to quench fluorescence of P680 by accepting its excited electron. PS II complex contains more than 20 proteins. One or more proteins contain bound manganese. Amesz (1983) named it manganese protein. It is a part of the inner side of the thylakoid membrane. Two proteins, D1 and D2 bind electron transfer prosthetic groups, such as P680, pheophytin and plastoquinone. Other proteins (CP43 and CP47) bind to chlorophyll, which present as a component of antenna pigments. Protein 33, 23 and 17kDa are involved in oxidation of water.

PS II transfers electrons from water to platoquinone (Fig. 5.4). The primary electron acceptor of PS II reduces plastoquinone with the uptake of protons form the matrix space. when this pigment system is excited, one chl-a molecule of the pair is released and goes to plastoquinone (Q_A) and forms plastosemiquinone (a semiquinone radical Q_{A-}). This electron is then transferred to Q_B and forms semiquinone Q_B. With this event, Q_A- is again converted into Q_A. Now a second electron is transferred from P680 to Q_A and again a second Q_{A-} is formed. QA- is again converted into Q_A by transferring its electron Q_{B-} and ultimately Q_{B2-} is formed. Thus semiquinone Q_{B-} is converted into fully reduced Q_{B2}-. Two protons from stromal side of the membrane goes to fully reduced Q_B and produces plastoquinol (Q_BH_2). Q_BH_2

dissociates from the PS II complex and diffuses into the lipid bilayer of the membrane. It functions as a mobile electron carrier.

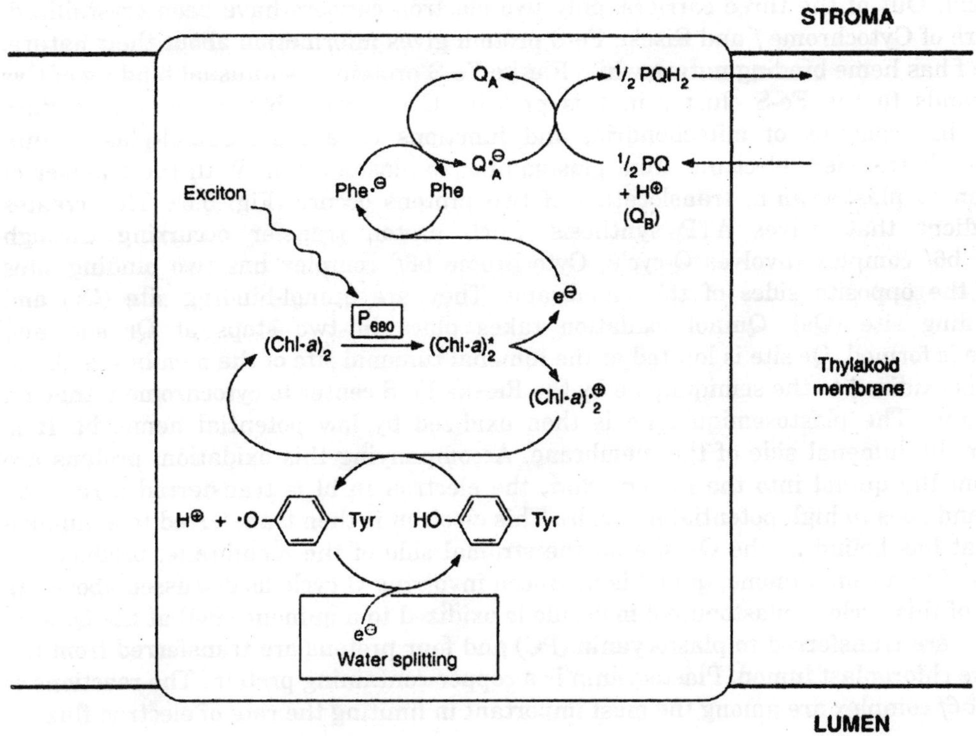

Fig. 5.4. Electron transport system in photosystem II

In oxygenic photosynthesis water acts as hydrogen donor and the electron deficit created during excitation of pigment is fulfilled by water. Thus PS II accepts electron from oxidation of water. The electron deficit, expressed as $(Chl\text{-}a)_{2+}$ radical has redox potential of about +1.1 volt, and it is a strong oxidant. According to Barber, (1984) manganese protein is probably involved directly in the first step of water oxidation. Manganese cations B and a tyrosine residue present in the protein of reaction center are involved in the transport of electrons from water to chlorophyll. (Chl-a)2+ radical withdraws an electron from tyrosine residue of this protein, and the electron deficit in the tyrosine is restored by oxidation of a manganese ion. PS II complex contains several manganese ions, and during electron transfer process, the Mn ion undergoes changes between the oxidation state Mn_{3+} and Mn_{4+}. The PS II complex contains several Mn ions, and to liberate one molecule of O_2 from water, the reaction center must withdraw four electrons thus, probably four electrons are released from Mn ions. A further advantage of many number of Mn ion is that it minimizes the formation of oxygen radicals. The latter is harmful for lipids and protein molecules. Mn cluster supplies four electrons one by one to the reaction center via tyrosine residue.

Role of Cytochrome b6f

Cytochrome b6*f* complex is an integral membrane protein. Cytochrome b6*f* complex contains three electron carriers; a high-potential c-type cytochrome (Cytochrome *f*), a high-potential 2Fe-2SFe-S protein (the Rieske Fe-S protein), and a b-type cytochrome (Cytochrome b6) with two b-type heme viz b_l and b_h. Heme bl is of low potential while b_h is of high potential. Out of the three carriers, only two electron carriers have been crystallized. The structure of Cytochrome *f* and Riseke Fe-S protein gives information about their nature. Cytochrome f has heme binding nature while Rieske Fe-S protein has unusual binding of Cys and His ligands to the Fe-S cluster in this protein. Cytochrome b6*f* complex resembles Cytochrome bc1 complex of mitochondria, and functions as a plastoquinol-plastocyanin reductase i.e., it transfers electrons from plastoquinol to plastocyanin. With the transfer of each electron to plastocyanin, translocation of two protons occurs (Fig. 5.5). This creates proton gradient that drives ATP synthesis. Such proton transfer occurring through Cytochrome b6*f* complex involves Q-cycle. Cytochrome b6*f* complex has two binding sites present on the opposite sides of the membrane. They are quinol-binding site (Q_P) and quinone-binding site (Q_B). Quinol oxidation takes place in two steps at Q_P site and semiquinone is formed. Q_P site is located at the lumenal tumenal site of the membrane. At Q_P site, quinol is oxidized to the semiquinone by the Rieske Fe-S center to cytochrome f, then on to plastocyanin. The plastosemiquinone is then oxidized by low potential heme, bl. It is located near the lumenal side of the membrane. Accompanying this oxidation, protons are released from the quinol into the lumen. Now, the electron in bl is transferred across the membrane and goes to high potential heme, b_h. This electron is then transferred to a quinone molecule that has bound at the Q_B site on the stromal side of the membrane, producing a semiquinone. From semiquinone, quinol is produced involving Q cycle as discussed above. As a net result of this cycle, a plastoquinol molecule is oxidized to a quinone (PQ) at the Q_P site, two electrons are transferred to plastocyanin (PC) and four protons are transferred from the stroma to the chloroplast lumen. Plastocyanin is a copper-containing protein. The reactions of cytochrome *b6f* complex are among the most important in limiting the rate of electron flux.

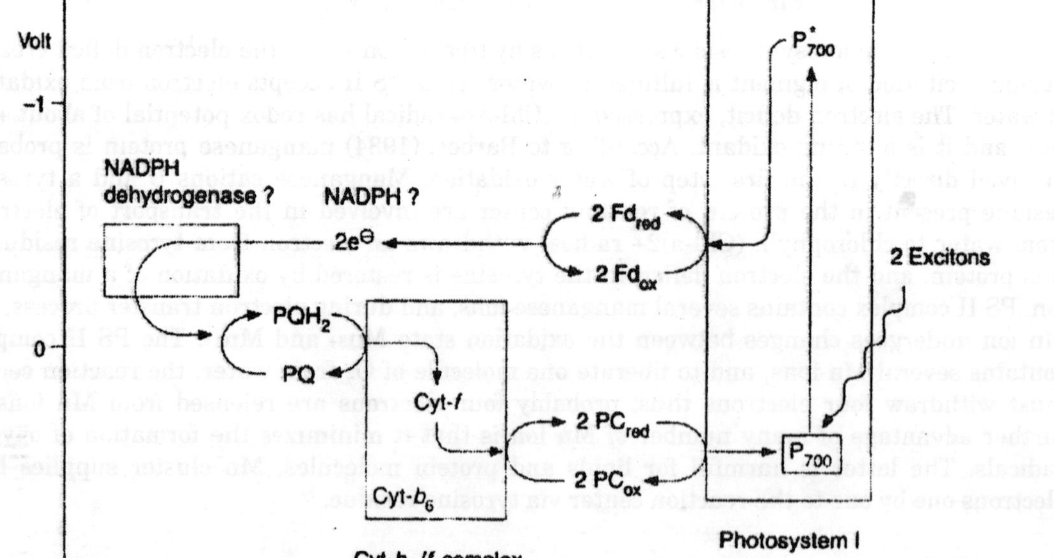

Fig. 5.5. Electron transport system between photosystem I and the cyt-b6lf

Physiology of PSI

PSI along with photosynthetic pigments, it also contains fifteen protein subunits. Out of them, PsaA and PsaB take part in binding the major electron transfer carriers. The major binding carriers of PSI are P700, chla acceptor molecule, phylloquinone (vitamin K1, the A1 acceptor), and the bound Fe-S center. PsaD binds with ferredoxin, and PsaF binds with plastocyanin. In 1957, Kok discovered P-700. When P-700 was titrated with mixture of ferro-ferricyanide, it showed that P-700 to have a midpoint potential of +430 mV and behave as a one-electron transfer agent. PSI functions as a light-dependent plastocyanin-ferredoxin oxidoreductase. Plastocyanin is the electron donor to PSI, thus it is an intermediate in the reducing system of PS I. Plastocyanin is a copper containing blue protein. Plastocyanin is bound loosely to the inside of thylakoid membranes next to channel. PS II reduces copper of plastocyanin ($Cu_2^+ \rightarrow Cu^+$). The reduced form can move along the membrane carrying an electron to PS I. PSI while receiving electrons from plastocyanin, oxides the latter ($Cu^{2+} \rightarrow Cu^+$). PSI accepts the electron and transfers it to ferredoxin. Ferredoxin is located in chloroplast stroma. It is a strong reductant in nature (redox potential -420mV), and reduces $NADP^+$. Electron transfer takes place through ferredoxin-$NADP^+$ redcutase (FNR) enzyme. FNR is an FAD-containing enzyme, and loosely associated with the thylakoid membrane. Ferredoxin and FNR form a complex through electrostatic interactions. FNR can accept two electrons. Addition of first electron, i.e., first reduction results in flavin semiquinone state and the second electron reduces FAD into fully reduced $FADH_2$. Reduced FNR ($FADH_2$) transfers its two electrons to $NADP^+$ (Fig. 5.6).

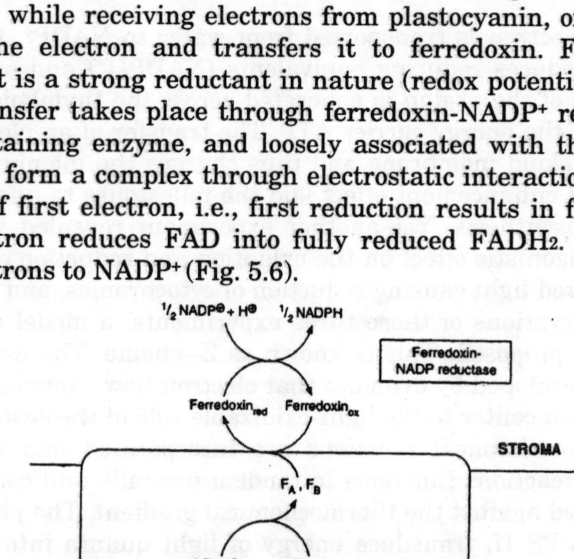

Fig. 5.6. Electron transport system in photosystem I

The ferredoxin

The strong reductant produced in PS I is X (chemically unidentified) passes its electron to ferredoxin through FRS (ferredoxin reducing substrate). Ferredoxin is an iron-sulphur protein with a redox potential of -0.43 V. SanPietro and Lang (1956) first extracted this protein and the partially purified protein was named as `photosynthetic pyridine nucleoide reductase' (PPNR). Later Tagawa and Arnon (1962) purified this protein and named it as ferredoxin, because of its similar properties to the ?iron-dedox protein? or `ferredoxin' isolated from nitrogen-fixing bacteria. Reduced ferredoxin is strongly autoxidazable. Reduced ferredoxin has affinity for the flavoprotein $NADP^+$, and reducing system is called ferrodoxin-$NADP^+$ reductase.

THE ELECTRON TRANSPORT SYSTEM

In photosynthesis, electron is transported from water to $NADP^+$. Such transfer serves two purposes: first it produces reducing equivalents ($NADPH^+$) and oxygen, the second is electrochemical potential of the proton is generated across the thylakoid membrane. This is used in the production of the energy carrier ATP. The transfer of an electron to the primary acceptor crosses the thylakoid membrane and thus charges the membrane electrically. The discovery of red drop and enhancement effect laid the milestones to understand the electron transport chain in photosynthesis. Yet another experiment revealed that red and far-red wavelengths had an antagonistic effect on the oxidation and reduction of cytochromes in the electron transport chain-red light causing reduction of cytochromes, and far-red causing their oxidation. With the observations of these three experiments, a model of electron transport chain in chloroplast was proposed. This is known as Z-scheme. The essential argument for the Z-scheme has been developed by evidence that electron flows between the light-reducible (Q) of the system 2 reaction center to the light-oxidizable side of the system 1 reaction center. In this scheme two photochemical reactions are incorporated and the electron flow is noncyclic. The two light reactions functions in tandem partially and collaboratively to allow for electron flow to proceed against the thermochemical gradient. The photochemically active chlorophylls of PS I and PS II, transduce energy of light quanta into the energy of redox couples. PS I and PS II are found in different regions of the thylakoid membrane, whereas the cytochrome b_6f complex is distributed equally between appressed and stroma-exposed membranes. This distribution expresses that mobile electron carriers connect the membrane complexes. Plastocyanin and plastoquinone are the diffusible electron carriers.

Reductional Potential and the Z-scheme

In photosynthesis, electron moves from water to $NADP^+$. It is difficult to oxidize H_2O as O_2 attracts electrons and protons strongly. Reduction of $NADP^+$ is easier thermodynamically. Standard reduction potential ($E_{o'}$) of oxidation-reduction occurring in photosynthesis is -0.815V, and for reduction of NADP+ it is -0.32 V. The overall process of electron transfer from H_2O to $NADP^+$ arrives at a change in $E_{o'}$ of -1.14V. This is expressed in the following equation

H_2O	$\rightarrow 1/2O_2 + 2e^- + 2H^+$	Eo' = -0.82V
$NADP^+ + 2e^- + H^+$	$\rightarrow NADPH$	Eo' = -0.32V
$H_2O + NADP^+$	$\rightarrow 1/2O_2 + NADPH^+ + H^+$	Eo' = -1.14V

The standard Gibbs free energy of the above reaction is approximately 220 kJ or 52.6 kcal. The highly positive value obtained demonstrates the strongly endergonic nature of photosynthesis (Fig. 5.7).

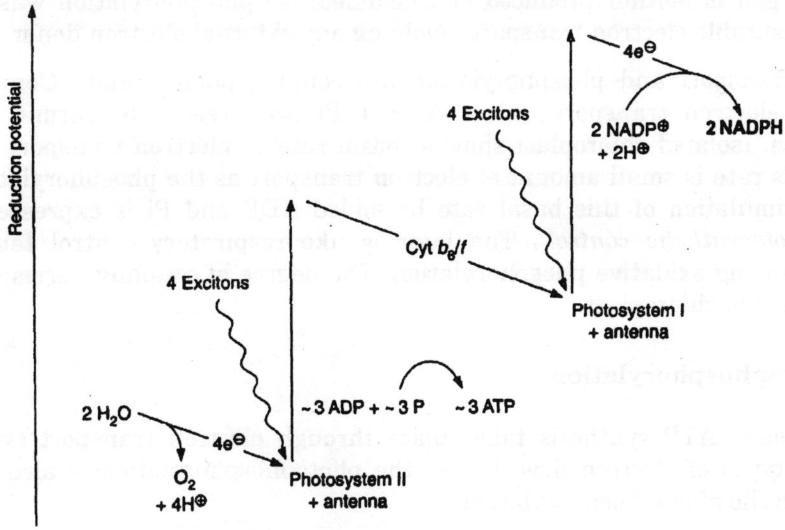

Fig. 5.7. The Z-scheme

Cyclic Electron Transport Chain

Cyclic electron flow in chloroplast was confirmed by application of dichlorophenyl-dimethyl-urea (DCMU) a inhibitor of PS II. In cyclic electron flow takes place in PS I and only ATP is produced. The cyclic pathway involves ferredoxin as a native cofactor. PSI reduces ferredoxin in light, i.e., electron ejected from PS I has no role in reduction of $NADP^+$. Reduced ferredoxin (Fdx_{red}) is proposed to interact with an Fdx-plastoquinone oxidoreductase that allows for the transfer of electrons into the quinone pool. Plastoquinol is oxidized by cytochrome b_6 complex, and proton translocates across the membrane. Transolcation of proton takes place possibly by Q cycle. The role of this pathway in vivo was a subject of discussion, however, it is established that this pathway provides ATP for CO_2 fixation.

PHOTOPHOSPHORYLATION

Photosynthetic electron transport system gives NADPH. Besides NADPH, ATP is also synthesized in this photoact. During photosynthesis, ATP synthesis is light dependent, hence it is known as photophosphorylation. The ATP produced by photophosphorylation is essential for CO_2 assimilation. Frenkel (1954) observed phosphorylation in cell free suspension of a bacterium, *Rhodospirillum rubrum*. Arnon *et al.* (1954) discovered ATP synthesis in isolated chloroplast in presence of light. Arnon *et al.* (1958, 1959) and Avron et al., (1958) discovered direct measurement of simultaneous, coupled electron transport and ATP synthesis in the cell-free milieu. In isolated chloroplast, photophosphorylation takes place in presence of some co-factors, *viz.* phenazine methosulfate (PMS), riboflavin phosphate and ferredoxin. ADP, Pi and Mg^{2+} are required for such photophosphorylation. Photophosphorylation shows some distinguishing features from other types of phosphorylations taking place in plant cells like (1) ATP formation occurred in chloroplast (lamellae) and was independent of other enzyme systems or organelles; (2) absorbed photon served as a source of energy; (3) oxygen is neither produced or consumed; (4) phosphorylation was not accompanied by a measurable electron transport involving any external electron donor or acceptor.

Electron transport and phosphorylation are coupled phenomenon. Conversely, there should be no electron transport unless ADP + Pi are present to permit simultaneous phosphorylation. Isolated chloroplast show a `basal rate' of electron transport in absence of ADP or Pi. This rate is small amount of electron transport as the phosphorylation system is absent. The stimulation of this basal rate by added ADP and Pi is expressed as a ratio, known as *`photosynthetic control'*. The later is like respiratory control taking place in mitochondria during oxidative phosphorylation. The degree of coupling varies depending on the physical state of chloroplast.

Types of Photophosphorylation

In chloroplast, ATP synthesis takes place through electron transport system, and as there are two types of electron flow, hence, the photophosphorylation is also of two types: cyclic and noncyclic photophosphorylation.

Cyclic photophosphorylation

Cyclic photophosphorylation is the only type of photophosphorylation in anoxygenic photosynthetic organisms. In presence of light and high concentrations of ferredoxin, isolated chloroplast carry out cyclic photophosphorylation.

In *Chlorella* sufficient ATP is synthesized by cyclic photophosphorylation, but for CO_2 fixation, both types of photophosphorylation are essential. The energy provided by additional ATP (produced by cyclic photophosphorylation) is consumed in synthesis of adaptive enzymes. Only PS I is involved in cyclic photophosphorylation, and it takes place at longer wavelength, greater than 680nm. Electron ejected from P700, goes to FRS or ferredoxin, and than moves downhill, and returns to PS I. The downhill movement of electron releases sufficient energy to synthesize ATP(•) Cyclic photophosphorylation may be expressed as following:

$$\text{Light (long wavelength)} + ADP + Pi \rightarrow ATP + H_2O$$

ATP is synthesized between cyt b6 and cyt *f*, as the potential gap is more than 32V. It is yet to confirm the site of phosphorylation and the role of cyclic photophosphorylation in CO_2 assimilation in higher plants.

Noncyclic photophosphorylation

During non-cyclic photophosphorylation, both types of pigment systems i.e., PS I and PS II are involved, and the flow of electron is unidirectional. Electron donated by PS II transports through plastoquinone, cytb6, cyt f, plastocyanin and goes to PS I. The electron received by PS I is transferred to ferredoxin and ultimately to NADP+. Thus the electron coming from PS II does not goes back to PSII. The loss of electron is furnished by water.

During the passage of excited electrons from PS II to PS I, the ATP synthesis is presumed to occur between cytochrome b6 and ctyochrome *f*. As the difference between the redox potentials of these two is approximately 0.33V, it is sufficient to synthesize ATP. Two molecules of water are oxidized with the yield of one molecule of oxygen, 2NADPH+ and 2 ATP molecules. The stoichiometry of non-cyclic photophosphorylation is one molecule of ATP for each molecule of NADPH produced, or a P/2e ratio of 1: 0. The overall reaction of noncyclic photophosphorylation is expressed as following equation :

$$2ADP + 2Pi + 2NADP^+ + 2H_2O \rightarrow 2ATP + 2NADPH{+}H^+ + O_2$$

Non-cyclic photophosphorylation is inhibited by substances that block oxygen evolution.

Pseudocyclic photophosphorylation

Arnon and his coworkers (1954) observed that if illuminated chlorophyll produces ATP even in the absence of CO_2 and $NADP^+$ and in presence of FMN or vit K and oxygen. This phenomenon is described as oxygen dependent FMN catalysed photophosphorylation and named it as pseudocyclic photophosphorylation. In this type of phosphorylation, reduction of FMN and oxidation of water takes place. Arnon explained the mechanism, water acts as the source of electron to reduce FMN. The reduced FMN is reoxidised by oxygen. FMN with the help of illuminated chloroplast, breaks and rejoins water molecule. ATP is synthesized during the process, which depends upon the reduction of FMN.

Physiology of photophosphorylation

ATP synthesis is energetically coupled to electron transport. The mechanism of ATP synthesis was explained in 1960s by Mitchell. He proposed chemiosmotic theory of phosphorylation. The basic principle of this theory is phosphorylation takes place in intact membrane system with low intrinsic permeability to protons, and electron flowing through carriers cross the membrane at defined site. This is discussed in details later in the chapter 2.

Lumen thylakoids in chloroplast contains regions of the electron transport chain, where energy conservation takes place. At these sites, proton accumulation is coupled with electron transport. One site is involved in accumulation of protons released during oxidation of water and the second site accumulates protons liberated during oxidation of plastoquinol. During electron transport, as the electron pass through a series of carriers. Protons are also translocated across the membrane at coupling sites. In chloroplasts, as electrons are transferred along the chain the protons move from the stroma to the lumenal side of thylakoid. The unidirectional flow of protons results in high concentration of protons at lumenal side. The resulting gradient supplies the proton motive force (pmf):

$$\Delta pmf = \Delta pH + D\Psi$$

where pmf is proton motif force, and <F128M>Y<F255D> is membrane potential. This pmf is the high energy state driving ATP synthesis. Mitchell (1966), McCarty (1976) gave arithmetic details of energy available from one mol equivalent of proton trapped inside. The equation is as following :

$$\Delta G' = 1.36\ (pH) + 1.36\ (\Delta\Psi/59)$$

A direct formula to understand pmf was also worked out.

$$Pmf = \Delta\mu\text{-}H+ /96.5 \text{ kJ } V^{-1} \text{ mol}^{-1}$$

Above equation shows interrelationship between the pH component and electrical potential component. pH component arises from a concentration difference of protons across the membrane (ΔpH), while the electrical potential component arises from a difference in charge across the membrane (ΔΨ). In chloroplast, ΔpH mainly contributes to Pmf. The latter is the driving force for ATP synthesis. The coupling of phosphorylation and electron transfer has a feed back system. As the electron transport chain moves protons into the lumen, the lumenal concentration of protons would be expected to increase, generating a back pressure that slows further proton pumping. The chemiosmotic model predicts a fixed stoichiometry between the number of protons transported across the membrane and the number of ATP molecules synthesized. In simple terms the so-called H^+/ATP ratio. In photophosphorylation the value is $4H^+$/ATP.

The ATP synthase

ATP synthase enzyme uses proton motive force to convert ADP to ATP. This enzyme has a hydrophilic head and a hydrophobic channel. The hydrophilic head is present towards the stromal side of the membrane and the hydrophobic channel allow H+ ions to diffuse spontaneously from the lumen to the stroma side. The stromal surface has CFo segments, which participates in translocating protons across the membrane to the catalytic portion of the enzyme ATP synthase. CF_o has subunits, I, II, III and IV. The thylakoid ATP synthase has a transmembrane segment, CF_1, which is actually involved in phosphorylation. CF_1 has two relative large subunits, α and β, and three smaller subunits γ, δ and ε. CF_1 contains three copies of each large subunits and one copy each of small subunits to give an overall sotichiometry of $\alpha_3\beta_3\gamma\delta\varepsilon$. The larger subunits bind ADP and phosphate and catalyze ATP synthesis. The δ subunit links CF_o to CF_1, and the γ subunit appears to control proton-grating through the enzyme, ε subunit blocks catalysis in the dark, preventing the breakdown of ATP.

Paul Boyer proposed the synthesis of ATP due confromational change mechanism occurring in binding sites in CF_1 portion of enzyme ATP synthase. There are distinct conformational states, designated as loose binding site (L), tight nucleotide binding (T) and an open binding sites (O). These sites are free of nucleotides. The CF_1 portion has three distinct nucleotide-binding sites. Each binding site can exist in one of the three conformational state i.e., either L, T or O. According to conformational mechanism, ADP and Pi initially bind to an unoccupied site in the open state. When proton moves from the lumen to the stromal site through CF_o channel, their is release of energy. This energy brings about rotation in the three nucleotide binding sites. The T site contains bound ATP. When ATP is released, T site is converted to an O site. L site, when containing the bound ADP and Pi is converted into a T site. Such rotation facilitates the ATP synthesis.

C3 plants

In dark reaction of photosynthesis, as is commonly known, CO_2 is reduced by reductant produced in photochemical reaction. It is a thermochemical stage of photosynthesis. In 1950s Melvin Calvin, Andrew Benson, and James A Bassham were the first to elucidate a multistep carbon fixation pathway called the *reductive pentose phosphate cycle*. This pathway is apparently ubiquitous in all photoautotrophic green plants (Norris et al., 1955). In these plants, a three carbon-compound, 3-phosphoglycerate is produced as the first `stable product', hence the name C3 plants. The C3 pathway has been thoroughly analyzed. Several techniques have helped to understand the reactions of dark reaction. Application of tracer technique has proved to be the most successful in understanding the path of carbon in this phase of photosynthesis. Two forms of isotopic carbon ^{11}C and ^{14}C were discovered, with half-lifetime of 20.5 seconds and 5720 years, respectively. The use of radioactive carbon, making it possible to tag the CO_2 ($^{14}CO_2$) which is incorporated into organic compounds during photosynthesis. Application of ^{14}C made it possible to follow the passage of the carbon tracer form a `tagged' substrate into one organic compound after another. The experimental photosynthesizing cells (algal cells) was exposed to $^{14}CO_2$ for seconds or minutes. The suspension of cells was collected and extracts was concentrated and two-dimensional chromatograms were prepared, and to identify the paper chromatogram, the filter paper was pressed against a X-ray film sheet. This sheet was developed after proper exposure. The radiation (β-particles), emanating from the radioactive carbon present in certain spots on filter paper, blackness the opposite spots on the X-ray film. Radio chromatograms serves the purpose of identifying each compound containing ^{14}C. An alternate method is to cut out the paper containing the radioactive compound and measure its radioactivity directly with a Geiger counter. Racker (1955) suggested another technique for the analysis of thermo chemical reaction which involves the isolation of the enzymes involved in each of the thermo chemical steps in photosynthesis.

Calvin and co-worker used Chlorella *pyrenoidosa* or *Scenedesmus obliques* for the experiments to identify the early stable intermediates in the path of carbon from CO_2 to plant metabolites. Experiments were carried out in a unique reaction vessel, a `lollipop', designed for the rapid removal samples at short time duration. The compounds were identified in the chromatographs both by their radioactive emissions and by chemical reactions. Calvin (1950, 1951, 1962) Bassham and Calvin (1962) and Bassham *et al.* (1955) observed that the radioactive carbon dioxide is rapidly spread throughout the various classes of compounds present in the plant cells. In cells killed immediately after a 1 second exposure to light the ^{14}C was found mainly in phosphoglyceric acid. In cells killed after 5 seconds of illumination, traces of phosphoglyceric acid, alanine, aspartic acid and malic acid containing ^{14}C appeared. Likewise cells were killed after 15 seconds, one minute, and five minutes of illumination, and compounds were identified containing ^{14}C. After 5 seconds in addition to the other compounds, faint traces of sucrose was detected and also some pentose, glucose phosphate and fructose phosphate, was detected. After a minute abundant radioactive glucose phosphate and fructose phosphate were detected. At the end of 5 minutes of illumination radioactivity may be able to reach the proteins.

Main Features of Calvin Cycle

Calvin cycle (Fig. 5.8) is also known as reductive pentose phosphate (RPP). The peculiarity of this pathway is carbohydrate synthesis as well as regeneration of phosphorylated pentose sugar. There are four main features of Calvin cycle :

1. Corboxylation : CO_2 joins with the ribulose 1, 5-bisphosphate (RuBP) to yield two molecules of 3-phosphoglycerate (PGA). This reaction is catalyzed by RuBP carboxylase, the details of this complex enzyme is discussed in another section of this chapter.

2. *Reduction* : PGA is reduced by NADPH+H⁺ to give rise to phosphotriose sugar.

3. *Regeneration* : Five molecules of triose phosphate are rearranged to regenerate 3 molecules of the CO_2-acceptor.

4. *Autocatalyst* : For every 3 molecules of CO_2 entering the pathway one triose phosphate molecule is produced. The latter can be converted to starch, sucrose, etc. It can also enter the regenerative phase of the pathway leading to an autocatalytic build-up of intermediates.

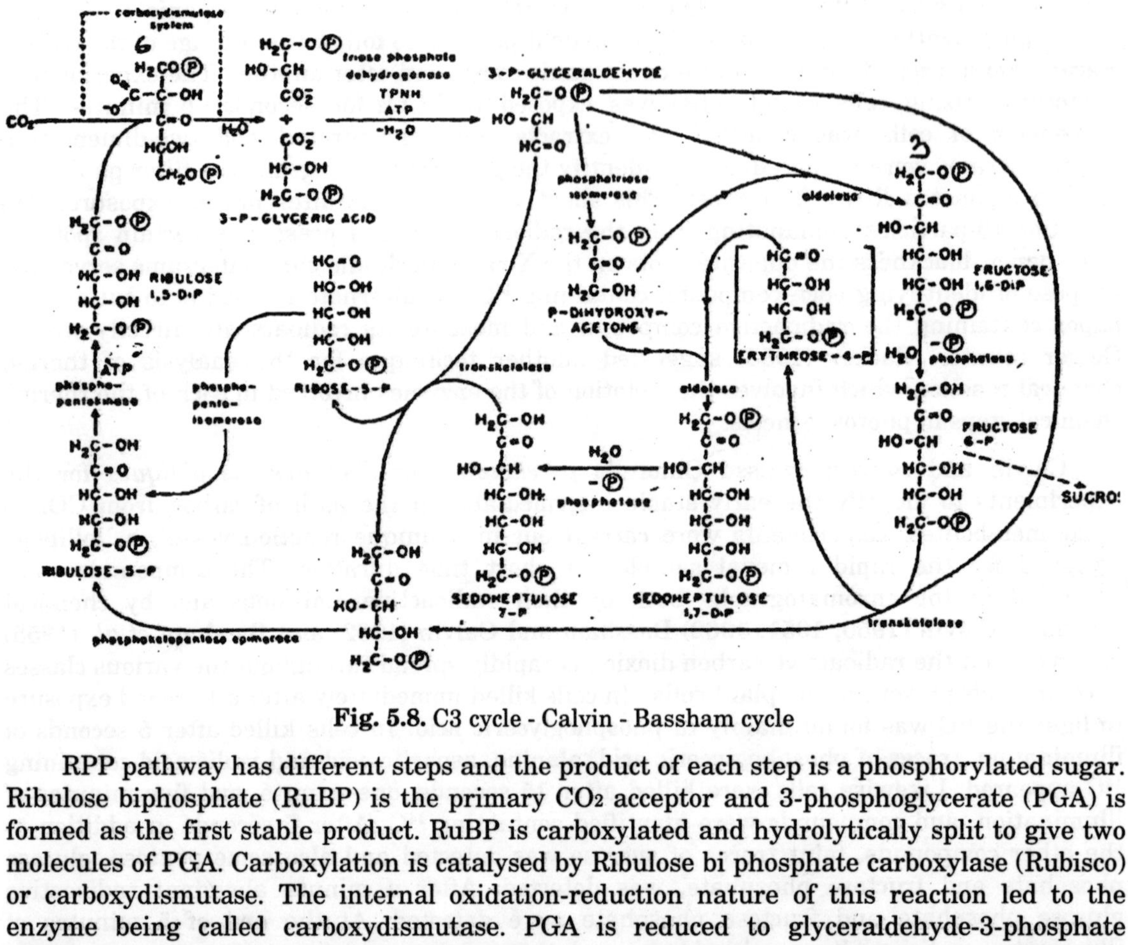

Fig. 5.8. C3 cycle - Calvin - Bassham cycle

RPP pathway has different steps and the product of each step is a phosphorylated sugar. Ribulose biphosphate (RuBP) is the primary CO_2 acceptor and 3-phosphoglycerate (PGA) is formed as the first stable product. RuBP is carboxylated and hydrolytically split to give two molecules of PGA. Carboxylation is catalysed by Ribulose bi phosphate carboxylase (Rubisco) or carboxydismutase. The internal oxidation-reduction nature of this reaction led to the enzyme being called carboxydismutase. PGA is reduced to glyceraldehyde-3-phosphate (GAP). The reduction of PGA to GAP occurs in two steps. First, the PGA is converted to the acyl phosphate in a reaction using ATP and mediated by PGA kinase and glycerate-1,3-bisphosphate (PPGA) is formed. The PPGA is reduced with NADPH+H⁺ and is catalysed by enzyme triose phosphate dehydrogenase. The redox potential of the couple carboxyl-carbonyl is about -0.5, and NADPH+H⁺ does not have enough reducing power. The redox potential of NADPH+H⁺ is Eo'=-0.35, which is not sufficient to reduce a carboxyl group to a carbonyl group and, therefore, this reaction is to be coupled with the hydrolysis of ATP. Some molecules of glyceraldehydes-3-phosphate take part in formation of hexose sugar and the rest regenerate the CO_2 acceptor. The conversion of triose phosphates and hexose phosphates to pentose phosphate, requiring changes in chain length, is initiated with transketolase which transfers C-1 and C-2 from Glyceraldehydes-3-phosphate which undergoes partial isomerization. Enzyme triose phosphate isomerase converts phsophoglyceraldehyde into dihydroxyacetone phosphate. In the next step enzyme aldolase catalyzes the formation of fructose 1,6 biphosphate (FBP). In this step, one molecule each of phsophoglyceraldehyde and dihydroxyacetone unites to give phosphorylated hexose sugar. FBP is then converted to fructose-6-phosphate(F6P) with enzyme fructose-1-6-bisphosphatase. One sixth molecules of this sugar are converted into glucose and ultimately into starch, and rest take part in regeneration of ribulose biphosphate. The conversion of triose phosphates and hexose phosphates to pentose phosphate, requiring changes in chain length, is initiated with transketolase which transfers C-1 and C-2 from F6P to GAP. In the regeneration pathway, fructose-6-phosphate and glyceraldehydes-3-phosphate react to form a four carbon, erythose-4-phosphate (E4P) and a five carbon sugar, xylulose-5-phosphate (Xu5P). This reaction is catalyzed by enzyme transketolase. The tetrose sugar erythose-4-phosphate combines with a molecule of dihydroxyacetone phosphate and forms sedoheptulose-1,7-diphosphate (SBP). Enzyme aldolase combines these sugars and then enzyme sedoheptulose-1,7 phosphatase removes one phosphate moiety to produce sedoheptulose-7-phosphate (S7P) and Pi. This enzyme is distinct from fructose-1-6-bisphosphatase. A second transketolase-mediated step follows in which C-1 and C-2 of S7P are transferred to GAP to give two pentose phosphates. Sedoheptulose-7-phosphate and glyceraldehyde-3-phosphate react to form two molecules of pentose sugars, viz. ribose-5-phosphate (R5P) and xylulose-5-phosphate (Xu5P). This completes the conversion of five GAP molecules to three pentose phosphate molecules. Ultimately both sugars isomerise into ribulose-5-phosphate (Ru5P). Ribose-5-phosphate is converted into ribulose 5-phosphate by enzyme phosphopentoisomerase, while enzyme epimerase converts xylulose-5-phsophate into ribulose-5- phosphate. In the last step of regeneration, one phosphate moiety comes from ATP and joins with ribulose-5-phosphate to give ribulose-1,5-biphosphate. This is catalyzed by enzyme phosphopentokinase.

The Calvin cycle shows a number of sugar intermediate hence, it becomes necessary to understand the balance sheet of carbon. This is presented in a simple manner by following equation :

(1) $RuDP + 6\ CO_2 \rightarrow 12\ PGA$

$12 \times (2H)$

(2) $12PGA \rightarrow 12C_3$ (phosphoglyceraldehyde)

12ATP

(3) $2C_3 \rightarrow 1\ C_6$ (net gain of hexose)

(4) $4C_3 \rightarrow 2C_6$

(5) $2C_6 + 4C_3 \rightarrow 2C_5 + 2C_7$

(6) $2C_7 + 2C_3 \rightarrow 4C_5$

6ATP

(7) $6C_5 \rightarrow 6RuDP$

Thus for the synthesis of one mole of hexose it needs assimilation of six molecules of CO_2, 18 moles of ATP and 12 moles of $NADPH+H^+$. According to Bassham *et al.* (1968); Stocking and Larson, (1969); Heber *et al.* (1970); Werden and Heldt, (1971); Bemberger *et al.*

(1975) three molecules of RuBP are carboxylated to give six molecules of PGA, and after reduction, six molecules of GAP. Only five of these GAP molecules are required to regenerate the three RuBP molecules. The sixth GAP molecule, equivalent in carbon, to the three carbon dioxide molecules fixed, can either be converted by a reverse of glycolysis to glucose phosphate for starch synthesis, or can be exported from the chloroplast to the cytoplasm for extrachloroplastic reactions.

Enzymes of RPP

The enzymes catalyzing biochemical reactions of RPP cycle are water soluble and are located in the stroma region of chloroplast. There are thirteen biochemical reactions in this pathway and out of them only three steps use up cofactors. Ribulose 5-phosphate is CO_2 acceptor and forms PGA, however, the latter is not formed directly. Fuller and Gibbs (1959) demonstrated the presence of enzyme RuBP in various photoautotrophic organisms. The CO_2 acceptor first forms a six carbon intermediate, which is immediately cleaved into two molecules of 3-PGA. As already mentioned this step is catalyzed by enzyme ribulose bisphosphate carboxylase/oxygenase (Rubisco). This enzyme is a soluble protein in chloroplast and consists of two types of subunits. These are eight large (L) subunit of 56-kDa and eight small subunit (S) of 14-kDa. L subunit contains the catalytic domain, and the role of S subunit is not clear. An enzyme Rubisco activase is specifically involved in the activation of Rubisco through carbamylation. This enzyme appears to promote the dissociation of sugar carbamylation. Rubisco also has an oxygenase activity that uses molecular oxygen as a substrate instead of CO_2. Reaction of O_2 with Rubisco produces 3-PGA and the two-carbon molecule, 2-phosphoglycolate. Studies of Rubisco from obligately anaerobic bacteria, indicate that the oxygenase activity is intrinsic to Rubisco and reflects that the ancestral enzyme evolved in an anaerobic atmosphere. CO_2 and O_2 compete for the same active site on the enzyme, and the activity of the enzyme toward these substrates is dictated by the relative amounts of O_2 and CO_2 in the environment. However, in air, the carboxylation reaction proceeds approximately three times faster than the oxygenation reaction.

Stoichiometry and energetics

In RPP cycle for fixation of each molecule of CO_2, one mol of ATP is required for the conversion of Ru5P to RuBP, the substrate of carboxylation. Two mol of ATP are required in the subsequent activation of the 2 mol of PGA formed, giving PPGA. The reduction of two mole of PPGA requires 2 mol of NADPH+H^+. To synthesize one mol of GAP, it requires 3 mol of CO_2, 9 mol of ATP and 6 mol of NADPH+H^+. The overall result can be expressed by the sum of the following two equations. The first equation represents the utilization of the cofactors from the light, and the second is for the conversion of CO_2, water, and Pi to GAP.

$$6(NADPH + O_2 + H^+ \rightarrow NADP+ + H_2O) \qquad \Delta G' = -315.5 \text{ Kcal}$$

$$9(ATP\ 4+ + H_2O \rightarrow ADP^{3-} + Pi^{2-} + H^+) \qquad \Delta G' = -68.8 \text{ Kcal}$$

$$\text{Total} \qquad \Delta G' = -384.3 \text{ Kcal}$$

$$3CO_2 + 3H_2O + Pi^{2-} \rightarrow GAP^{2-} + 3O_2 \qquad \Delta G' = +350.4 \text{ Kcal}$$

The balance of subtracting the energy stored from the energy expended, is $\Delta G' = -33.9$ Kcal as the driving energy for one turn of the cycle. The efficiency would be 350/384 = 91%.

Regulation in Calvin cycle

The enzymes of Calvin cycle like aldolase, transketolase, glyceraldehydes- 3-phosphate dehydrogenase catalyze reversible reactions in glycolytic pathway. Changes in stromal pH and Mg^{++} concentration are important regulators. Stromal Mg^{++} concentration is an important factor in the activation of enzyme Rubisco. Evidences indicates that Mg^{++} concentrations in the stroma increases from approximately 1-3 mM to approximately 3-6 mM in the light. As light drives H^+ into the thylakoid lumen, Mg^{++} is released from the lumen to compensate for the influx of positive charges. Other enzymes are also activated in the light in response to change in the pH of the stroma.

A redox-dependent regulatory system involving ferredoxin, thioredoxin and enzyme ferredoxin-thioredoxin reductase plays an important role in light activation of certain chloroplast enzymes. Thioredoxin is a low-molecular-mass disulfide containing protein. Reduced ferredoxin is produced by the light reactions of photosynthesis and reacts with oxidized thioredoxin *f* or *m* in a reaction catalyzed by ferredoxin-thioredoxin reductase (FTR). The reduction results in altering the target's conformation an modulating its enzymatic activity. In the dark, the regulatory sulfhydryl groups of the enzyme become oxidized. Reduced thioredoxin are activated by reduction in the light and deactivated by oxidation in dark. This links the light reactions to CO_2 fixation. This regulatory mechanism has been demonstrated for enzymes Rubisco, NAD+-glyceraldehyde-3-phosphate dehydrogenase, sedoheptulose-1,7 bisphosphatase, phosphoribulokinase, as well as ATP synthase.

According to Kanazawa (1970), there are rate limiting steps to keep in balance the concentrations of intermediate compounds as the physiological needs and rates of photosynthesis of the cells changes. The rate limiting steps in the light are the formation and the carboxylation of RuBP and the conversions of FBP and SBP to F6P and S7P, respectively. Thus the rate of carboxylation reaction relative to the bisphosphate reaction determines the rates of formation and utilization of triose phosphates.

Sucrose Synthesis

The principle end product of photosynthesis is sucrose, which is synthesized in the cytosol of the mesophyll cells. It can be easily transported to the other parts of the plant. Glucose-1-phosphate reacts with UTP and glucose-UDP is formed and pyrophosphate is released. This reaction is catalyzed by enzyme UDP-glucose pyrophosphorylase. The formation of UDP-glucose from glucose-1-phosphate and UTP has a small negative free energy change ($\Delta G^{o'}$=-2.88 KJ/mol). The glucose residue is transferred from glucose-UDP to fructose-6-phosphate by enzyme sucrose phosphate synthase. In this reaction sucrose-6-phosphate is produced. The free energy of the formation of sucrose-6-phosphate by sucrose phosphate synthase is ($\Delta G^{o'}$=-5.7 KJ/mol). The is dephosphorylated by enzyme sucrose phosphate phsophatase. The hydrolysis of sucrose-6-phosphate sucrose plus phosphate has large negative free energy change ($\Delta G^{o'}$=-16.5 KJ/mol). Sucrose phosphate synthase and sucrose phosphate phsophatase form a complex. The overall free energy of sucrose synthesis is highly negative ($\Delta G^{o'}$=-25 KJ/mol). This favours the formation of synthesis. Sucrose synthesis is regulated by the activity of sucrose-phosphate synthase. the phosphorylation-dependent down-regulation of sucrose-phosphate synthase is enhanced by 14-3-3 proteins in a manner similar to regulation of nitrate reductase (Fig. 5.9).

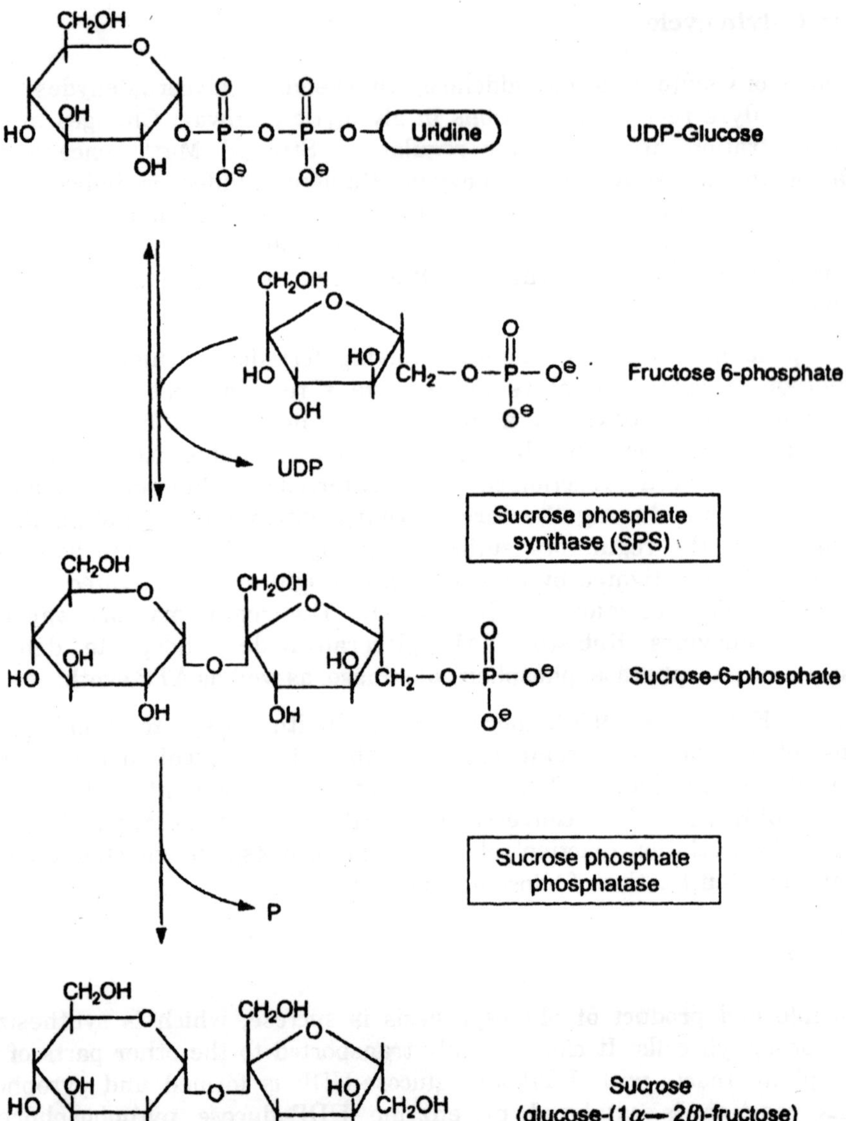

Fig. 5.9. Synthesis of sucrose

In nonphotosynthetic tissues contain enzyme sucrose synthase. This enzyme catalyzes breakdown of sucrose. It is present in high concentration in sucrose utilizing tissues. Sucrose reacts with UDP to form UDP-glucose and fructose. UDP-glucose thus formed is utilized in starch synthesis, which is stored in amyloplast.

Starch Synthesis

Starch is synthesized and stored in plastids. In chloroplast, when the export of sucrose from the cell cannot keep pace with photosynthesis, the excess fixed carbon enters the hexose

phosphate pool and is finally stored in the chloroplast as starch. Starch is stored in amyloplast for longer period. Cytosol provides carbon source to amyloplast for the synthesis of starch.

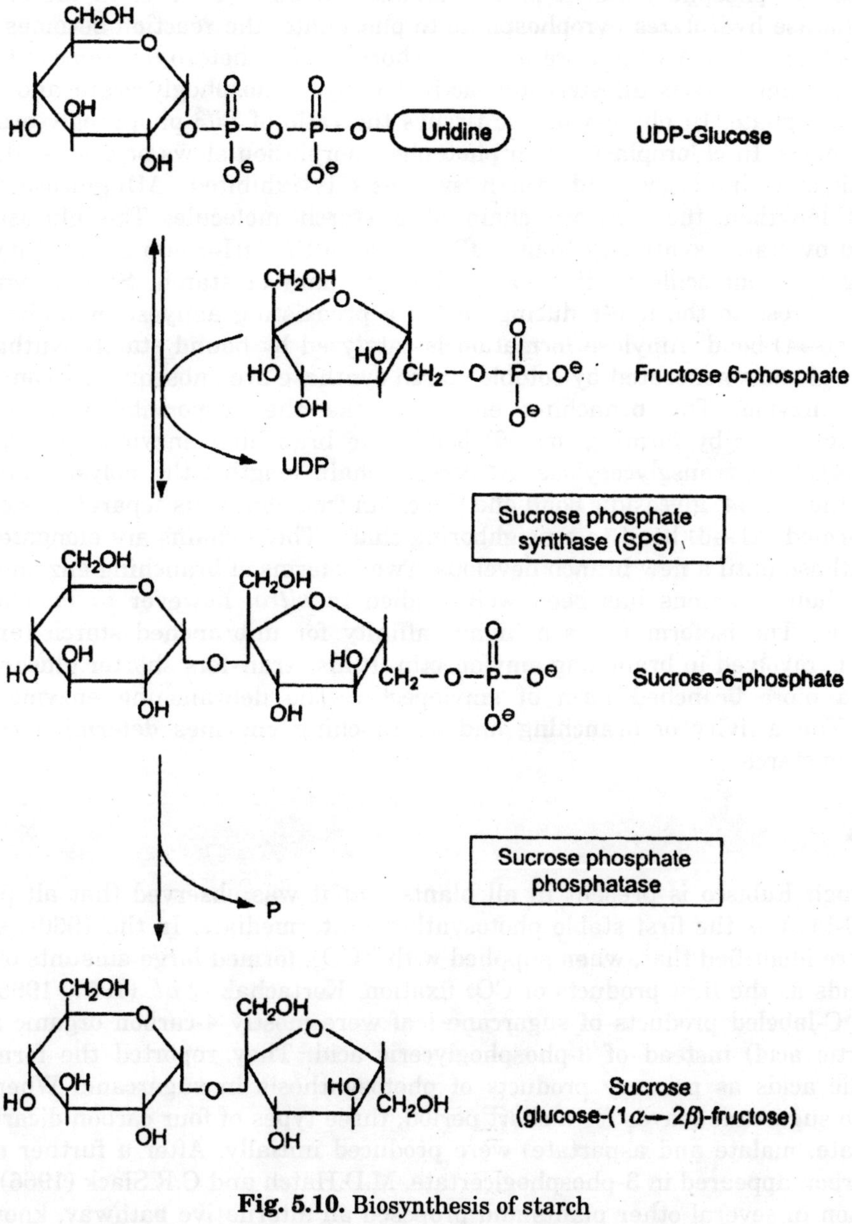

Fig. 5.10. Biosynthesis of starch

Synthesis of starch from hexose phosphate is catalyzed by three enzymes, ADP-glucose pyrophosphorylase, starch synthase and starch branching enzyme (Fig. 5.10). Beside this another important enzyme is pyrophoshatase. Starch synthase has a number of isoenzyme. The first biochemical reaction of starch synthesis pathway is conversion of GAP to F6P. Enzyme hexose phosphate isomerase converts F6P into glucose-6-phosphate (G6P). The next

step is catalyzed by phsophoglucomutase and the phosphate moiety of 6th position is transferred to 1st position and glucose-1-phosphate is produced. G1P reacts with ATP to give ADPglucose and inorganic pyrophosphate is released. This reaction is catalyzed by ADPglucose pyrophosphorylase. It is a reversible reaction. However in the stroma, as the pyrophosphatase hydrolyzes pyrophosphate to phosphate, the reaction becomes irreversible. Jack Preiss found that ADP-glucose pyrophosphorylase is a hetero tetramer of two large and two small subunits. It is allosterically activated by 3-phosphoglycerate and inhibited by inorganic phosphate. In photosynthetic tissues the ratio of Pi/3-phopshoglycerate regulates starch synthesis. In chloroplast, when photophosphorylation slows or ceases, the phosphate concentrations is increased and starch synthesis is inhibited. ADPglucose transfers its glucose to lengthen the amylose chain of a starch molecule. The glucose residue is transferred by starch synthases from ADP-glucose to the OH-group in the 4-position of the terminal glucose molecule in the polysaccharide chain of starch. Starch synthases adds individual glucose to the non-reducing end of a preexisting amlyose or amlyopectin chain forming, (α1$\rightarrow$4) bond. Amylose formation is catalyzed by bound starch synthase, whereas the (α1$\rightarrow$4) glucans generated by soluble starch synthase are subsequently converted by the branching enzyme. The branching enzymes take the responsibility of branching of synthesizing starch by forming (α1$\rightarrow$6) bond. The branching enzyme, are also known as amylo-(1$\rightarrow$4),(1$\rightarrow$6)-transglycosylase. At certain chain lengths, the polysaccharide chain is cleaved at the (α1$\rightarrow$4) glycosidic bond and the chain fragment thus separated is connected via a newly formed (α1$\rightarrow$6) bond to a neighboring chain. These chains are elongated further by starch synthase until a new branch develops. Two isoforms of branching enzyme I and II are reported. Their functions has been well studied *in vitro*, however *in vivo* it still needs conformation. The isoform I has a higher affinity for unbranched starch (amylose). The isoform II is involved in branching amylopectin. It also transfers shorter glucan chains, and produces a more branched form of amylopectin. The debranching enzyme cleaves the branches. The activity of branching and debranching enzymes determine the degree of branching in starch.

C4 Plants

Although Rubisco is present in all plants, but it was observed that all plants do not generate 3-PGA as the first stable photosynthetic intermediate. In the 1960s, several plant species were identified that, when supplied with $^{14}CO_2$, formed large amounts of four-carbon organic acids as the first products of CO_2 fixation. Kortachak *et al.* (1957, 1965) found that the first ^{14}C-labeled products of sugarcane leaf were mostly 4-carbon organic acids, (malic and aspartic acid) instead of 3-phosphoglyceric acid. They reported the formation of C_4 dicarboxylic acids as primary products of photosynthesis in sugarcane. When $^{14}CO_2$ was supplied to sugarcane leaves for a short period, three types of four carbon dicarboxylic acids (oxaloacetate, malate and aspartate) were produced initially. After a further exposure the labeled carbon appeared in 3-phosphoglcertate. M.D.Hatch and C.R.Slack (1966) studied this phenomenon in several other plants and proposed an alternative pathway, known as Hatch and Slack cycle.

Hatch *et al.* (1967), Slack and Hatch (1967), Johnson and Hatch (1968) found that C4 photosynthesis occurs in a variety of higher plants. Downton (1975) Raghvendra and Das (1978) has listed over 1500 species 120 genera of plants belonging to 18 families possessing C4 mechanism of CO_2 fixation. Out of the 18 families, 15 were dicotyledons while the remaining 3 were monocotyledons. At least 11 genera had both C3 and C4 species. All these plants have some features in common : similar anatomical structure ("Kranz" type of leaf anatomy) and capability to grow successfully in environments with low levels of CO_2 in the air. C4 are generally of tropical or subtropical origin. Maize, sugarcane, numerous tropical grasses, and some dicotyledonous plants (e.g., *Amaranthus*) are among the species that demonstrate this C4 labeling pattern.

The Pathway

All C4 carbon-fixation pathway begin in the mesophyll, where PEP carboxylase converts phosphoenolpyruvate to oxaloacetate (Fig. 5.11). This enzyme acts on HCO_3- and not on CO_2 , thus HCO_3- acts as a substrate of this enzyme. The aqueous equilibrium of these two species strongly favours HCO_3- ion over gaseous CO_2, allowing for a more efficient initial fixation step. The C_4 pathway of CO_2 fixation is based on a complex interaction between mesophyll and bundle sheath cells. The pathway involves an initial HCO_3- fixation in the outer mesophyll cells. The organic acid formed is transferred from mesophyll to bundle sheath cells, and release of CO_2 for subsequent refixation in the Calvin cycle.

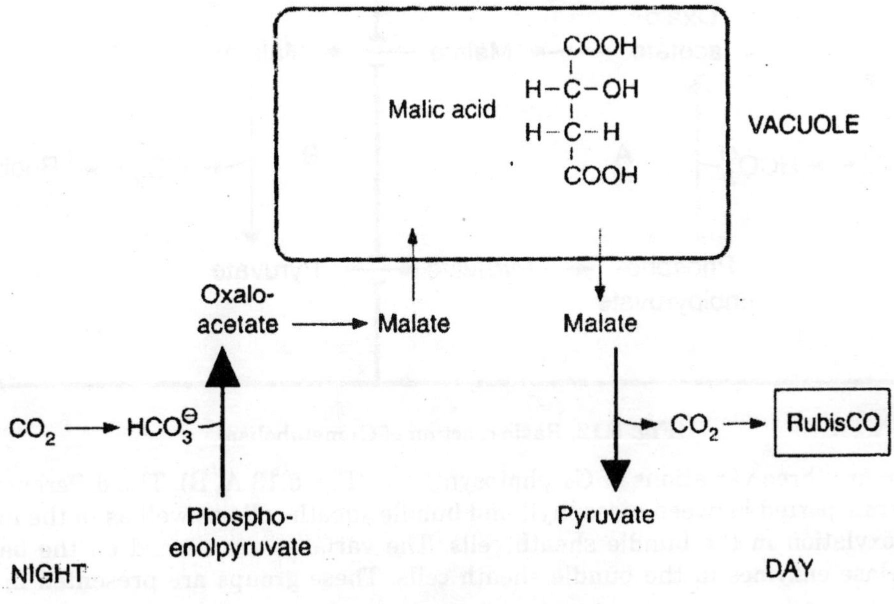

Fig. 5.11. CAM pathway

The first step of CO_2 fixation is catalyzed by enzyme PEP-carboxylase (Phosphoenolpyruvate carboxylase). Phosphoenolpyruvate is the primary CO_2 acceptor molecule. It is carboxylated and a four carbon product oxaloacetate is produced. The subsequent metabolism of oxaloacetate differs among species, but in all cases, a C4 acid is transported to the bundle sheath cells. This product is unstable and the decarboxylation breaks down it readily to malate (MDH enzyme) and aspartate through transmination (GOTenzyme). These C4 acids are enzymatically degraded to yield free CO_2 and pyruvate. The latter 3-carbon intermediate compound is phosphrylated to produce PEP. Thus C4 cycle can continue in this way as the new CO_2 enters the xylem.

The scheme proposed by Hatch and Slack was modified and a major difference is the compartmentation of pathway in mesophyll cells and bundle sheath cells (Fig 5.12). Slack (1969) and Slack *et al.* (1969) isolated chloroplasts from mesophyll and bundle sheath cells and found indications that PEP carboxylase and RuBP carboxylase activity was enriched in mesophyll cell chloroplasts and in bundle sheath cells, respectively. Edwards and Black (1971a) observed in crabgrass (*Digitaria sanguinalis*) that PEP carboxylase and NADP+-malic dehydrogenase were localized in the mesophyll cells while RuBP carboxylase, ribose-5-phosphate isomerase, ribulose-5-phophate kinase, and malic enzyme were localized in bundle sheath. Coombs and Baldry (1972), proposed a scheme in which the C3 is presented as being in the mesophyll cells and the bundle sheath chloroplasts act simply as amyloplasts. There are difference of opinion about the idea of bundle sheath cell chloroplasts carrying out the C3 cycle by either directly fixing CO_2 from the air and/or using the C4 acids as an alternate pathway to supply their CO_2.

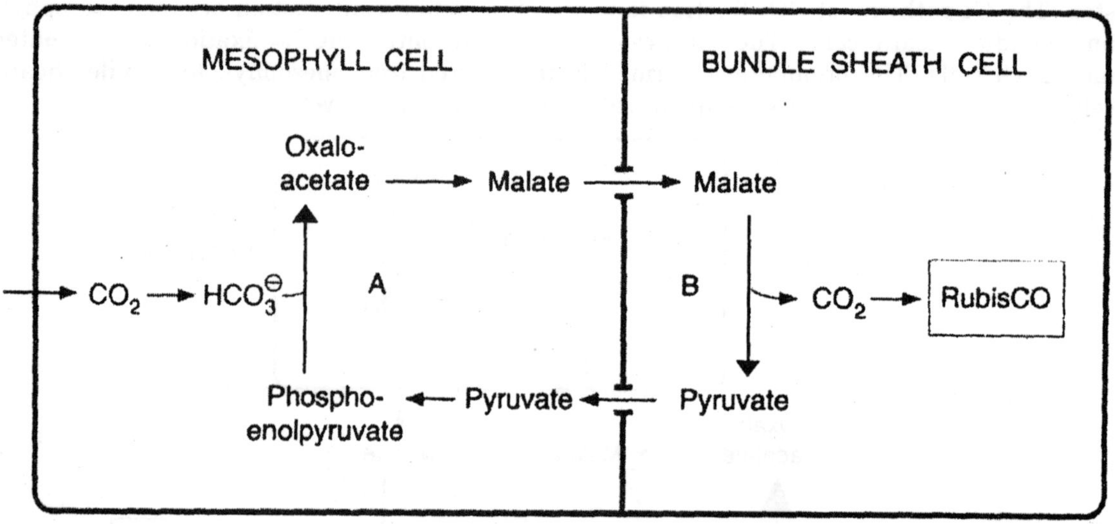

Fig. 5.12. Basic reaction of C4 metabolism

There are three variations of C4 photosynthesis (Fig. 5.13 A, B). The difference is in the C4 acids transported between mesophyll and bundle sheath cells as well as in the mechanism of decarboxylation in the bundle sheath cells. The variants are named on the basis of the decarboxylase enzymes in the bundle sheath cells. These groups are presented in following table

Major BSC decarboxylase	Major substrate moving from		Examples
	MC to BSC	BSC to MC	
NADP+ malic enzyme	Malate	Pyruvate	*Zea mays; Digitaria sanguinalis*
NAD+ malic enzyme	Aspartate	Alanine/Pyruvate	*Artiplex spongiosa Portulaca oleracea*
PEP carboxykinase	Aspartate	PEP	*Panicum maximum*

It is believed that CO_2 released from C4 acid and enters the RuBP carboxylase system and further fixed in the same manner as in Calvin cycle. The CO_2 is reduced to carbohydrates. In summary C4 plants have an additional CO_2 fixing sequences which is coordinated with the C3-cycle and RuBP-carboxylase. Thus CO_2 is fixed by PEP carboxylase and is finally handed over RuBP carboxylase within the leaf.

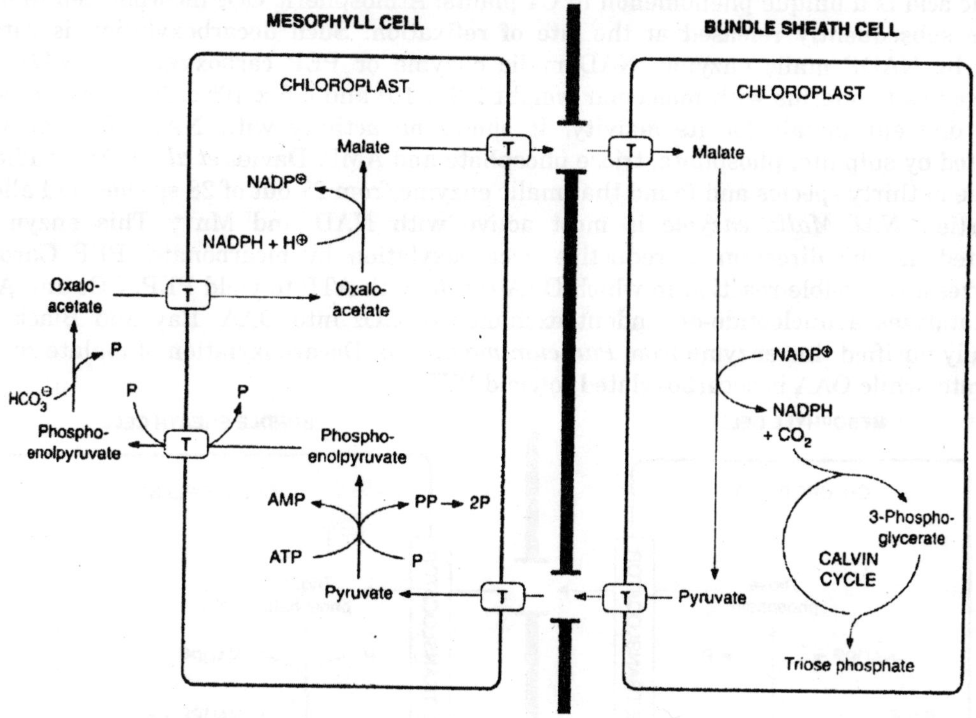

Fig. 5.13A. C4 Cycle in maize type

Enzymes of C4 Pathway

In C4 plants photosynthetic carbon reduction requires the cooperative interaction of the following sequence of reactions: (a) carboxylation (b) reduction or transamination (c) decarboxylation and (d) substrate regeneration. The details of the individual reactions taking part in biochemical pathway differ within various groups of C4 plants. In this section properties of some enzymes of C4 pathway is discussed. *PEP Carboxylase* catalyzes the initial carboxylation, and requires Mg^{++}. This enzyme has molecular weight of around 4 x 10^5 daltons, and is found in a number of distinct forms. In plant tissues, this enzyme is found in heat-stable form. It is located in chloroplast of mesophyll cells. This enzyme is activated by sugar phosphates and inhibited by organic acids. *NADP⁺Malate Dehydrogenase* : Johnson and Hatch, 1970) first observed this enzyme in C4 plants. It is present in chloroplast of mesophyll cells and reduces oxaloacetic acid. This enzyme can be isolated in several forms of differing molecular weight ranging from 3.9 x 10^4 to 1.8 x 10^5 dalton. Sulphite inhibits the activity of this enzyme, it also favours the appearance of a low molecular weight form. *NAD⁺Malate Dehydrogenase* reduces OAA to malate in the same manner as $NADP^+$malate dehydorgenase. Its activity was demonstrated in the species of C4 plants having active NAD malic enzyme. The molecular weight varies from 1.7x10^4 to 7x10^5 dalton. One form of this enzyme with molecular weight 1.27x10^5 dalton is stabilized in the presence of Mg^{++}.

Aspartate Aminotransferase has two isoenzymes, one present in the mesophyll cell and other in bundle sheath cell. The isoenzyme of mesophyll cell converts OAA to aspartate, and the second isoenzyme present in bundle sheath catalyzes reverse reaction. Decarboxylation of organic acid is a unique phenomenon in C4 plants. Atmospheric CO_2 incorporated in organic acid is subsequently released at the site of refixation. Such decarboxylation is catalyzed either by NADP malic enzyme, NAD malic enzyme or PEP carboxykinase. *NADP-Malic Enzyme* has two forms with molecular weight 2.3 x 10^5 and 4.6 x 10^5 dalton, respectively. It needs divalent metals for its activity. It shows no activity with NAD. This enzyme is inhibited by sulphite, phosphate, triose phosphate and AMP. Davies *et al.* (1974) studied this enzyme in thirty species and found that malic enzyme from 27 out of 28 species had allosteric properties. *NAD-Malic enzyme* is most active with NAD and Mn^{++}. This enzyme was inhibited, in the direction of reductive decarboxylation by bicarbonate. PEP Caroxylase catalyzes a reversible reaction in which OAA reacts with ATP to yield PEP, CO_2 and ADP. It also catalyzes a nucleotide-dependent exchange of CO_2 into OAA. Ray and Black (1976) partially purified this enzyme from *Panicum maximum*. Decarboxylation of malate results in pyruvate, while OAA is decarboxylated to yield PEP.

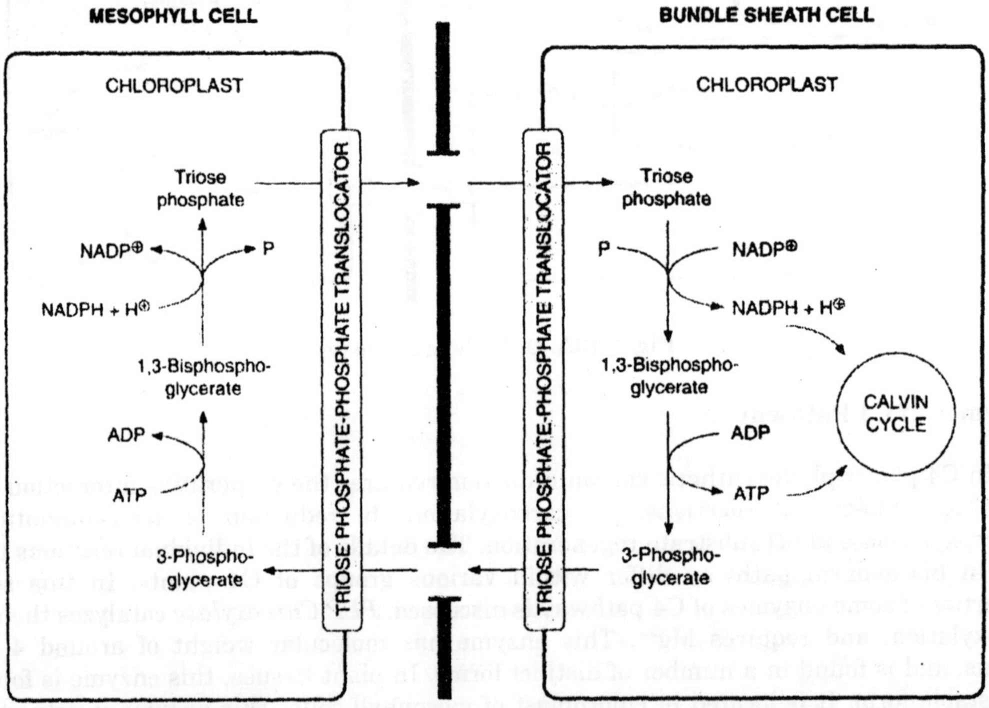

Fig. 5.13B. C_4 Cycle in maize type transfer of NADPH and ATP

Anatomical features of C4 plants

The leaves of these plants demonstrate an unusual anatomy called `Kranz Anatomy'. The 19th century German botanists who initially described this feature called it Kranz anatomy The leaves have two different types of chloroplast-containing cells: mesophyll cells and bundle sheath cells. The anatomical features of C4 leaf in light microscopy shows large green bundle sheaths cells flanked by a layer of green mesophyll cells. There are dense concentration of chloroplasts, mitochondria, and peroxisomes in bundle sheath cells compared to mesophyll cells, the presence of numerous plasmodesmata between mesophyll and bundle sheath cells and the thick-walled nature of grass sheath cells. Electron Microscopic examination of Crabgrass (*Digitaria sanguinalis*) shows mesophyll chloroplasts which are strikingly different in ultrastructure from bundle sheath chloroplasts. The mesophyll cells surround the bundle sheath cells, which in turn surround the vascular tissue. Electron micrographs of bundle sheath and mesophyll chloroplasts show structural differences that reflect distinct function. The bundle sheath chloroplasts have no stacked membranes and demonstrate little PSII activity, whereas mesophyll chloroplasts have retained both PSII and PSI activities. The presence of plasmodesmata connecting the two cell types led to the speculation that photosynthetic intermediates are transported through plasmodesmata.

Kranz anatomy is critical in the biochemistry of CO_2 fixation in these C4 plants. Biochemical fractionation analyses have shown that enzymes unique to the Calvin cycle are located only in the bundle sheath chloroplasts, whereas oxalaoacetate is generated form HCO3- and phosphoenolpyruvate (PEP) in the cytosol of mesophyll cells. As mentioned earlier, there are three groups of C4 plants. These groups show some general correlations of in their chloroplast ultrastructure. The $NADP^+$-malic enzyme plants have reduced grana development in bundle sheath chloroplast. Contrary to it the NAD^+-malic enzyme plants have good grana development of their chloroplasts, which are located centripetally within bundle sheath cells. PEP carboxykinase also have good grana development of their chloroplast but it is distributed randomly or centrifugally within the bundle sheath cells.

C4 plants-environmental regulation

Environmental factors are also important in considerations of C4 photosynthesis.

C4 plants show hyperbolic response of leaf photosynthesis to increasing light intensity approaching values near those observed in full sunlight. In *Tidestromia oblongifolia* there is a linear response of leaf photosynthesis even at the high light intensities.

Hesketh (1963) believed that C4 plants tend to require lower levels of CO_2 in the atmosphere as compared to C3 plants. Now it is proved that air CO_2 concentration are close to saturation for C4 photosynthesis while C3 plants in air respond linearly to increase in CO_2. O_2 does shift the flow of carbon from organic acids into the C4 cycle.

Some enzymes of C4 cycle shows temperature sensitivity. The cold sensitivity of C4 photosynthesis has been related to the cold sensitivity of a key enzyme, pyruvate Pi kinase which generate PEP from pyruvate. Several C4 enzymes have high temperature optima for maximum activity. Shifts in temperature also alter Rubisco's affinity for each of its substrates. As a result, the ratio of carboxylation to oxygenation of Rubisco declines as the temperature warms, and the ratio of photorespiration to photosynthesis increases.

C_4 plants have evolved mechanisms to improve photosynthetic efficiency and decrease water loss in hot, dry environments. Their biochemical and anatomical modifications result in improved CO_2-trapping efficiency, allowing them to reduce stomatal conductance and conserve water without diminishing rates of carbon fixation. C4 plants concentrate CO_2 in bundle sheath cells, where Rubisco is located, effectively suppressing that enzyme?s oxygenase activity and eliminating photorespiration almost completely.

CAM PLANTS

In 1804, de Saussure observed that *Opuntia* produced oxygen in the presence of light and absence of CO_2. Members of family Crassulaceae show unusual metabolism known as Crassulacean acid metabolism (CAM). All the cacti, epiphytes growing in tropical rain forests have this metabolic pathway. To understand this pathway and enzyme system, experiments are worked on *Kalancho daigremontiana* and *Bryophyllum blossfeldianum*. Pineapple and agave are commercially significant plants showing CAM. CO_2 fixation takes place in night and produce organic acid primarily malic acid, which is stored until the following day. In the day the acid is decarboxylated and refixed by rubisco and it enters Calvin cycle. Malate is synthesized (in night) and decaroxylated (in day), and CO_2 fixation by rubisco takes place in the same cell. CAM is generally found in succulent and tropical epiphytic plants, stomata remain open during the cool and relatively humid night. This permits CO_2 entry while sustaining minimal water loss. In CAM plants, stomata remain closed in day and photosynthesis takes place in closed stomata. Thus, the water requirement for CO_2 assimilation is 5 to 10% as compared to C3 plants. It is not essential that the plants having CAM should be succulent. CAM plants have poorly developed palisade parenchyma, and most of the leaf or stem cells are spongy mesophyll.

The Metabolic Pathway

In dark hours, CO_2 fixation is catalyzed by enzyme PEP carboxylase and malic acid is produced. The mechanism that inhibit PEP carboxylase activity in C4 plants do not operate in CAM plants. In light (Fig. 5.14 A,B), enzyme rubisco becomes active and catalyzes refixation of CO_2. In night, starch breaks down to phosphoenolpyruvate (PEP). The enzymes of glycolysis catalyze these reactions. Starch located in chloroplast is degraded to triose phosphate, which is then released to the cytosol via the triose phosphate-phosphate translocator. In cytosol it is converted to PEP. The latter accepts CO_2 in the form of HCO_3^- and oxaloacetate is synthesized. This reaction is catalyzed by PEP carboxlase. In the next reaction, enzyme NAD-malate dehydrogenase takes part and oxaloacetate is reduced to malate. The NADH required for this reaction is provided by the oxidation of triose phosphate in the cytosol. The malic acid thus formed is pumped into vacuoles and is stored up to daytime. The pumping process is an energy-dependent step supported by H+-V-ATPase. The latter is located in vacuolar membrane. The uptake of malate is through a malate channel and is driven by the proton potential. This system is different from the guard cell, and protons are not exchanged for potassium ion. At certain osmotic pressure, three times malate can be stored as malic acid. In daytime, malic acid is released from the vacuole and decarboxylated by one of the mechanism as it takes place in C4 plants. Release of malate is by a regulated efflux through the malate channel. Decarboxylation of malate shows variations in pattern in different plants. The various ways are: via NADP-malic enzyme, NAD-malic enzyme or also via PEP caroxykinase. Malate is transported to chloroplast by a specific transporter. In chloroplast NADP-malic enzyme catalyzes decaroxylation of malate and produce pyruvate, CO_2 and NADPH. The CO_2 released by decarboxylation is refixed by rubisco and it enters Calvin cycle. Pyruvate formed is converted to phosphoenolpyruvate

(PEP) by enzyme pyruvate phosphate dikinase. In chloroplast PEP is not converted into 3-phosphoglycerate, hence it is transported with the help of translocators. In exchange, 3-phosphoglycerate enters chloroplast through two different translocators. In CAM plants, like C4 plants chloroplast have triose phosphate-phosphate translocators, which takes the responsibility of transporting in a counter-exchange triose phosphate, phosphate, and 3-phophoglycerate. In addition there is PEP-phosphate translocator which catalyzes counter-exchange for phosphate. The triose phosphate formed is primarily for resynthesis of the starch consumed during the previous night.

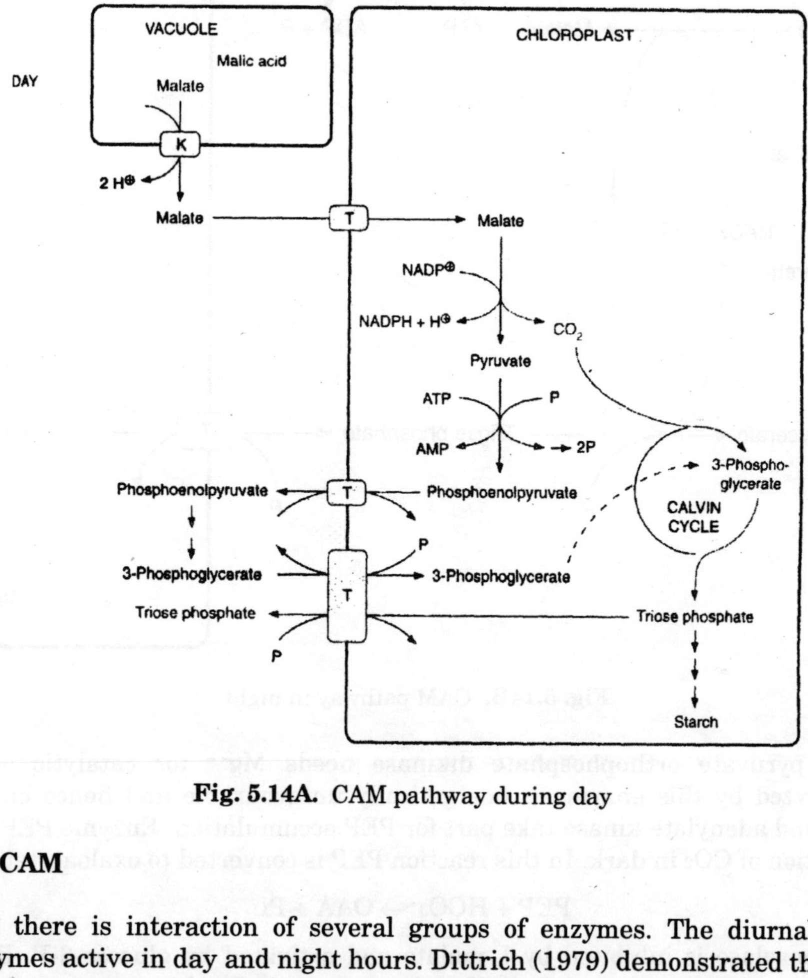

Fig. 5.14A. CAM pathway during day

Enzymes of CAM

In CAM there is interaction of several groups of enzymes. The diurnal cycle show different enzymes active in day and night hours. Dittrich (1979) demonstrated three enzymes in *Kalancho daigremontiana* which breaks down starch into glucose. These enzymes are α- and β-amylase in concert with α-glucosidase. Enzyme phosphorylase converts glucose into glucose-1-P. Carbon entry into gluconogenic sequence takes place either by phosphorylation of pyruvate or phosphorylative decarboxylation of oxaloacetate. Term gluconeogenic is used here is because the enzyme is considered to divert pyruvate from breakdown within the kreb's cycle to synthesis of carbohydrates. Pyruvate is converted to PEP by enzyme pyruvate orthophosphate dikinase as shown in the following equation:

$$\text{Pyruvate} + \text{ATP} + \text{Pi} \rightarrow \text{PEP} + \text{AMP} + \text{PPi}$$

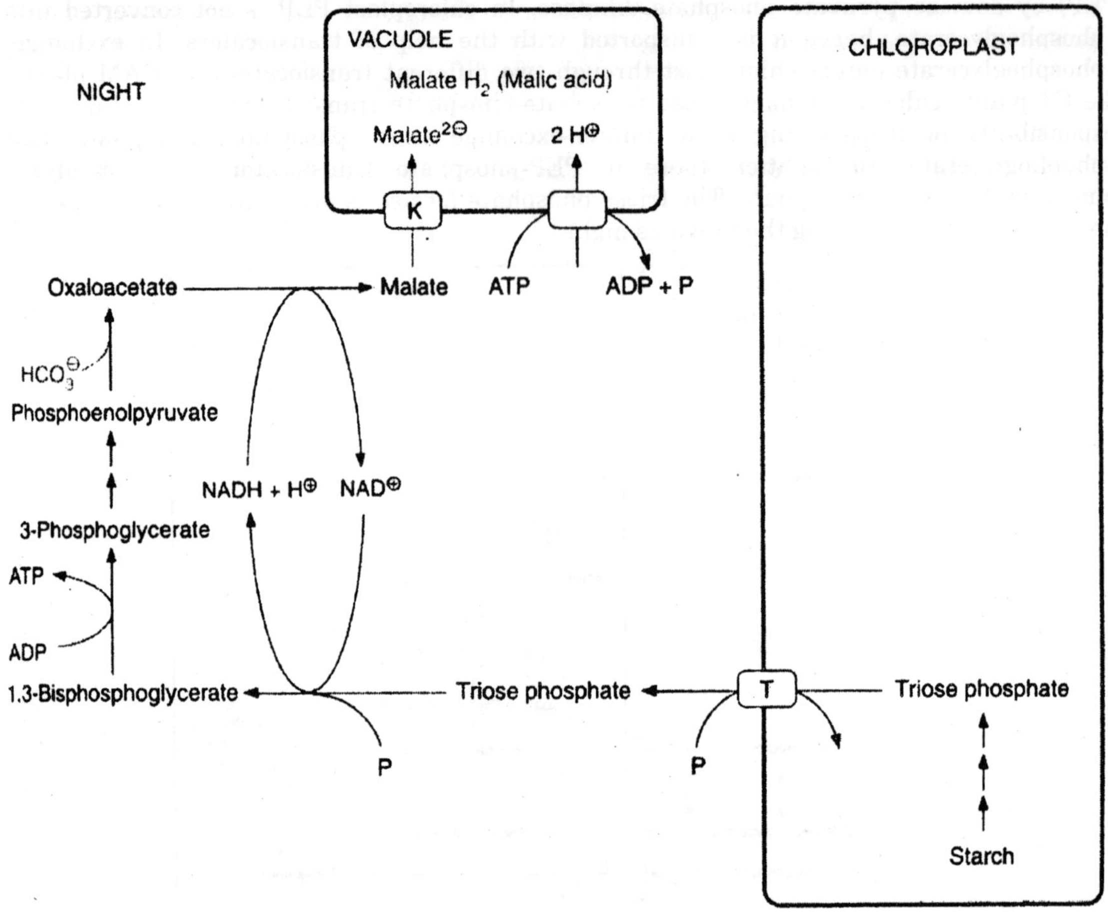

Fig. 5.14B. CAM pathway in night

Enzyme pyruvate orthophosphate dikinase needs Mg^{++} for catalytic activity. The reaction catalyzed by this enzyme is energetically unfavourable and hence enzymes pyrophosphatase and adenylate kinase take part for PEP accumulation. Enzyme PEP carboxylase catalyzes fixation of CO_2 in dark. In this reaction PEP is converted to oxaloacetate (OAA):

$$PEP + HCO_3^- \rightarrow OAA + Pi$$

PEP carboxylase is inhibited by L-malate and activated by glucose-6-P. Badger *et al.* (1975) rubisco in a comparative study in K. *daigremontiana*. The kinetic properties of enzyme rubisco isolated form this plant exhibit similarities with enzyme rubisco of C_3 and C_4 plants. Anderson *et al.* (1952) described NADP-dependent malic enzyme, which takes part in decarboxylation of malic acid. This enzyme is present in cytoplasm and requires Mg^{++} or Mn^{++}. NAD-dependent malic enzyme is present in mitochondria and it is specific for Mn^{++} and does not operate in Mg^{++}. This enzyme shows significant activities only after activation with CoA, Acetyl-CoA or SO_4^{2-}. Malic enzyme CAM plants use both cytosolic NADP-malic enzyme and mitochondrial NAD-malic enzyme to decarboxylate malate. Enzyme PEP caroxykinase is present in those plants which lack adequate activities of both malic enzymes. This enzyme catalyzes following reaction :

$$OAA + ATP \rightarrow PEP + ADP$$

The decarboxylation of OAA by PEP carboxykinase not only fulfils the demand for CO_2 by RuBP carboxylase, but also caters for the process of gluconeogenesis.

In CAM, a major part of triose phosphate synthesized in day is utilized in the resynthesis of starch, consumed during the previous night, hence only a small surplus amount of triose phosphate is available for synthesis of carbohydrate which can be utilized by the plant. Much more, the storage capacity for malate is limited in vacuole, thus again there is limitation of CO_2 assimilation. It is due to all these reasons, the daily increase in biomass in CAM plants is usually very low.

Chapter - 6

RESPIRATION

Respiration is a catabolic process. In this process, there is breakdown of inorganic or organic compounds with the supply of energy. It is a series of 50 or more component reactions, each catalyzed by a different enzyme. The gradual, stepwise breakdown of large molecules provides a means for converting energy into ATP. The intermediates are precursors for biosynthesis of cell material. Respiration is of two types: aerobic and anerobic.

Aerobic respiration involves the controlled oxidation of reduced organic substrates (carbohydrate, proteins, fats, and organic acids) to CO_2 and H_2O.

Respiration releases a large amount of free energy, which is conserved in the acid anhydride linkages of ATP molecules. The primary pathways of respiration provide metabolic intermediates that serve as substrates for the synthesis of nucleic acids, amino acids, fatty acids, and many secondary metabolites.

The Respiratory Quotient

When the respiratory substrates are completely oxidized, the volume of O_2 taken up exactly balances the volume of CO_2 released from the cells. This ratio of CO_2/O_2, is known as respiratory quotient or RQ.

The three stages of respiration- glycolysis, the citric acid cycle, and electron transfer/ oxidative phosphorylation - take place in different sub cellular locations. Gycolysis involves a series of soluble cytosolic enzymes that oxidize sugars to organic acids.

GLYCOLYSIS

The biochemical pathway of breakdown of hexose sugar into pyruvic acid is known as glycolysis (Gr. Glyco = sweet: lysis = splitting) or `glycolytic' pathway. The biochemical reactions of this pathway were worked out by Embden Meyerhof and Paranas and hence it is also known as EMP pathway. Glycolysis is usually depicted as a linear sequence of reactions in which glucose is converted to pyruvate and a small net gain in ATP and NADH (Fig. 6.1). The overall reaction may be expressed in a single formula :

$$Glucose + 2NAD^+ + 2ADP + 2H_2PO_4^- \rightarrow 2pyruvate + 2NADH + 2H^+ + 2ATP + H_2O$$

Free hexose are produced by the degradation of sucrose and starch. In glycolysis one molecule of hexose sugar (glucose /fructose) is catalyzed into two molecules of pyruvic acid. It completes in ten biochemical reactions. Out of these reactions the first five reactions form preparatory phase where the hexose sugar is phosphorylated with ATP. The remaining five reactions form payoff phase where energy is released and is utilized in phosphorylation of ADP to ATP. All the intermediates of this pathway are phosphorylated compounds. These intermediates are biologically important as (i) they are negatively charged and hence they are impermeable through cell membrane, (ii) The intermediates can very easily form enzyme-substrate complex, (iii) The phosphorylated intermediates pass their phosphate moiety to ADP to give a molecule of ATP.

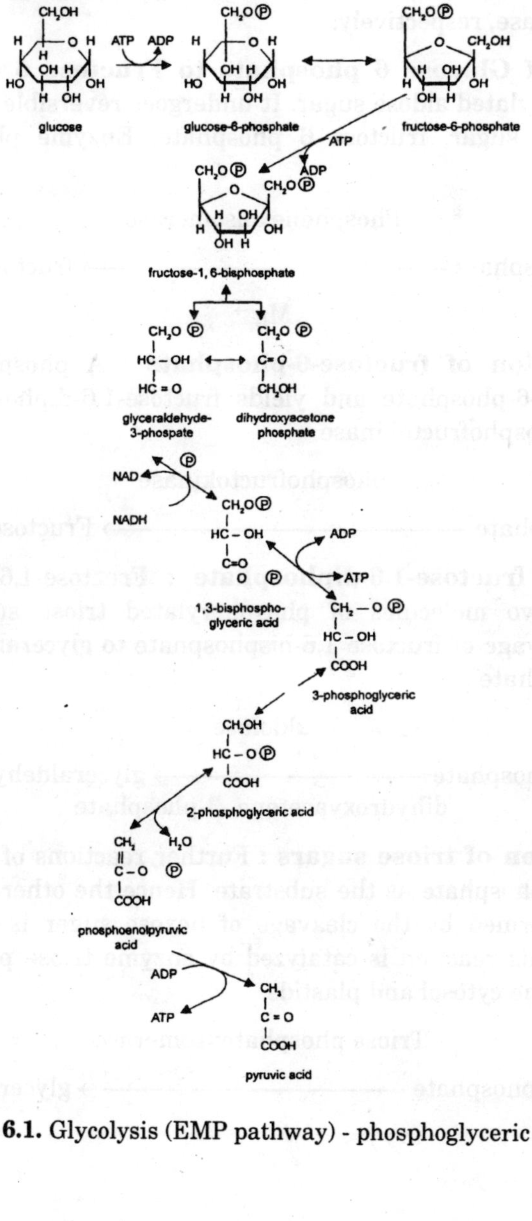

Fig. 6.1. Glycolysis (EMP pathway) - phosphoglyceric acid

The detailed reactions of the pathway are as following —

(1) Phosphorylation of Glucose : In first reaction glucose is converted into glucose 6 phosphate with the help of enzyme hexokinase. This enzyme is known as kinase as it catalyzes the transfer of the terminal phosphate group from ATP to the substrate glucose (hexose sugar). Mg ions act as cofactor. ATP supplies phosphate moiety.

$$\text{Glucose} + \text{ATP} \xrightarrow{\text{Hexokinase } Mg^{++}} \text{Glucose 6 phosphate} + \text{ADP}$$

Plants contain multiple forms of hexokinase. ATP is preferred phosphoryl donor for all these isoenzyme, which differ in cellular location and affinity for particular hexose substrates. The hexokinase which are specific for glucose or fructose should be called glucokinase or fructokinase, respectively.

(2) Conversion of Glucose 6 phosphate to Fructose 6 phosphate : Glucose 6 phosphate is a phosphorylated aldose sugar. It undergoes reversible isomerization and forms a phosphorylated Keto sugar, fructose 6 phosphate. Enzyme phosphohexose isomerase catalyzes this reaction.

$$\text{Glucose-6-phosphate} \xrightarrow[Mg^{++}]{\text{Phosphohexoisomerase}} \text{fructose-6-phosphate}$$

(3) Phosphorylation of fructose-6-phosphate : A phosphate moiety of ATP is transferred to fructose-6-phosphate and yields fructose-1,6-diphosphate. This reaction is catalyzed by enzyme phosphofructokinase.

$$\text{Fructose-6-phosphate} \xrightarrow{\text{phosphofructokinase}} \text{Fructose-1,6-diphosphate}$$

(4) Cleavage of fructose-1,6-diphosphate : Fructose-1,6-diphosphate undergoes cleavage and forms two molecules of phosphorylated triose sugars. Enzyme aldolase catalyzes the aldol cleavage of fructose 1,6-bisphosphate to glyceraldehyde 3-phosphate and dihydroxyacetone phosphate :

$$\text{Fructose-1,6-diphosphate} \xrightarrow{\text{aldolase}} \text{glyceraldehyde-3-phosphate} + \text{dihydroxyacetone -3-phosphate}$$

(5) Interconversion of triose sugars : Further reactions of glycolysis are continued with glyceraldehyde-3-phosphate as the substrate. Hence the other triose sugar (dihydroxy-acetone-3-phosphate) formed by the cleavage of hexose sugar is converted into glyceral-dehyde-3-phosphate. This reaction is catalyzed by enzyme triose phophate isomerase. This enzyme occurs in both the cytosol and plastids.

$$\text{Dihydroxyacetone-3-phosphate} \xrightarrow{\text{Triose phosphate isomerase}} \text{glyceraldehyde-3-phosphate}$$

(6) Oxidation of glyceraldehyde-3-phosphate : Glyceraldehyde-3-phosphate is oxidized into 1,3-diphosphoglycerate.The aldehyde group is oxidized to a carboxylic acid anhydride with phosphoric acid. This type of anhydride is called as acyl phosphate. This reaction is catalyzed by enzyme glyceraldehyde-3-phosphate dehydrogenase.

$$\text{Glyceraldehyde-3-phosphate} + \text{iP} + \text{NAD}^+ \xrightarrow{\text{Glyceraldehyde 3 phosphate dehydrogenase}} \text{1,3-diphosphoglyceric acid} + \text{NADH} + \text{H}^+$$

The mixed acid anhydride product, 1.3-bisphophoglycerate, is a high-energy compound capable of donating its C-1 phosphate to ADP. It is an example of substrate-level phosphorylation.

(7) Transfer of phosphate : The phosphate moiety present at 1 position of 1,3-diphosphoglycerate is transferred to ADP and ATP is formed. The phosphorylation is referred as substrate level phosphorylation. This reaction is catalyzed by enzyme phosphoglycerate kinase.

$$\text{1,3-diphosphoglycerate} + \text{ADP} \xrightarrow{\text{Phosphoglycerate kinase}} \text{3-phosphoglycerate} + 2\text{ATP}$$

(8) Formation of 2-phosphoglycerate : The phosphate moiety of 3-phosphoglycerate is transferred to C-2. Enzyme mutase (3-phosphoglycerate mutase) catalyzes this reaction. Mg^{++} is required for this enzyme. Enzyme mutase is a subclass of isomerase. These enzymes transfer functional group from one position to another in the same molecule.

$$\text{3-phosphoglycerate} \xrightarrow{\text{mutase}} \text{2-phosphoglycerate}$$

(9) Dehydration of 2-phosphoglycerate : A molecule of water is removed from 2-phosphoglycerate and phsophoenolpyruvate is formed. This reaction is catalyzed by enzyme enolase (2-phosphoglycerate hydrolyase). This enzyme is a hydratase.

$$\text{2-phosphoglycerate} \xrightarrow{\text{enolase}} \text{2phosphoe nolpyruvate} + 2H_2O$$

(10) Synthesis of pyruvate : Phosphate moiety of 2phosphoenolpyruvate is transferred to ADP and the later is converted into ATP. Pyruvate is synthesized in this reaction. This reaction is catalyzed by enzyme pyruvate kinase. This enzyme needs Mg^{++} and K^+ for its function. Mn^{++} may replace Mg^{++}.

$$\text{2 phosphoenol pyruvate} + \text{ADP} \xrightarrow{\text{Pyruvate kinase}} \text{Pyruvate} + \text{ATP}$$

In phosphoenol pyruvate, the enol form is stabilized by the phosphoester bond. Removal of the phosphate group allows the resulting enol form of pyruvate to convert to the more stable keto form, removing the product from the reaction and driving the equilibrium toward product formation.

Table I. Energetics of Glycolysis (per glucose / Fructose molecule).

Reaction	Enzyme	ATP/NADH+H^+ Consumed
1. Glucose → Glucose-6-phosphate	Hexokinase	1 ATP Consumed
Fructose → Fructose-6-phosphate		
2. Fructose-6-phosphate → Fructose-1,6 diphosphate	Phospho-fructokinase	1 ATP Consumed
3. Glyceraldehyde – 3-phosphate → 3 phospho-glyceric acid	Glyceraldehyde -3-phos phate dehydrogenase	2 NADH +H^+ produced
4. 1,3 diphosphoglyceric acid → 3 phospho-glyceric acid	Phospho glycerate kinase	2ATP
5. Phosphoenol pyruvate → Pyruvic acid	Pyruvate kinase	2ATP

FORMATION OF ACETYL CoA

Before entering the citric acid cycle, pyruvate is oxidized and decarboxylated by the pyruvate dehydrogenase enzyme complex to form CO_2, acetyl-CoA, and NADH. Enzyme pyruvate dehydrogenase complex catalyzes the oxidative decarboxylation of pyruvate. It requires the bound cofactor thiamine pyrophosphate, lipoic acid, and FAD as well as free coenzyme A (CoASH) and NAD^+. The complex contains three separate enzymes : pyruvate dehydrogenase, dihydrolipoyl transacetylase, and deihydrolipoyl dehydrogenase. The pyruvate dehydrogenase subunit contains thiamine pyrophosphate (TPP) as the prosthetic group. The reactive group of TPP is the thiazole ring. Due to the presence of a positively charged N-atom, the thiazole ring contains an acidic C-atom. After dissociation of a proton, a carbaninon is formed, which is able to bind to the carbonyl group of the pyruvate. The positively charged N-atom of the thiazole ring enhances the decarboxylation of the bound pyruvate to form hydroxethyl-TPP. The hydroxethyl group is now transferred to lipoic acid. Lipoic acid is the prosthetic group of the dihydrolipoyl transacetylase subunit. It is covalently bound by its carboxyl group to a lysine residue of the enzyme protein via an amide bond. Lipoic acid contains two S-atoms linked by a disulfide bond. When the hydroxyethyl residue is transferred to the lipoic acid residue, lipoic acid is reduced to dihyrolipoic acid and the hydroxyethyl residue is oxidized to an acetyl residue. The latter is attached to the dihyfrolipoic acid by a thioester bond.

Acetyl CoA is also called active acetic acid and was discovered by Feodoe Lynen. It is biologically significant intermediate and takes part in a number of biosynthetic pathways. In aerobic respiration acetyl CoA enters in mitochondria where it is completely oxidized via Kreb's cycle.

The mechanism of formation of acetyl-CoA from pyruvate is as following –

$$\text{Pyruvate + TPP} \xrightarrow{\text{Pyruvate dehydrogenase}} \text{Hydroxyethyl-TPP} + CO_2$$

$$\text{Hydroxyethyl-TPP + lipoate} \xrightarrow{\text{dyhydrolipoamide transacetylase}} \text{6-S-acetyl-dihydrolipoate + TPP}$$

Dihydrolipoamide transacetylase

Acetyl-dihydrolipoate + CoA ⟶ acetyl-CoA + dihydrolipoate

Dihydrolipoamide dehydrogenase

Dihydrolipoate + NAD^+ ⟶ lipoate + $NADH+H^+$

KREB'S CYCLE

In 1937, Hans Kreb's proposed a cycle for catabolic oxidation of pyruvic acid and all the intermediates of this cycle were acids. He studied the effect of the anions of various organic acids on the rate of oxygen consumption during pyruvate oxidation. In his studies he used suspensions of minced pigeon-breast muscle. The first intermediate of this cycle is citric acid a tricarboxylic acid, hence this cycle is also known as citric acid cycle or TCA. In 1948, Eugene Kennedy and Albert Lehninger observed that isolated mitochondria can also show oxidative catabolism of pyruvic acid. The citric acid cycle moves electrons from organic acids to the oxidized redox cofactors NAD+ and FAD, forming NADH, $FADH_2$, and carbon di oxide.

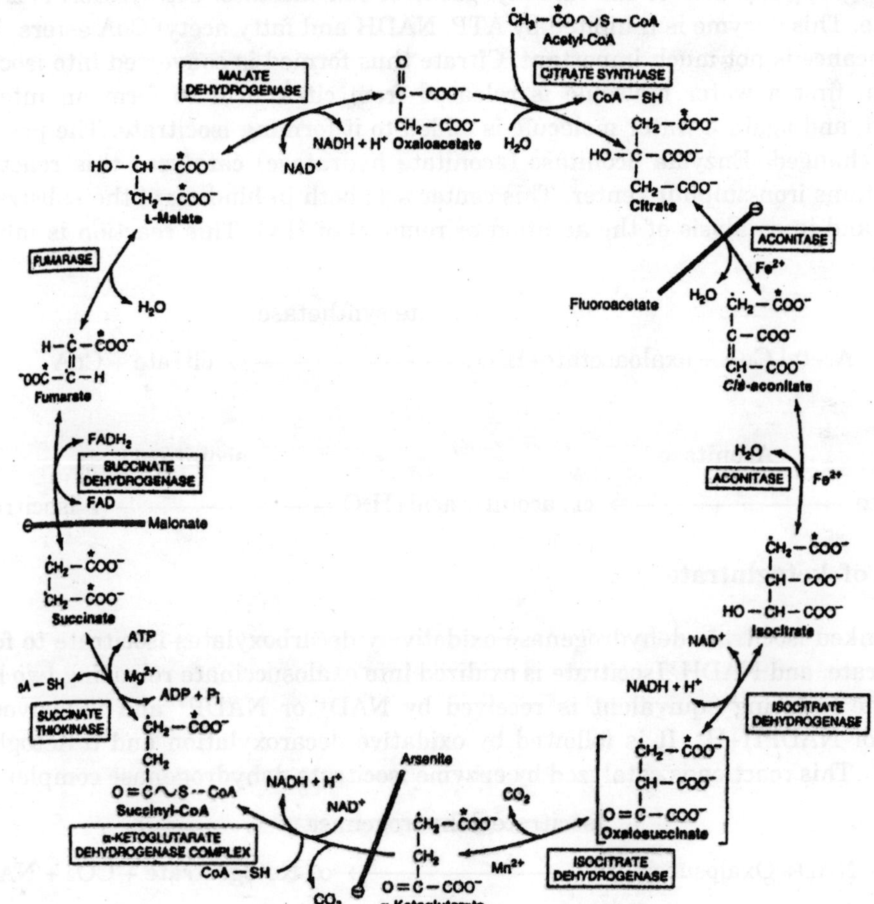

Fig. 6.2. Kreb's cycle

Pyruvic acid formed by the catabolism of glucose molecule in the biochemical pathway of glycolysis is further activated and is converted into acetyl CoA. Acetyl CoA (CH_3-CO~S-CoA) is an ester of coenzyme A. CoA contains pantothenic acid which is a vitamin. The later is an important metabolite and has significant role in many biological processes. It acts as a bridge between glycolysis and Krebs cycle. The biochemical reactions of TCA (Fig. 6.2) are as following:

Citrate and isocitrate synthesis

The first intermediate of TCA is citric acid. Acetyl CoA condenses with a dicarboxylic acid (oxaloacetic acid) forming a tricarboxyilic acid (citric acid) and releases CoA due to hydrolysis of thioester bond of CoA. The acetyl moiety of acetyl CoA and oxaloaectic acid combines and produces citric acid. This reaction is catalysed by enzyme citrate synthetase. The thioester group promotes the removal of a proton of the acetyl residue, and the carbanion thus formed binds to the carbonyl carbon of oxaloacetate. In this reaction the methyl carbon of the acetyl group is joined to the carbonyl group of oxaloacetate. Citroyl-CoA is a transient intermediate. This enzyme is inhibited by ATP, NADH and fatty acetyl CoA esters. However, there significance is not much important. Citrate thus formed is converted into isocitrate. In this reaction first a water molecule is released from citric acid to form an intermediate aconitic acid, and again a water molecule is added to it forming isocitrate. The position of H and OH is changed. Enzyme aconitase (aconitate hydratase) catalyzes this reaction. This enzyme contains iron-sulphur center. This center acts both in binding of the substrate at the active site and in catalysis of the addition or removal of H_2O. This reaction is inhibited by fluroacetate.

$$\text{Acetyl CoA + oxaloacetate} + H_2O \xrightarrow{\text{Citrate synthetase}} \text{citrate + CoA}$$

$$\text{Citrate} \xrightarrow{\text{Aconitase}} \text{cis aconitic acid} + H_2O \xrightarrow{\text{aconitase}} \text{isocitrate}$$

Formation of ketoglutrate

NAD-linked isocitrate dehydrogenase oxidatively decarboxylates isocitrate to form CO_2, α-ketoglutarate, and NADH. Isocitrate is oxidized into oxalosuccinate releasing two hydrogen ($2H^+2e^-$). The reducing equivalent is received by NAD^+ or $NADP^+$ and is converted into $NADH+H^+$ or $NADPH+H^+$. It is followed by oxidative decaroxylation and α Ketoglutrate is synthesized. This reaction is catalyzed by enzyme isocitrate dehydrogenase complex.

$$\text{Isocitrate + NAD+ Oxalosuccinate} \xrightarrow{\text{Isocitrate dehydrogenase}} \alpha\text{- Ketoglutrate} + CO_2 + NADH+H^+$$

Isocitrate dehydrogenase occurs in two forms (isozymes). These two enzymes require NAD^+ and $NADP^+$, respectively.

Formation of succinyl CoA

Succinyl CoA is formed by oxidation of α ketoglutrate. CoA takes part in this reaction and NADH+H^+ is formed due to oxidation reduction reaction. This reaction is catalyzed by Succinate thiokinase (succinyl-CoA synthetase). It includes three enzymes, analogous to E1, E2 and E3 of the pyruvate dehydrogenase. The mechanism of this reaction is similar to that of formation of acetyl CoA from pyruvic acid.

$$\alpha \text{ Ketoglutrate} + \text{CoA} + \text{NAD} \xrightarrow{\alpha \text{ Ketoglutrate dehydrogenase}} \text{Succinyl CoA} + CO_2 + \text{NAD.2H}$$

Deacylation of succinyl CoA

Succinyl CoA releases CoA forming succinic acid. A molecule of GTP is produced in this reaction. Enzyme Succinate synthetase catalyzes this reaction.

In eukaryotes, this enzyme is tightly bound to the inner membrane of mitochondria while in prokaryotes, it is bound to the plasma membrane. This enzyme is inhibited by malonate.

$$\text{Succinyl CoA} + \text{GDP} + \text{Pi} \xrightarrow{\text{Succinate synthetase}} \text{Succinate} + \text{CoA} + \text{GTP}$$

Oxidation of succinate to fumerate

Succinate is oxidized to fumarate. A flavoprotein enzyme, succinate dehydrogenase catalyzes this reaction. FAD is converted into $FADH_2$ by accepting reducing equivalent released from succinate.

$$\text{Succinate} + \text{FAD} \xrightarrow{\text{Succinate dehydrogenase}} \text{Fumarate} + FADH_2$$

Hydration of Fumarate

A molecule of water is added to fumaric acid and produce malate. This reaction is catalyzed by enzyme fumarate hydratase.

$$\text{Fumarate} + H_2O \xrightarrow{\text{Fumarate hydratase}} \text{Malate}$$

Oxidation of malate

Malate is oxidized into oxaloacetic acid and thus the later is recovered. NAD^+ receives the reducing equivalents forming NADH+H^+. This reaction is catalyzed by enzyme malate dehydrogenase.

$$\text{Malate} + NAD^+ \xrightarrow{\text{Malate dehydrogenase}} \text{Oxaloacetate} + NADH + H^+$$

It is important to note that four components of vitamin B complex take part in Kreb's cycle. They are (a) ribofalvin, (b) niacin, (c) thiamin and (D) pantothenic acid.

ENERGETICS OF KREBS CYCLE

Table II. Energetics of Kreb's Cycle (per glucose molecule)

Reaction	Enzyme	$NADH+H^+/FADH_2/$ GTP produced
1. Isocitrate → oxalosuccinate	Isocitrate dehydrogenase	2 $NADH+H^+$
2. α-ketoglutarate → succinyl CoA	α-ketoglutarate dehydrogenase	2 $NADH+H^+$
3. Succinyl CoA → succinate	Succinate thiokinase	2 GTP
4. Succinate → fumarate	Succinate dehydrogenase	2 $FADH_2$
5. L. Malate → Oxaloaxcetate	Malate dehydrogenase	2 $NADH+H^+$

MITOCHONDRIAL ELECTRON TRANSPORT

In all mitochondria, the respiratory electron transfer chain consists of a series of membrane-bond redox centers that catalyze the multi step transfer of electrons from NADH and $FADH_2$ to oxygen, forming water, proton translocates from the matrix to the inter-membrane space. This endergonic proton pumping is driven by the exergonic transfer of electrons from strong reducing agents of strong oxidant. The $2e^-$ reduction of *1/2* O_2 by NADH involves a reduction potential difference of 1.14 V. This translates to 219.2 KJ of free energy released for every mole of NADH oxidized.

COMPONENTS OF ELECTRON TRANSPORT SYSTEM (ETS)

The mitochondrial electron transfer chain consists of four multiple subunit protein complexs, commonly referred to as Complex I to IV. Youssef Hatefi (1962) working with beef heart mitochondria, succeeded in isolating four different complexes, which he termed complexes I-IV. Fo-F1-ATP synthase, which does not have electron transfer activity, is occasionally referred to as Complex V.

Complex I The NADH dehydrogenase complex oxidizes the NADH generated in the mitochondrial matrix by citric acid cycle activity, regenerating NAD^+ and reducing ubiquinone. It feeds the respiratory chain with the electrons from NADH, formed from the degradation of substrates in the matrix. These enzymes contain nicotinamide rings resembling pyridine as coenzyme. This complex includes as many as 9 distinct iron-sulfur centers in addition to a flavoprotein. It has the most complicated structure of all the mitochondrial electron transport complexes. It is made up of more than 40 different subunits. One part is embedded in the membrane and a peripheral part protrudes into the matrix space. The peripheral part contains the binding site for NADH, a bound flavin mononucleotide and at least three Fe-S-centers. The membrane contains a further Fe-S center, as well as the binding site for ubiquinone.

Complex II: In plants complex II (succinate dehydrogenase), oxidizes succinate to fumarate. Like complex I, complex II transfers electron to ubiquinone. This complex consists of several polypeptides, two of which compose succinate dehydrogenase, a membrane-bound FAD-containing enzyme that catalyzes a key reaction of the TCA cycle. This complex has seven subunits. It contains a flavin adenine nucleotide (FAD) as the electron acceptor; several Fe-S center as redox carriers: and one cytochrome-b, the function of which is not known. Complex II provides a pathway for feeding "lower-energy" electrons from succinate to FAD to ubiquinone. Electron transfer through complex II is not accompanied by proton translocation.

Complex III (UQH2-cytochrome c oxidoreductase or Cytochrome bc1) complex III catalyzes the transfer of electrons from reduced ubiquinone to cytochrome c, which is a peripheral membrane protein and not an integral part of any of the respiratory-chain complexes. It is calculated that four H^+ are translocated across the membrane for every pair of electrons transferred through ubiquinone to complex III. Complex III contains approximately 10 polypeptide, one of which is encoded by the mitochondrial genome, it includes several hemes and an iron-sulphur center.

At Complex III, transfer of one electron pair from ubiquinol to cytochrome c is accompanied by the translocation of four protons across the inner membrane. This mechanism, called the Q-cycle, presents the only case in which proton translocation by the electron transport chain is largely understood. During the Q-cycle, ubiquinone is reduced to ubiquinol on the matrix side of the inner membrane. Three electron transfer chain sites can reduce ubiquinone : one each at Complex I and II, and Center N on Complex III. Ubiquinol then diffuses to the outer face of the membrane and binds to Center P, a ubiquinol oxidation site on Complex III. Once bound to Center P, the ubiquinol is oxidized; one electron reduces the Rieske Fe-S center, and the other reduces cytochrome b566. The oxidation of ubiquinol to ubiquinone also releases two protons into the intermembrane space. The electron on the Rieske Fe-S center is subsequently transferred to cytochrome c1 and then to cytochrome C. the electron residing on cytochrome b566 is transferred to cytochrome 560, which donates electrons to ubiquinone at Center N.

Complex IV (cyt-a/a3complex) or cytochrome oxidase contains 13 different subunits, three of them-subunits I, II and III- are the three largest polypeptides of the complex and contain all of its redox centers. They are encoded in the mitochondrial genome. The complex has a large hydrophilic region that protrudes into the intermembrane space and contains the binding site for cyt-c.

The mechanism of electron transfer through complex IV has been intensively studied. A major challenge to investigators is to explain how carriers that are only able to transfer single electrons can reduce a molecule of O_2 to two molecules of H_2O, a process that required four electrons. In addition of acting as an agent of O_2 reduction, cytochrome oxidase is a proton pump.

All four redox centers are located within mitochondrion-encoded subunits I and II.

The Machinery for ATP Formation

During the early 1970s, Humberto Fernandez-Moran of Massachussetts discovered a layer of spheres attached to the inner side of the inner membrane of mitochondria, projecting from the membrane and attached to it by stalks. A few years later, Efraim Racker of Cornell University isolated the inner membrane spheres, which he called coupling factor1, or simply F1. Racker discovered that the F1 spheres behaved like an enzyme that hydrolyzed ATP, that is, an ATPase. At first glance this seems like a peculiar finding. If one considers that the hydrolysis of ATP is the reverse reaction from its formation, the function of the F1 sphere becomes more evident; it contains the catalytic site at which ATP formation normally occurs.

The structure of ATP synthase

The enzyme ATP synthase is composed of two principal components: the F1 head piece and a basal section, called Fo, embedded in the inner membrane. The two portions are connected by a narrow stalk.

The F_1 headpiece has a molecular weight of about 360,000 daltons and contains five different polypeptides. In the mitochondrial enzyme, the homologous polypeptides are encoded by the nuclear DNA, synthesized in the cytosol, and imported post-translationally. In mitochondria, the Fo base piece is embedded in the inner mitochondrial membrane and forms a channel through which protons are conducted from the inter membrane space into the F1 headpiece.

The presence of channel in the Fo base piece is best demonstrated in experiments where the inner mitochondrial membrane is broken into fragments that form membrane vesicles, called submitochondrial particles. Intact submitochondrial particles are capable of oxidizing substrates, generating a proton gradient, and synthesizing ATP. However, if the F1 spheres are removed from the particles by treatment with urea, the vesicle membrane can no longer maintain a proton gradient despite continuing substrates oxidation and electron transport. Protons that are translocated across the membranes simply cross back through the ATP synthase, and the energy is dissipated.

The mechanism of ATP formation

Movement of electrons along the respiratory chain leads to the establishment of a proton gradient and that the movement of protons back across the membrane through the ATP synthase acts to drive the formation of ATP.

Paul Boyer and his colleagues proposed a hypothesis, according to which the movement of protons down an electrochemical gradient through the channel in the Fo base and into the F1 head induces conformational changes that alter the affinity of the protein for ADP, Pi and ATP.

The cellular reactions occurring in an aqueous environment in which the concentration of water is 55 M and the reactants and products are simply dissolved in the medium. Under these conditions, energy is required to drive the formation of the covalent bond linking ADP with inorganic phosphate to form ATP. It has been demonstrated, however, that once ADP and Pi are tightly bound within the catalytic site of the ATP synthase, the two bound reactants readily condense to form a bound ATP molecules without the input of additional energy. In other words, even though the reaction

$$\text{Soluble ADP + Soluble Pi} \longrightarrow \text{soluble ATP} + H_2O$$

may require the input of considerable energy (7.3 Kcal/mol under standard conditions), the reaction

$$\text{Enzyme-bound ADP + Enzyme-bound Pi} \longrightarrow \text{enzyme-bond ATP} + H_2O$$

Has an equilibrium constant close to 1 ($\Delta G^o = 0$) and thus can occur spontaneously without the input of energy. The energy is required for the binding of reactions to the active site or the release of the product form the active site, or both, rather than the phosphorylation event itself. This process is thought to occur in the following way.

According to the Boyer "binding-change" hypothesis, each F1 headpiece contains three catalytic sites that shift progressively through three different conformations as reflected in their affinity for nucleotides. These three conformations can be described as the `tight' or T conformation in which nucleotides (ADP + Pi or ATP) are tightly bound: the "loose", or L, conformation in which ADP and Pi are loosely bound; and the "open" or O, conformation, because it has a very low affinity for sites for nucleotides, it is considered to be empty. In step I, ADP and Pi bind loosely to a catalytic site in the L conformation. Protons then bind to the enzyme at allosteric sites facing the inter-membrane space where the H^+ concentration sis high. Once bound, the protons induce conformational changes that increase the affinity of the catalytic site for ADP and Pi causing these reactants to bind much more tightly to the enzyme. The reactants then spontaneously condense to form ATP, which remains tightly bound to the active site to the T conformation. The enzyme then undergoes a conformational change that express the allosteric, proton-binding sites to the matrix side of the membrane, where the H^+ concentration is low. The dissociation of the protons from the enzyme into the matrix changes the conformation of the catalytic site, decreasing its affinity for the bound ATP and causing the product to be released into the matrix. The overall net reaction would lead to the synthesis of ATP at the expense of protons moving down their gradient from the inter-membrane space to the matrix.

The changes in conformation of the active site are thought to be accompanied by the rotation of the F1 headpiece relative to its stalk, bringing successive catalytic sites in position to interact with protons coming through the Fo channel.

Pentose Phosphate Pathway

In 1950s plant physiologists observed alternate pathway (other than glycolysis and citric acid cycle) by which plants were able to obtain energy by oxidation of sugars. Plant physiologists like Martin Gibbs, Bernard Axelrod and Harry Beevers were the pioneers to observe this catabolic pathway in which the intermediates were phosphorylated pentose sugars hence this pathway was named as Pentose Phosphate Pathway. This pathway is also known as Hexose Mono Phosphate Pathway or Phosphogluconate Pathway. Some intermediates of this pathway are same as that of Calvin cycle. However, in this pathway, the pentose sugars are degraded and do not synthesize other sugars. Like glycolysis, this pathway also occurs in cytosol. One typical difference with glycolysis is that in this pathway $NADP^+$ acts as electron acceptor instead of NAD^+.

In this pathway (Fig. 6.3), pyruvate is not produced directly. Enzyme starch phosphorylase and phosphoglucomutase take the responsibility to breakdown starch and form glucose-6-phosphate. Otherwise also glucose may be phosphorylated to produce glucose-6-phosphate, as it happens in glycolysis. The latter is dehydrogenated to 6-phosphogluconolactone.

Enzyme glucose-6-phosphate dehydrogenase catalyzes this reaction and $NADP^+$ is reduced to $NADP+H^+$. 6-phosphogluconolactone (lactone) is hydrolysed by enzyme lactonase and 6-phosphogluconate is formed. The dehydrogenase enzymes involved in the initial reactions of this pathway are highly specific for $NADP^+$ and not for NAD^+. The enzyme Glucose-6-phosphate dehydrogenase present in cytosol has no allosteric regulatory properties. However, the enzyme is strongly inhibited by one of its products, NADPH; accordingly, the $NADP^+$/NADPH ratio appears to determine the activity of the enzyme. Enzyme 6-phosphogluconate dehydrogenase converts 6-phosphogluconate into 3-keto-6-phosphogluconate. The regulatory properties of 6-phosphogluconate dehydrogenase have been observed. 3-keto-6-phosphogluconate is decarboxylated to yield a pentose sugar ribulose-5-phosphate. It is epimerized to xylulose-5-phosphate and isomerises into ribose-5-phosphate. Enzyme transketolase and transaldolase catalyze further reactions. Enzyme transketolase transfers 2-carbon compound (glycoaldehyde group) while transaldolase transfers 3-carbon compound (dihydroxyacetone group). The mechanism of this pathway is cyclic and glucose-6-phosphate is regenerated. In this pathway three molecules of CO_2 and one molecule of glyceraldehyde is produced. Six molecules of $NADP^+$ are reduced to $NADPH+H^+$. All of the reaction of the nonoxidative part of the pentose phosphate pathway are readily reversible.

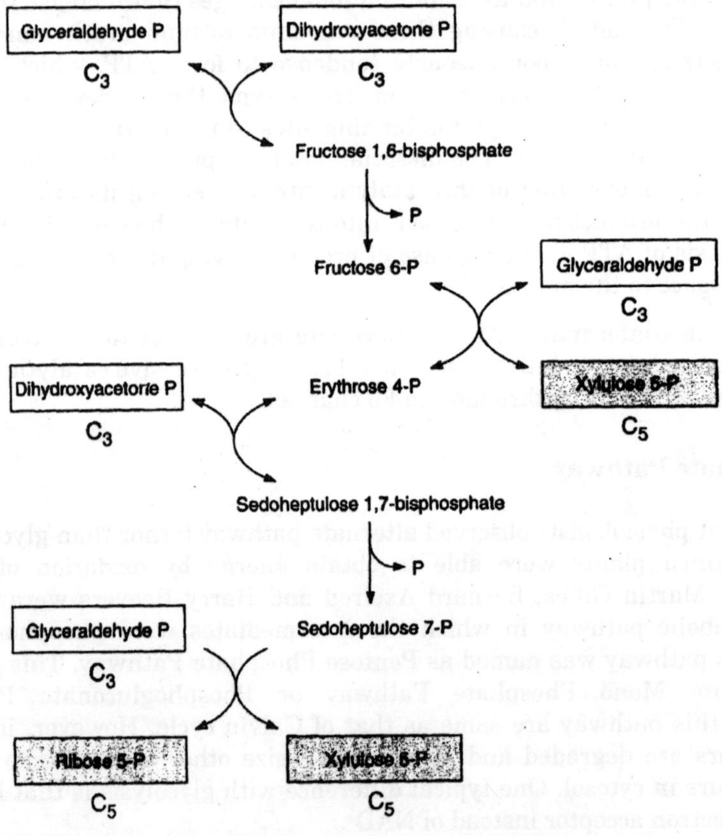

Fig. 6.3. Pentose phosphate pathway

PPP has three significance, they are (1) production of NADPH, (2) production of erythose-4-phosphate, and (3) production of ribose-5-phosphate.

SUMMARY

Respiration is a catabolic process. Respiration releases a large amount of free energy, which is conserved in the acid anhydride linkages of ATP molecules. The three stages of respiration- glycolysis, the citric acid cycle, and electron transfer/oxidative phosphorylation - take place in different subcellular locations. The biochemical pathway of breakdown of hexose sugar into pyruvic acid is known as glycolysis. Before entering the citric acid cycle proper, pyruvate is oxidized and decarboxylated by the pyruvate dehydrogenase enzyme complex to form CO_2, acetyl-CoA, and NADH. Hans Krebs proposed a cycle for catabolic oxidation of pyruvic acid and all the intermediates of this cycle were acids. In all mitochondria, the respiratory electron transfer chain consists of a series of membrane-bond redox centers that catalyze the multi step transfer of electrons from NADH and $FADH_2$ to oxygen, forming water. Martin Gibbs, Bernard Axelord, Harry Beevers observed alternate pathway in which the intermediates were phosphory lated pentose sugars.

Chapter - 7

PHOTORESPIRATION

Photorespiration is evolution of carbon di oxide from photosynthetic tissues in the presence of light. Jackson and Volk (1970) defined photorespiration to describe "all respiratory activity in the light regardless of the pathways by which CO_2 is released or O_2 consumed". Beevers (1971) described photorespiration as "light-dependent O_2 and/or CO_2 output. Otto Warburg (1920) observed that in algae photosynthesis is inhibited in the presence of O_2. This phenomenon was termed *Warburg Effect*. This effect was also observed in C3 plants. This is explained in the next paragraph. Rabinowitch (1945) described this phenomenon as dark respiration occurring in photosynthetic tissues in light. Decker and Tió (1959) first clearly described the existence and importance of photorespiration. Zelitch (1964) observed that glycolate was the primary substrate of photorespiration. In 1971, he worked on the details of this phenomenon. Tolbert (1981), Huang *et al.* (1983) and Ogren (1984) discovered that photorespiration occurs in chloroplasts, peroxisomes, and mitochondria in a cooperative way.

In the natural environmental conditions, rubisco shows carboxylase to oxygenase activity in the ratio of 3:1. Due to oxygenase activity of rubisco, ribulose bisphosphate is converted into 2-phosphoglycolate. Plants are not able to use this compound and it is metabolized photorespiratory carbon oxidation cycle. This cycle is also known as *C2-oxiadative photosynthetic carbon cycle*. Two molecules of phosphoglycolate are converted to one CO_2 and one molecule of 3-phosphoglycerate. Under normal conditions the rate of photorespiration of most of the higher C3 plants ranges from 3 to 10 mg CO_2 $dm^{-2}h^{-1}$. It rates are 1.2 to 4.0 times higher as compared with dark respiration. In the submerged aquatic plants rate of photorespiration is less than dark respiration.

Photorespiration is undetectable in C4 plants and takes place in higher rate in C3 plants. To explain this, it is believed that carbons 1 and 2 of the ribulose bisphosphate were the precursors to glycolic acid. Much more, rubisco and other enzymes taking part in Calvin cycle are present only in the bundle sheath cells, and the CO_2 concentration in those cells is maintained too high for O_2 to compete, with CO_2. Ribulose bisphosphate is a substrate of CO_2 fixation in Calvin cycle and the reaction is catalyzed by rubisco. During photophosphorylation, ribulose bisphosphate and O_2 react to form 3-PGA and phospoglycolic acid. This reaction is catalyzed by rubisco. Molecular oxygen and carbon di oxide compete for the same substrate ribulose bisphosphate, and the same enzyme rubisco catalyzes both the reactions. The competition between O_2 and CO_2 explains the greater inhibition of photosynthesis in C3 plants at low than at higher CO_2 level. In general rubisco fixes about one-third to one-fourth as much O_2 as CO_2. Oxygen fixation seems to represent about two-thirds of the total O_2 absorbed during photorespiration. The remaining O_2 comes from oxidation of phosphoglycolate.

Mechanism of Photorespiration

Glycolic acid is synthesized as an early product of CO_2 assimilation and it disappears due to its oxidation by glycolate oxidase system. Experiments performed with inhibitors of glycolate synthesis show that the rate of glycolate synthesis may be as high as 50% of the net CO_2 assimilation during photosynthesis. Photorespiration involves glycolate synthesis and oxidation. This pathway completes in three organelles : chloroplast, peroxisomes and mitochondria.

Biosynthesis of Glycolate

The kinetics using $^{14}CO_2$ show that glycolic acid arises from the C-1 and C-2 of a sugar phosphate. It was observed that the transketolase type of reactions among phospho-pentose sugars result in the production of active glyceraldehydes during Calvin cycle. Transketolase catalyzes transfer of two carbon moiety from the donor to the acceptor sugars. Enzyme transketolase contains thiamine pyrophosphate and the donor and acceptor are phospho-pentose sugars. In the absence of an acceptor, "active glycoladehyde" is synthesized on the position-2 of the thiazolium ring of thiamine. Active glyceraldehydes is oxidized with isolated enzymes to yield glycolate in the presence of ferricyanide. According to Bradbeer and Racker (1961); Goldberg and Racker (1962); Holzer and Schrter (1962); two molecules of ferricyanide is reduced for the synthesis of each mole of gylcolate. Griffith and Byrerrum (1959) working with tobacco leaves supported the view that glycolic acid may be derived from C-1 and C-2 of pentose sugars. Richardson and Tolbert (1961) gave a different opinion about the synthesis of glycolate. He discovered an enzyme active phosphatase in green plants. This enzyme was found to be highly specific for phosphoglycolate and catalyzes the hydrolysis of phospho-glycolic acid to glycolic acid. It was presumed that phosphoglycolate would be derived from the oxidation of a two-fragment from C-1 and C-2 of intermediate phospho-sugars of Calvin cycle. Kearney and Tolbert (1962) working with labeled CO_2 found that phosphoglycolate as one of the radioactive products excreted by isolated spinach chloroplasts during photosynthesis. Phosphoglycolate is the source of CO_2 released in photorespiration. A specific phosphatase removes the phosphate moiety, and glycolic acid is liberated. Tolbert et al., (1968) separated cell organelles using sucrose density gradient technique and found that phosphoglycerate phosphatase activity is found in both chloroplast and mitochondria fractions.

Glycolate may be synthesis by more than one mechanism. Merrett and Goulding (1967); Goulding and Merrett (1967) observed that in *Chlorella pyrenoidosa* in presence of light acetate is converted into glycolate. In their experiments, they used acetate- ^3H and acetate-2-^{14}C. It was observed that both ^3H and ^{14}C was incorporated into glycolate before Calvin cycle intermediate were labeled. Thus, they concluded that glycolate was not synthesized from a two-carbon fragment derived from the cycle. When $3H/^{14}C^-$ was added together with the labeled acetate, the $^3H/^{14}C$ ratio decreased with time. This indicates that at least there are two independent pathway of glycolate biosynthesis, one from CO_2 and another from acetate. Plamondon and Bassham (1966) on the basis of increased ratio of $^3H/^{14}C$ in glycolate than in phosphoglyceric acid concluded glycolate was continuously synthesized by reductive formation of an unlabeled carbon compound as well as from $^{14}CO_2$. The unlabeled compound was glyoxylate.

The glycolate ion is excreted from chloroplast. It is because, glycolate oxidase system is absent in chloroplast.

Glycolate Oxidase and Glyoxylate Reductase

Glycolate oxidase plays a quantitative role in the production of O_2 uptake and CO_2 release in the light. Zelitch and Ochoa (1953) purified this enzyme from spinach leaves and found that it contains a prosthetic group riboflavin phosphate (FMN). Frigerio and Harbury (1958) further purified, crystallized and studied the properties this enzyme from spinach leaves. This flavoprotein can utilize quinines, like p-benzoquinone, as electron acceptors in addition to O_2 and certain dyes, and the corresponding hydroquinone is then produced in the reaction. Glycolate oxidase transfers electron from glycolate to O_2, reducing the O_2 to hydrogen peroxide. Though this enzyme is most active on glycolate, but it can act on other substrates like phenyllactate and p-hydroxylactate.

Zelitch (1953) reported an enzyme from spinach leaves extract which could reduce glyoxylate into glycolate in the presence of NADH. This enzyme, was named as glyoxylate reductase. Zelitch (1955) purified and crystallized this enzyme from tobacco leaves. It is specific for NADH and does not react with pyruvate.

The Enzymatic Activities in Peroxisomes

NADH-glyoxylate reductase activity takes place in peroxisomes (Fig. 7.1 A,B). Peroxisomes are microbodies with several oxidative enzymes. Five major enzymes present in peroxisomes, are glycolate oxidase, catalase, NADH-glyoxylate reductase, NAD+-malate dehydrogenase, and transaminase. The enzyme may vary in their complement according to cell type and the state of differentiation in the cells.

The following reactions show a glycolate-glyoxylate cycle :

$$\text{Substrate} + NAD^+ \xrightarrow{\text{dehydrogenase}} \text{Oxidized substrate} + NADH + H^+$$

$$NADH + H^+ + \text{glyoxylate} \xrightarrow{\text{glyoxylate reductase}} \text{glycolate} + NAD^+$$

$$\text{Glycolate} + O_2 \xrightarrow{\text{glycolate oxidase}} \text{glyoxylate} + H_2O_2$$

$$H_2O_2 \xrightarrow{\text{catalase}} H_2O + 1/2\ O_2$$

$$\text{Substrate} + 1/2\ O_2 \rightarrow \text{oxidized substrate}$$

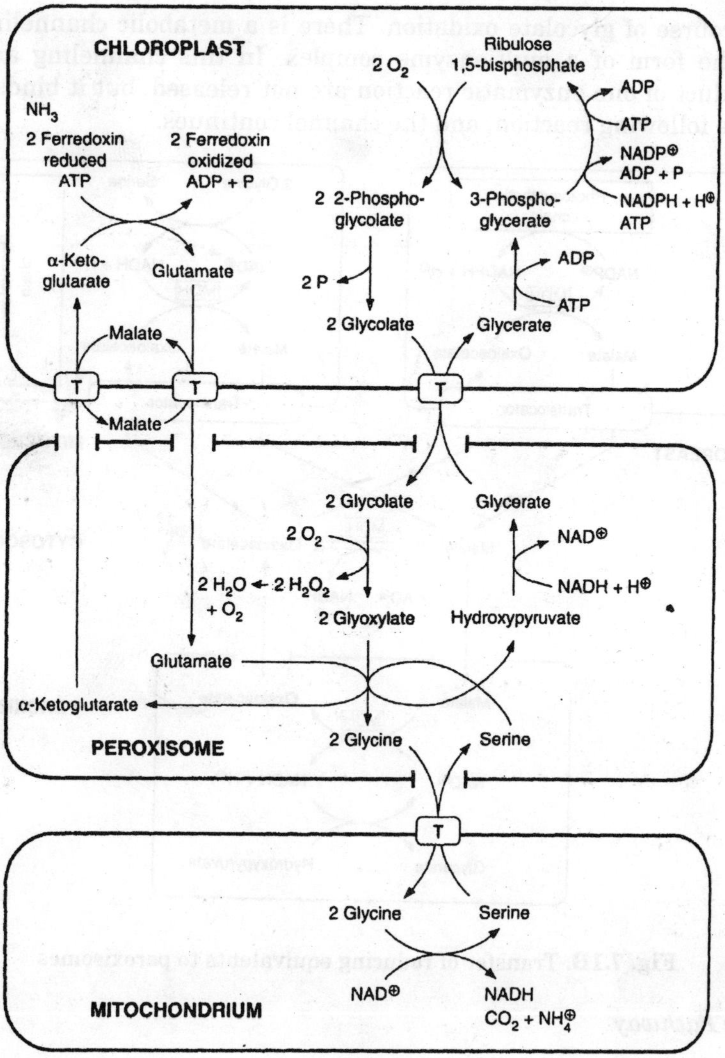

Fig. 7.1A. The cell organelles involved in photo respiration and the biochemical reaction occuring in them.

Glyoxylate reductase along with NAD linked dehydrogenase oxidizes a substrate. NAD acts as electron carrier. Glycolate takes up a molecule of oxygen and forms glyoxylate and hydrogen peroxide. Both of these products are toxic compounds. Enzyme catalase breaks H_2O_2 into water and oxygen, while glyoxylate is converted to amino acid glycine.

In peroxisomes, there is compartmentalization, where glyoxylate is converted to glycine, thus eliminating its toxic effects. The compartmentalization is not due to membrane boundaries, but due to specific properties of peroxisomal matrix. If the boundary membrane of peroxisome is disrupted, it is observed that peroxisomal matrix proteins remain aggregated in the form of particle size similar to peroxisomes. This gives the compartmentalization of the peroxisomal reaction is maintained. The intermediates of peroxisomal metabolism (glyoxylate, H_2O_2, and hydroxypyruvate) are not released from these particles in the course of glycolate oxidation. There is a metabolic channeling, the enzymes are present in the form of a multienzyme complex. In this channeling and multienzyme complex, the product of one enzymatic reaction are not released, but it binds immediately to the enzyme of the following reaction, and the channel continues.

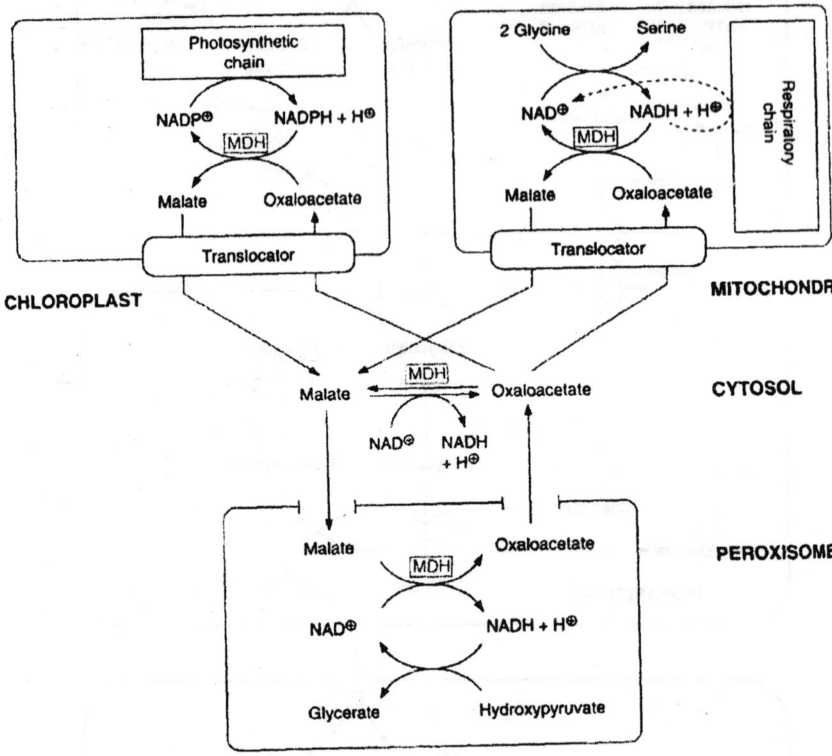

Fig. 7.1B. Transfer of reducing equivalents to peroxisomes

Fate of Glycolate Pathway

Glycine and serine are the early products of glycolate metabolism. The glyoxylate is converted to the amino acid glycine by two different reactions proceeding in the peroxisome simultaneously. Most of the glyoxylate undergoes transamination reaction. Amino acid glutamate donates its amino group and glyoxylate is converted to glycine (Fig. 7.2). This reaction is catalyzed by enzyme glutamate-glyoxylate aminotransferase. Analine may also donate its amino group to glyoxylate. The second type of transamination reaction is catalyzed by enzyme serine glyoxylate aminotransferase, and serine is synthesized. Two molecules of this glycine molecule enter inside the mitochondria. It is believed that glycine is transported to mitochondria via a specific translocator. In mitochondria, two molecules of glycine are oxidized yielding one molecule each of serine and CO_2. A NH_4^+ ion is also released in this reaction, which is reincorporated in the synthesis of glycine. The oxidation of glycine is catalyzed by the glycine decarboxylase-serine hydroxymethyltransferase complex. This enzyme is a multi-enzyme complex and shows similarities with pyruvate dehydrogenase. There are four subunits in this complex viz. H-protein (having prosthetic group lipoic acid),

P-protein (contains pyridoxal phosphate), T-protein (has tetrahydrofolate) and L-protein (also called dihydrolipoate-dehydrogenase). Serine hydroxymethyltransferase of this complex catalyzes the transfer of the formyl residue to another molecule of glycine to form serine. Serine molecule has different fate. Amino moiety of this amino acid is released and phosphate moiety is added to it. Enzyme serine-glyoxylate aminotransferase converts serine to hydroxypyruvate. The latter is reduced to glycerate (Fig. 7.3). This reaction is catalyzed by enzyme hydroxypyruvate reductase, NAD H + H^+ provides reducing equivalents. Glycerate is released from peroxisomes and glycolate-glycerate translocator transfers it to the chloroplast. Here, ATP donates a phosphate moiety to glycerate, which is converted to 3-phosphoglycerate by enzyme glycerate kinase.

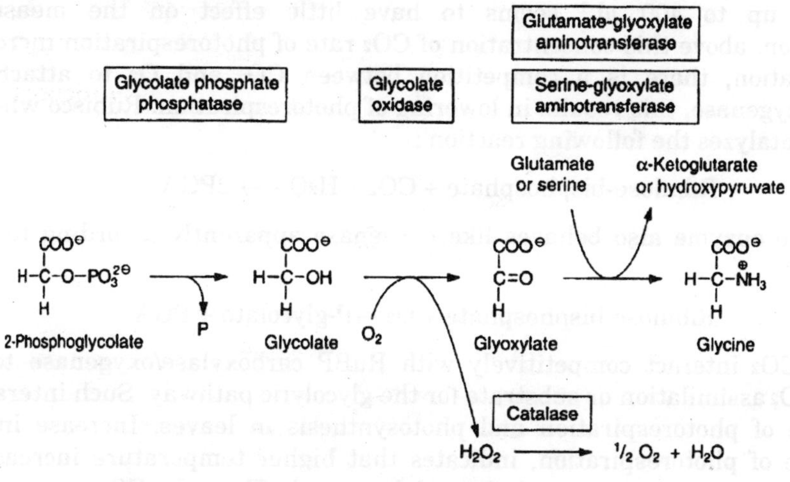

Fig. 7.2. Conversion of 2-phosphoglycolate to glycine

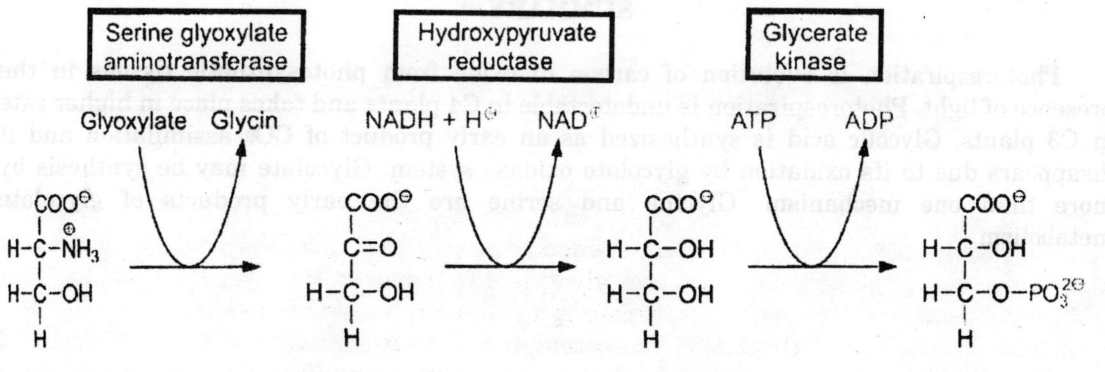

Fig. 7.3. Conversion of serine to 2-phosphoglycerate

Factors Affecting Photorespiration

Photorespiration occurs at greater rate than dark reaction. It is completely dependent on photosynthesis, which provides substrate to photorespiration. There is a direct effect of

CO_2 and O_2 concentrations on photorespiration. The latter increases with the increase in temperature (upto 30°C). Biosynthesis of glycolate in the light was about 29% less in CO_2 free air than in normal air. As the light intensity increases, photorespiration also increases and exceeds dark respiration. At low light intensities, the rate of CO_2 evolution in light is less than that in darkness. Very high light intensities, the rate of photorespiration do not increases, however, the photosynthesis continues to increase.

Increasing concentration of O_2 stimulates photorespiration. It causes an increased rate of synthesis of glycolate in the light. The O_2 requirement for glyxolate synthesis is possibly related to an oxidative step in the synthesis of a two-carbon compound at the glycolaldehyde oxidation level than is derived from the photosynthetic carbon reduction cycle. CO_2 concentration up to 400 μl^{-1} seems to have little effect on the measured rate of photorespiration, above this concentration of CO_2 rate of photorespiration increases. At high CO_2 concentration, there is a competition between CO_2 and O_2 to attach with RuBP carboxylase/oxygenase, this results in lowering of photorespiration. Rubisco when behaves as carboxylase, catalyzes the following reaction :

$$\text{Ribulose-bisphosphate} + CO_2 + H_2O \dashrightarrow 2PGA$$

The same enzyme also behaves like oxygenase apparently according to the following equation :

$$\text{Ribulose-bisphosphate} + O_2 \rightarrow \text{P-glycolate} + PGA$$

O_2 and CO_2 interact competitively with RuBP carboxylase/oxygenase to yield either products of CO_2 assimilation or substrate for the glycolytic pathway. Such interaction governs both the rate of photorespiration and photosynthesis in leaves. Increase in temperature increases rate of photorespiration, indicates that higher temperature increases affinity of RuBP carboxylase/oxygenase towards O_2 and decreased affinity for CO_2.

SUMMARY

Photorespiration is evolution of carbon di oxide from photosynthetic tissues in the presence of light. Photorespiration is undetectable in C4 plants and takes place in higher rate in C3 plants. Glycolic acid is synthesized as an early product of CO_2 assimilation and it disappears due to its oxidation by glycolate oxidase system. Glycolate may be synthesis by more than one mechanism. Glycine and serine are the early products of glycolate metabolism.

Chapter - 8

NITROGEN AND SULPHUR ASSIMILATION

Plants absorb nitrogen and sulphur primarily through interaction with the soil solution. In plants a major portion of nitrogen is present in the form of proteins. In a plant cell, it is present in oxidized state (e.g., R-NH_2, NH_3, NH_4^+, N_2, N_2O, NO, NO_2 and NO_3). Sulphur constitutes about one fifth of nitrogen content. It is a constituent of many molecules, including proteins. Plants absorb these elements from soil in the oxidized form and reduce them by energy-dependent processes.

The Nitrogen Cycle

In environment, nitrogen is present in many forms and undergoes physical and biological inter-conversion. Such processes in environment are known as nitrogen cycle. Though it is present in 78% by volume, still it cannot be used by living organisms. Nitrogen enters in leaf via stomata with air, but as there are no enzymes to fix it, it goes back to the environment through gas exchange. Plants utilize nitrogen in the form of nitrate or nitrite. Nitrogen is fixed either by physical or biological processes. The physical methods of nitrogen fixation are rain, volcanic activity and forest fires. Nitrite arises from oxidation of N_2 or ozone in the presence of lightning or ultraviolet radiation. Ocean is another source of NO_3^+ Wind-whipped white caps produce minute droplets of water are called aerosols, from which the water evaporates, leaving ocean salts suspended in the atmosphere near coastline. These salts can be brought to the land in rainwater. They are called cyclic salts, because they eventually cycle through streams back to the oceans. After intake of nitrogen plants convert it into numerous nitrogenous compounds. This nitrogen goes back to environment via manure and dead plants, microorganisms, and animals. Soil also contains various amino acids. These amino acids are produced largely by microbial decay of organic matter. They are also by excretion from living roots. Though they may be absorbed by plants, their importance is by high merit of reservoir. There is a huge amount of organic nitrogen in soil. The conversion of organic nitrogen to the more mobile, inorganic state is known as nitrogen mineralization. It is conducted by many micro-organisms like bacteria, fungi and actinomycetes. Conversion of organic nitrogen to NH_4^+ by soil microbes is called ammonification. A number of ammonifying micro-organisms are present in all types of soil and water conditions which oxidize proteins. The amino acids formed as a result of breakdown of proteins, is hydrolysed to form lower fatty acid and ammonia; or an alcohol, CO_2 and ammonia or an aldehyde, lower acids and ammonia. This phenomenon is known as hydrolytic deamination.

In warm, moist soils with near natural pH ammonium is further oxidized, this process is *Nitrification*. Ammonium salts are oxidized into nitric acid (nitrates). Plants are able to uptake nitrogen in this form. S. Winogradsky (1890) observed for the first time that nitrification is associated with the metabolism of certain chemoautotrophic bacteria. The bacteria, which take part in nitrification, are known as nitrifying bacteria. These bacteria have two types of nature one which derives energy by oxidation of ammonium and the second derive energy by the oxidation of nitrite. Nitrification is a strictly aerobic process and does not occur at redox potential lower than +200 mV. Alkaline pH (7-8) favours nitrification. The biochemical reactions of nitrification can be studied in two phases.In the first phase ammonia is oxidized into nitrous acid through the removal of electrons.

$$NH_3 \rightarrow NH_2OH \dashrightarrow (HNO) \dashrightarrow NO_2$$

In the second phase nitrous acid is oxidized into nitric acid and it takes place in the following manner.

$$HO\text{-}N{=}O \xrightarrow{+H_2O} HO\text{-}N{=}(OH)_2 \xrightarrow{2H} HNO_3$$

Along with this process carbon di oxide is reduced due to chemosynthesis.

The process in which, nitrates are reduced is known as *denitrification*. The anaerobic decomposition of protein is known as *putrifaction*. The end products are generally amines and related compounds and hydrogen sulphides. Denitrifying bacteria use NO_3^- rather than O_2 as an electron acceptor during respiration, thus obtaining energy for survival.

Biological Nitrogen Fixation

The process by which N_2 is reduced to NH_4^+ is called nitrogen fixation. It is carried out only by prokaryotic microorganisms. The nitrifying bacteria convert nitrogen into NO_2^- and NO_3^-. In turn NO_3^- may enter metabolism through reduction to NH_4^+. Nitrogen fixation is an anaerobic process.

Hellriegel and Wilfarth established the fact that nitrogen is fixed by certain bacteria living in root nodules of the leguminous plants. Beijerinck (1888) isolated pure culture of root nodule forming bacteria and named it as *Bacillus radicicola*. This bacterium was later named as *Rhizobium*. In soil, *Rhizobium* is aerobic bacteria with rod shaped cells and persists saprophytically until they infect a root hair or damaged epidermal cell (Hubbell, 1981; Pierce and Bauer, 1983). The root cap cells secrete large amounts of polysacchride and other materials. Globulos globules of material may be formed at the root hair tips. This material forms the `mucigel' which attracts the rhizobia, and bacteria forms an infection thread in plant body. The infection thread consists of an infolded and extended plasma membrane. The bacteria multiply extensively inside the thread, which extends inwardly and penetrate through and between the cortex cells. Prazmonski (1889) studied the mechanism of root infection by the legume organism. According to him the nodule forming bacteria may be three forms: (1) those, which develop normally by the path of infection threads. This type is found in most of the leguminous plants. (2) Those, which develop through the infection. In the second stage bacteria changes into motile form and (3) those, developing from the bacteria with long infection. This is inside the plant tissues, the cells are transformed into bacteroids. The bacteriods usually occur in the cytoplasm in groups, each group surrounded by a membrane called the peribacteriod membrane. The space between the peribacteroid membrane and the bacteriod group is a region called the peribacteroid space. Thus the bacteria are separated from the cytoplasm of the host cell in a so-called symbiosome. Outside the peribacteroid space in the plant cytoplasm is a protein called the leghemoglobin. Nitrogen fixation in roots occurs directly within the bacteriods. The nodules are thought to be relatively simple organs. They are connected with the root via vascular tissues, which supply them with substrates formed by photosynthesis. Nodules contain perhaps dozens to a few hundred nodule-specific plant synthesized proteins or nodulins. Many of these nodulins are involved in metabolizing carbon and nitrogen, transporting compounds across the symbiosome membrane, or establishing a low-oxygen environment.

The rhizobia form species-specific nodulation factors (Nod factors). The Nod factors of rhizobia are lipochito-oligosaccharides that acquire a high structural specificity. Lipochito-oligosaccharides are derivatized oligomers of chitin (β-1,4-linked N-acetylglucosamine). Many of the enzymes that catalyze derivatizations are encoded by so-called host-range genes. Direct biochemical assays indicate that the lipochito-oligosaccharides Nod factors are products of the known nodulation genes. Very low concentration of Nod factor (10^{-9} or less) can provoke profound morphological changes in the plants. Nod factors induce expression of many plant genes, which response to rhizobial inoculation. These factors bind to specific receptor kinase of the host, which are part of signal transduction chains. In this way the "key" induces the root hair of the host to curl and the root cortex cells to divide, forming the nodule primordium.

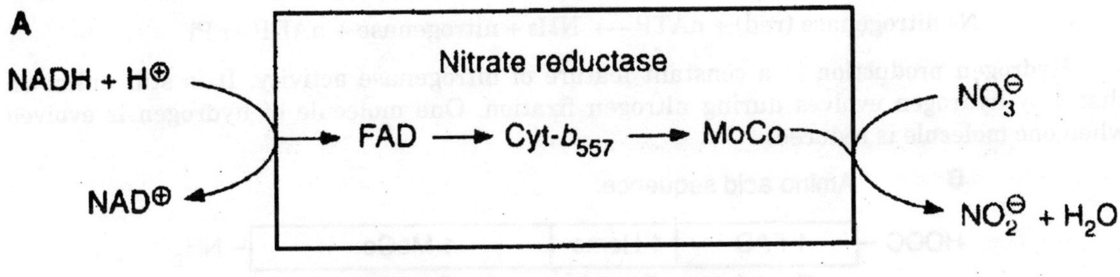

Fig. 8.1. Role of Nitrate reductase

Enzyme nitrogenase complex catalyzes nitrogen fixation. It is a highly complex system with nitrogenase reductase and nitrogenase as the main components. Former is a 64-kDa homodimer (Fig. 8.1), while the later is a 240-kDa heterotetramer. At least a total of 16 molecules of ATP are required for the fixation of one molecule of N_2. The overall chemical reaction for nitrogen fixation (reduction) is as following:

$$N_2 + 6 \text{ electrons} + 8H^+ + 16ATP \text{ -- } 2NH_4^+ + \sim16 \text{ ADP} + \sim16 \text{ Pi}.$$

The original source of electrons and protons is carbohydrate translocated from the leaves. Sucrose formed by photosynthesis reaches up to the roots from its site of synthesis i.e., leaf. This is the main source of energy for this reaction. In host cell malate is formed from sucrose and is accepted by bacteroids. Rhizobia possess a respiratory chain like that of mitochondria. This path branches at the cyt-bc1 complex of the respiratory chain and conducts electrons to another terminal oxidase. Nitrogenase reductase receives electrons from NADH via soluble ferredoxin. NADH is formed in the citrate cycle. Nitrogenase reductase is a one-electron carrier, consisting of two identical subunits, which together form a 4Fe-S cluster and contain two binding sites for ATP. Nitrogenase couples ATP hydrolysis to ADP with electron transfer from a reduced electron donor, probably a ferredoxin or a flavodoxin, to reduce N_2 to NH_3. After reduction of nitrogenase reductase, two molecules of ATP bind to it. This brings conformational change in protein, raising its redox potential from -0.25 to -0.40V. When nitrogenase receives two electrons, two ATP molecules bound to the protein are hydrolyzed to ADP and phosphate. Thus, protein is free from bound ATP and comes back to its conformation having lower redox potential. In this way enzyme regains the position to accept one electron from ferredoxin. The redox activity is brought about by coenzyme ferredoxin (non-haem Fe compound) or flavodoxin (FMN containing protein). Nitrogenase is a $\alpha_2\beta_2$ heterotetramer. This enzyme complex comprises of two subunits α and β, with similar size and folds. The tetramer contains two pair of unusual metal clusters. All nitrogenases contain two identical P-clusters, 8Fe-7S iron-sulphur complex that are integral components of the enzyme. In the reduced state a P cluster resembles two distorted 4Fe-4S cubes that share one sulfur at the corner. A second pair of metal clusters is held in nitrogenase by only cysteinyl sulphur and histidine nitrogen. It is an iron molybdenum cofactor. Molybdenum is essential element for plant associated nitrogen fixation (Fig. 8.2). Both molybdenum and iron become reduced and then oxidized as nitrogenase accepts electrons from ferredoxin and transfers them to N_2 to form NH_4^+. The mechanism of electron pathway to reducible substrate through the nitrogenase complex is as following.

$$\text{Fd (ox)} + \text{nH} \longrightarrow \text{Fd (red)}$$

$$\text{Fd (red)} + \text{nitrogenase} + \text{nitrogen} \longrightarrow \text{Fd (ox)} + N_2\text{-nitrogenase (red)}$$

$$N_2\text{-nitrogenase (red)} + \text{nATP} \longrightarrow NH_3 + \text{nitrogenase} + \text{nADP} + \text{nPi}$$

Hydrogen production is a constant feature of nitrogenase activity. It is still unknown that why hydrogen evolves during nitrogen fixation. One molecule of hydrogen is evolved when one molecule is reduced.

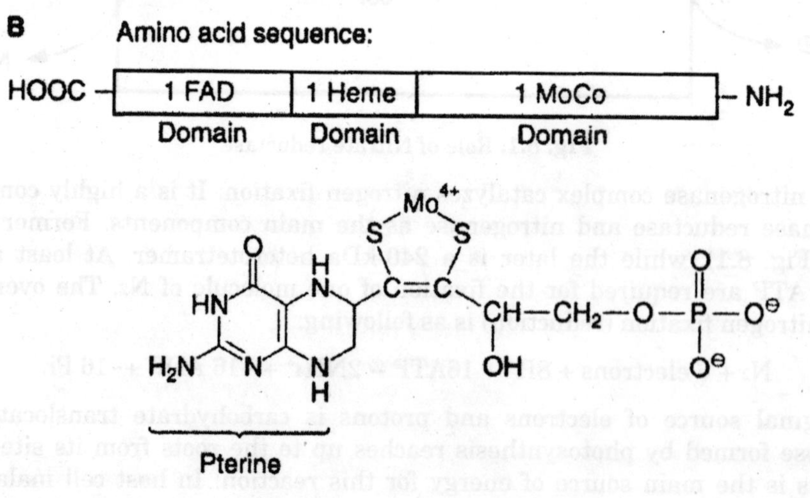

Fig. 8.2. The molybdenum cofactor (MoCo)

The low oxygen concentration within the nodule is a key element in regulating bacteroids expression of nitrogenase. The oxygen supply to bacteroid is excluded due to presence of leghaemoglobin around it. The latter is an oxygen-binding plant protein. It plays an active role in regulating and delivering oxygen in the infected cells. In some rhizobial species, an oxygen-sensitive hemo-protein kinase, FixL, controls a regulatory cascade that activates transcription of nitrogen fixation genes. FixL phosphorylates FixJ. Once phosporylated, FixJ activates transcription of other regulatory proteins. Two of these, NifA and FixK, control expression of diverse *nif* and *fix* genes. Bacteriods of developed nodule with low-oxygen concentration will express nitrogenase and the associated proteins required for reduction of dinitrogen to ammonia.

Ammonia formed by the bacteroids is transferred to host cells. In cytosol, it is converted into organic form, glutamin, glutamic acid, and asparagines. In many species, ammonium is metabolized into nitrogen-rich ureides. The formation of ureides in the host cells requires a complicated synthetic pathway. First, inosine monophophate is formed. It is an intermediate of purine biosynthetic pathway. The inosine monophosphate thus produced is degraded via xanthine and ureic acid to the ureides. Leguminous plants of tropical origin contain two types of urides, allantoin ($C_4N_4H_6O_3$) and allantoic acid ($C_4N_4H_8O_4$). Leguminous plants of temperate origin have high asparagines content in root nodules. These compounds have relatively high CN ratios. They move from bacteriod containing cells into pericycle cells adjacent to the vascular bundles that surround the nodule proper. These compounds are ultimately secreted into the conducting xylem cells. There is a cycle of nitrogenous compounds where these compounds are first transported to lead through xylem and then excess nitrogen moves back to roots via the phloem.

Assimilation of Nitrate

Plants unable to fix nitrogen utilize nitrogen in the form of NO_3^- and NH_4^+. Synthesis of organic nitrogen from inorganic source may occur either in root or in shoots (Fig. 8.3). Crop plants and many native species absorb most nitrogen as NO_3^-, while climax communities of conifers and of grasses absorb it as NH_4^+. According to Rice (1974); Haynes and Goh (1978) nitrification is inhibited either by low soil pH or by tannins and phenolic compounds.

Nitrate performs two functions i.e., as nutrients as well as a signal. The assimilation of nitrate begins with its uptake into the cell. Nitrate is stored in roots, stems, and leaf midribs. Root cells contain several nitrate transporters in their plasma membrane. It can be also loaded into xylem for long-distance transport to the shoot. The relative amounts of NO_3^- and organic nitrogen in the xylem depend upon environment conditions. The transport of nitrate into the root cells proceeds as symport with two protons. H^+-P-A-ATP generates a proton gradient across the plasma membrane.

Nitrate is reduced to ammonium so that it can be incorporated into organic compounds. The reduction of NO_3^- to NH_4^+ is a two step process. Nitrate is reduced to nitrite by nitrate reductase and nitrite is reduced to ammonium by nitrite reductase. Nitrate reductase (NR) transfers two electron, and NADH (or in some species NADPH as well as with NADH) acts as reductant. NR is a large and complex enzyme with two subunits. Each subunit contains an electron transport chain consisting of one FAD, one heme of the cytochrome-b type (cyt-b557), and one cofactor moloybdenum. This cofactor is pterin with side chain. In this side chain molybdenum is attached by two sulfur bonds. It is called the molybdenum cofactor (MoCo).

All the redox carriers are covalently bound to the protein subunit of the enzyme. These carriers become consecutively reduced and oxidized as electrons are transported from NADH in the nitrogen atom in NO_2^-. This reaction occurs in cytosol. The reaction may be expressed as following

$$NO_3^- + NADH + H^+ \xrightarrow{NR} NO_2^- + NAD^+ + H_2O$$

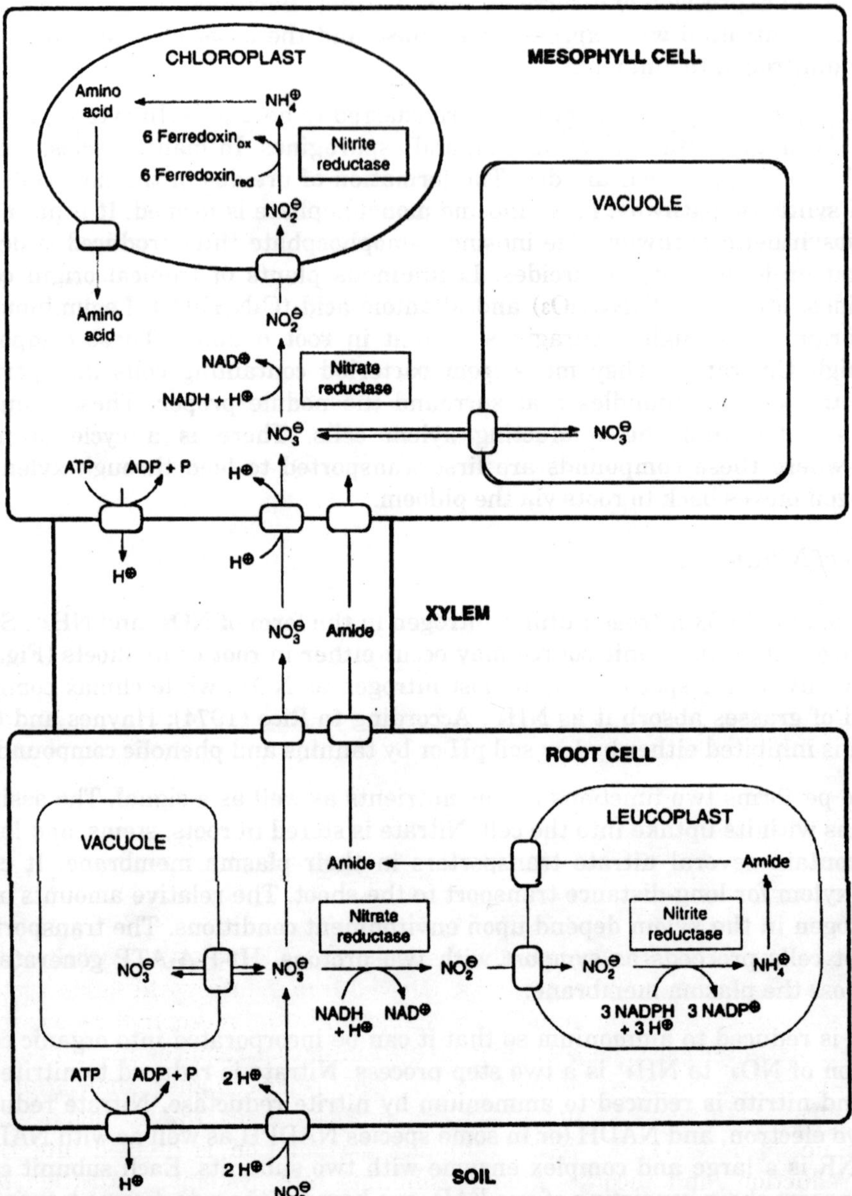

Fig. 8.3. Nitrate assimilation in the rest and mesophyll cell, mode of transport through xylem.

Nitrite produced by nitrate is transported into chloroplasts in leaves or into proplastids in roots. In these organelles nitrite reductase catalyzes reduction of nitrite to NH_4^+. In leaves, reduction of NO_2^- and NH_4^+ requires six electrons. These electrons are derived from H_2O by the chloroplast noncyclic electron transport system. Electrons are transported from H_2O to ferredoxin in the presence of light. Later, reduced ferredoxin provides the six electrons for the reduction of NO_3^- as following mechanism

$$3H_2O + 6Fd\ (Fe^{3+}) + light \longrightarrow 15O_2 + 6H^+ + 6\ Fd\ (Fe^{2+})$$

$$NO_2^- + 6Fd\ (Fe^{2+}) + 8H+ \rightarrow NH_4^+ + 6Fd\ (Fe^{2+}) + 2\ H_2O$$

$$NO_2^- + 3H_2O + 2H^+ + light \rightarrow NH_4^+ + 15O_2 + 2H_2O$$

In leaf, reduced ferredoxin acts electron donor. However, it is still not clear about the electron donor in roots.

Nitrite reductase contains a covalently bound 4Fe-4S cluster, one molecule of FAD, and one siroheme. They form an electron transport chain. Siroheme is a cyclic tetrapyrrole with Fe in the center. It shows difference with heme due to the presence of additional acetyl and propionyl residues. The electrons are transferred from ferredoxin to nitrite.

SULPHUR CYCLE

Sulphur is an important element of a cell. It is a component of S-containing amino acids (cystine, cysteine, homocysteine and methonine). It is contained in the detoxifying agent glutathione, in various iron sulphur redox clusters, in peroxiredoxins, and in thioredoxins. It is also a building element of vitamin B (thiamine, biotin and lipoic acid). In lithosphere sulphur is present as sulphate and sulphide. It is absorbed in the form of $SO_4\ ^{2-}$, which provides the necessary sulphur for plant growth. Roots contain a specific translocators which take up sulphur. Sulphur reaches up to leaves by xylem vessels. Along with three protons sulphur moves in the mesophyll cells by symport mechanism. Transport occurs due to proton gradient maintained across the plasma membrane by proton-pumping ATPase. Its surpulus quantity is stored in vacuole and is deposited there. The rate of sulfate transport across the tonoplast depends on an electrochemical gradient. Sulphate is reduced to sulphide by uptake of two electrons. Uptake of one more electron produces hydrogen sulphide. Reduction occurs in root as well as shoot.

The SO_4^{2-} assimilation

Sulphate assimilation takes place primarily in chloroplasts. During assimilation, reduction takes place and the reductants are derived from photosynthesis. In roots, it occurs in plastids. The basic scheme for sulphate assimilation in the mesophyll cells corresponds to that of nitrate assimilation (Fig. 8.4). Sulphate assimilation requires much energy, i.e., the ATP consumption of sulphate assimilation is four times higher than that of nitrate assimilation. The reduction of sulphate to sulphide consumes 732 kJ mol^{-1}, while nitrate assimilation needs only 347 kJ mol^{-1}. Sulphate assimilation is divided into three steps: activation, reduction, and incorporation of sulphide into cysteine. *Activation of sulphate* proceeds via the formation of an anhydride bond with the phosphate residue of AMP. SO_4^{2-} is exchanged by enzyme ATP sulfurylase. It catalyzes the first reaction in which, SO_4^{2-} reacts with ATP, producing adenosine-5-phosphosulfate (acronym APS) and pyrophoshpate is released. APS contains a high-energy phosphoric acid-sulfuric acid anhydride bond that potentiates the sulfuryl moiety for subsequent metabolic reactions. APS is central branch-point intermediate that feeds two pathways sulfate reduction and sulfation. The pyrophosphate breaks down into two inorganic phosphates in the presence of enzyme pyrophosphatase. At least two plant enzymes, APS kinase and APS reductase metabolize APS. These two enzymes; are now known to be the same as APS sulfotransferase. APS sulfotransferase are tetramer of 49-50-kDa subunits and have similar kinetic constants. Second step of assimilation is *reduction*, in which sulphate is reduced to sulphite (Fig. 8.5). In chloroplast, the sulphur of APS is reduced by ferredoxin. In case of reduction in proplastids, probably NADPH acts as reductant. Two hypothesis are proposed to explain the mechanism of reduction. The first theory is named as carrier-bound pathway and takes place in higher plants and eukaryotic algae. The second hypothesis explains assimilation in microorganisms. According to the first hypothesis, sulfate is transferred form APS to a reduced thiol compound by APS sulfotransferase. Thiol compound has not been identified. The reduced compound is glutathione. Thiosulfonate is produced in this reaction. Another enzyme, thiosulfonate reductase acts on thiosulfonate and produces thiosulfide.

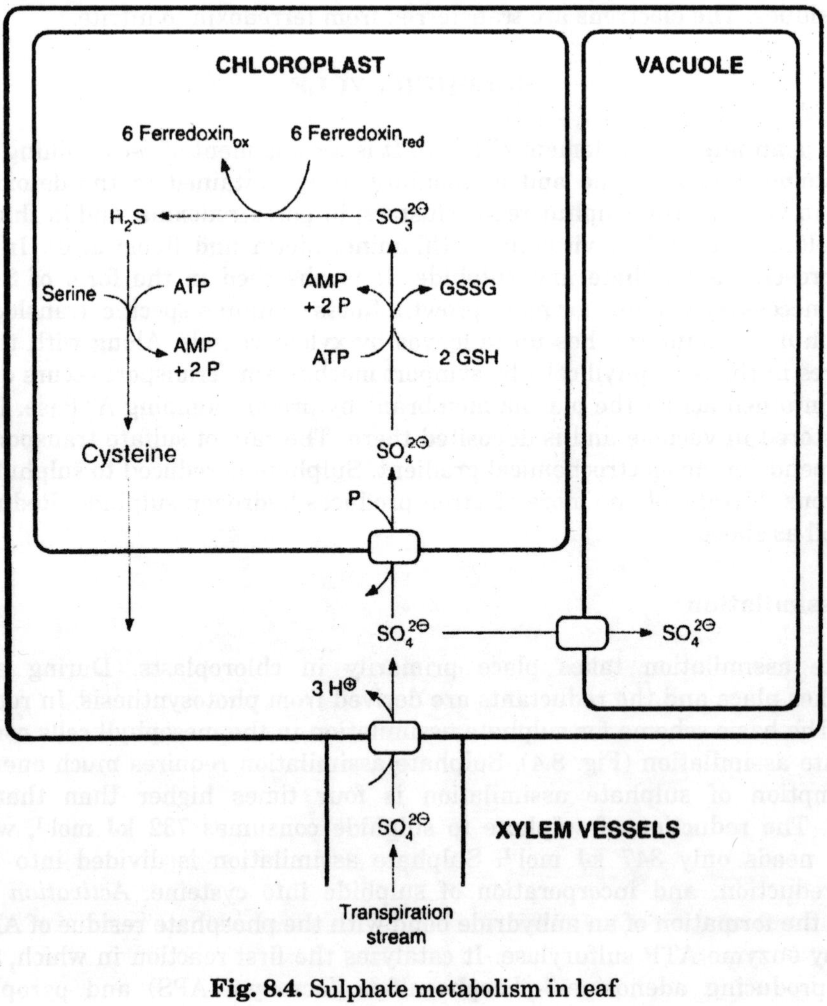

Fig. 8.4. Sulphate metabolism in leaf

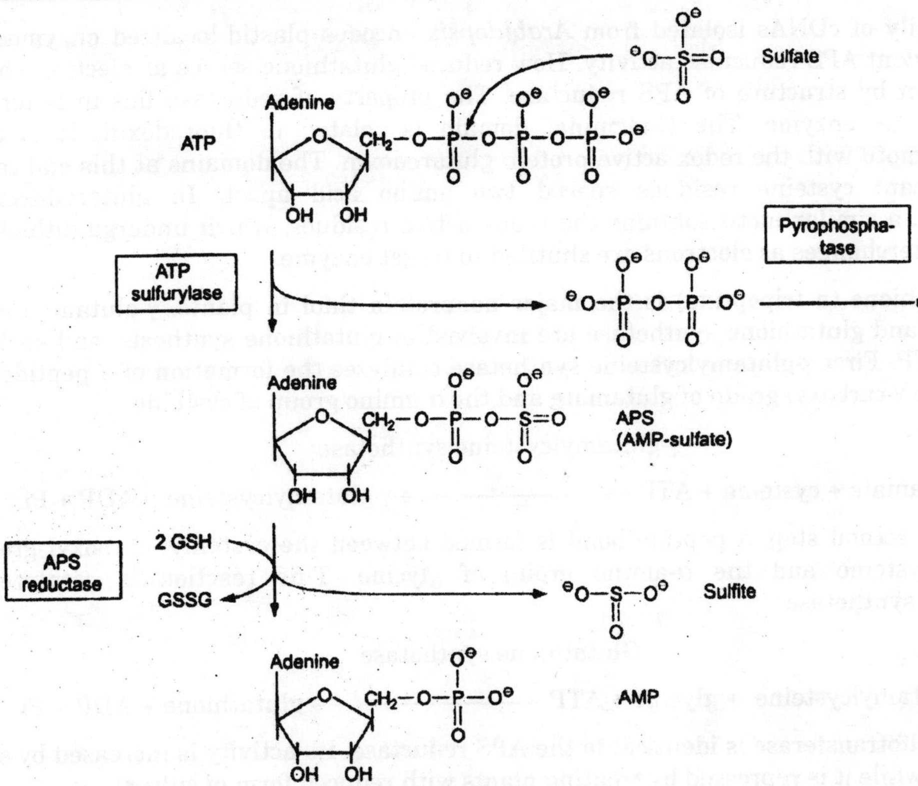

Fig. 8.5. Reduction of sulfate to sulfite

$$\text{APS + thiol compound}_{red} \xrightarrow{\text{APS sulfotransferase}} \text{thiosulfonate + 5'AMP}$$

$$\text{Thiosulfonate + 6 ferredoxin}_{red} \xrightarrow{\text{Thiosulfonate reductase}} \text{thiosulfide + 6 ferredoxin}_{ox}$$

APS sulfotransferase and Thiosulfonate reductase have proved difficult to characterize.

The mechanism of second hypothesis is as following

$$\text{APS + ATP} \xrightarrow{\text{APS Kinase}} \text{PAPS + ADP}$$

$$\text{PAPS + thioredoxin}_{red} \xrightarrow{\text{PAPS reductase}} SO_3^{2-} + \text{thioredoxinox + PAP}$$

$$SO_3^{2-} + \text{6 ferredoxin}_{red} \xrightarrow{\text{Sulfite reductase}} S_2^{-} + \text{6erredoxin}_{ox}$$

APS kinase and sulfite reductase are present in plants but PAPS reductase is not reported in plants.

A family of cDNAs isolated from *Arabidopsis* encodes plastid-localized enzymes with thiol-dependent APS reductase activity. How reduced glutathione serves as electron donor is made known by structure of APS reductase. The property of reductase lies in N-terminal domain of the enzyme. The C-terminal domain is related to thioredoxin. It shares a disulphide motif with the redox active protein glutaredoxin. The domains at this end contain two important cysteine residues spaced two amino acid apart. In glutaredoxin and thioredoxin, a similar motif contains the redox active residues, which undergo dithiol ←→ disulfide interchanges as electrons are shuttled to target enzyme.

Glutathione (a tripeptide) is the major nonprotein thiol in plants. γ-glutamylcysteine synthetase and glutathione synthetase are involved in glutathione synthesis, and each step requires ATP. First γ-glutamylcysteine synthetase catalyzes the formation of a peptide bond between the γ-carboxyl group of glutamate and the α-amino group of cysteine.

$$\text{Glutamate + cysteine + ATP} \xrightarrow{\gamma\text{-glutamylcysteine synthetase}} \gamma\text{-glutamylcysteine + ADP + Pi.}$$

In the second step, a peptide bond is formed between the cysteinyl carboxyl group of γ-glutamylcysteine and the α-amino group of glycine. This reaction is catalyzed by glutathione synthetase.

$$\gamma\text{-glutamylcysteine + glycine + ATP} \xrightarrow{\text{Glutathione synthetase}} \text{glutathione + ADP + Pi}$$

APS sulfotransferase is identical to the APS reductase. Its activity is increased by sulfur starvation, while it is repressed by treating plants with reduced form of sulfur.

Sulfite is reduced to sulfide by enzyme sulfite reductase. The latter resembles nitrite reductase. This enzyme adds six electrons to free sulfite. These electrons are donated by ferredoxin. Sulfite reductase is a homooligomeric hemeprotein. It is composed of two or four identical 64 to 71-kDa subunits, depending on the source. Each subunit of C-terminal half contains one siroheme and one iron-sulphur cluster. This cluster is like nitrite reductase (4Fe-4S). The homology point out that both reductase belong to a superfamily of anion redox enzymes. Bacteria and yeast contain NADPH-dependent sulfite reductases. These enzymes are heteromeric enzymes and they have one flavoprotein and one hemoprotein subunit. In the last step, *synthesis of cysteine* takes place. sulfide condenses with O-acetylserine (OAS) to produce cysteine. Serine and acetyl-CoA react to form OAS and CoA is liberated. This reaction is catalyzed by enzyme serine acetyltransferase.

$$\text{Serine + acetyl-CoA} \xrightarrow{\text{Serine acetyltransferase}} \text{O-acetylserine + CoA.}$$

A-acetylserine(thiol)lyase catalyzes next reaction in which, OAS and sulfide ion react to form cysteine. This enzyme contains pyridoxal phosphate as a prosthetic group. If the carrier-bound pathway is not present, instead of sulfide thiosulfate reacts with OAS.

$$\text{O-Acetylserine} + S^{2+} \xrightarrow{\text{O-Acetylserine(thiol)lyase}} \text{cysteine + acetate}$$

Both the enzymes catalyzing above reactions are present in a complex. This complex may function to regulate serine acetyltransferase activity. The enzymes are fully active when isolated free of complex.

SUMMARY

Plants absorb nitrogen and sulphur primarily through interaction with the soil solution. The inter-conversion of nitrogen in physical and biological form completes nitrogen cycle. Fixation of N_2 in the form of NH_4 is known as nitrogen fixation. Plants unable to fix nitrogen utilize nitrogen in the form of ENO_3^- and ENH_4^+. Synthesis of organic nitrogen from inorganic source may occur either in root or in shoots. Nitrate performs two functions i.e., as nutrients as well as a signal. Nitrate is reduced to ammonium so that it can be incorporated into organic compounds. Sulphur is an important element of a cell. In lithosphere sulphur is present as sulphate and sulphide. It is absorbed in the form of ESO_4^{2-}, Sulphate assimilation takes place primarily in chloroplasts. Enzyme sulphite reductase converts sulfite in to sulphide

Chapter - 9

NUCLEIC ACID THE MOLECULAR APPROACH

DNA is the genetic material. In 1944, O.T. Avery, C.M. Macleod and M. McCarty performed an experiment with bacterium *Diplococcus pneumoniae* and observed that the transforming principle is DNA. The nonvirulent strains of this bacterium were converted into virulent strains by receiving transformation principle from virulent cells. This principle was purified and was found as a molecule of DNA.

In 1952 A.D. Hershey and M. Chase experimented with bacteriophage T2, and observed that genetic information necessary for viral reproduction was present in DNA. Bacteriophage was the experimental tool and labeled phosphorus and sulphur (^{32}P and ^{35}S) were used to identify the genetic material. During replication of such bacteriophage, it was found that DNA containing ^{32}P was present inside the cells. As the progeny viruses are produced inside the cell. The progeny viruses were shown to contain some of the ^{32}P, but none of the ^{35}S of the parental phage.

F. Miescher (1869) named "nuclein" to the material isolated from nucleus of the cell. It contained phosphorus. He also observed that the number of purine nucleotides is always equal to number of pyrimidine nucleotides (A+G = C+T). Moreover, the number of A residues is always equals the number of T residues and likewise, number of G residues is equal to number of C residues. Franklin and Wilkins studied the X-ray diffraction patterns of fibers of DNA, which exhibited that DNA is made up of more than one helical polynucleotide chain. The bases are stacked with a spacing of around 3Å. The detailed structure of DNA and RNA is described in chapter cell biochemistry. In this chapter only the molecular aspect of nucleic acid is discussed.

STRUCTURE OF DNA

DNA is composed of polynucleotide chains twisted around each other. J.D. Watson and F.H.C. Crick (1953) proposed a double helix model of DNA structure. The double helix is right handed and composed of two polynucleotide chains that are held together by weak, noncovalent bonds between pairs of bases.

DNA has a fibrous nature. The molecule is made up of long and unbranched chain. The backbone is made up of alternatively arranged deoxyribose sugar and phosphate groups. These are joined by phosphodiester bonds in 3'→5' direction. Nitrogenous base is attached to sugar. The physicochemical analysis involving sedimentation, diffusion, light scattering, and viscosity measurement techniques have indicated that DNA is a very asymmetrical molecule with 20 Å in diameter and thousands Å long (Fig. 9.1). The molecular weight is between 5 X 10^6 and 10^7. There are two chemical chains in the fiber. X-ray crystallographic studies of sodium salt of DNA shows that the fibers are highly birefringent, having marked ultraviolet and infrared dichroism. Sodium salt of DNA shows two structures. Structure-À is crystalline with 75 percent relative humidity and contains approximately 30 percent water. At higher humidity the fibers take up more water, increase in length by about 30 per cent. It is paracrystalline. Structure B has a very strong 3.4 Å reflexion on the meridium. Structure A is shortens by about 30 per cent. It is about 2.5 per cent Å apart axially. All these evidences show that there are two polynucleotide chains. The two strands have the same helical geometry but base pairing holds them together with the opposite polarity. There is an approximately two folds axis of symmetry that relates the two sugars and all four base pairs can be accommodated within the same arrangement without any distortion of the overall structures of the DNA.

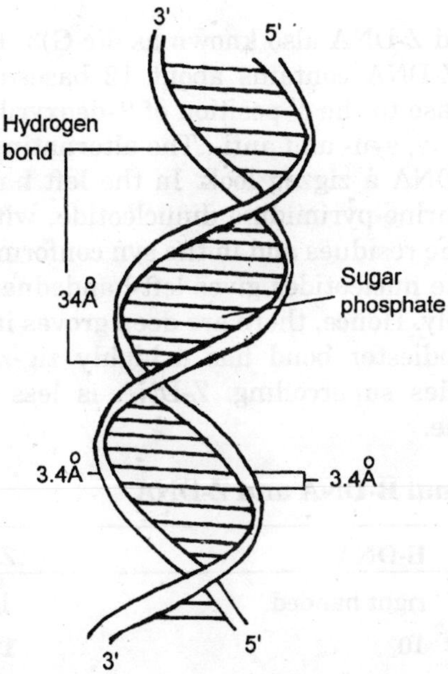

Fig. 9.1. Double Helical Structure of DNA

DNA molecule is a three dimensional crystal. The two polynucleotide chains are complementary to each other and are coiled round a common axis and joined together by hydrogen bonds. Both the chains follow right handed helices, but the sequences of the atoms in the phosphate-sugar backbones run in opposite directions. The phosphate and sugar are on the outside of the helix whilst the bases are on the inside. The distance of a phosphorus atom from the fiber axis is 10 Å. The two chains show a turn at each 34 Å distance. There are ten base pairs in each turn. Thus, each base pair has distance of 3.4 Å. The base pairs are joined by hydrogen bonds. The pairing is very specific, with one purine and one pyrimidine nitrogenous base forming the pair. Adenine always pairs with thymine while cytosine pairs with guanine. There is no restriction to the neighboring base pair. The base pair are flat and are arranged one above the other roughly perpendicular to the direction of the helical axis. Electron cloud interactions (π-π) between bases in the helical stacks contribute significantly to the stability of the double helix. This specificity of base pairing results in a complementary relationship between the sequence of bases on the two interwined chains and gives DNA its self-encoding character. Thus, if one chain has sequence of 5'-ATGTC-3', the complementary chain will have sequence 3'-TACAG-5'. The hydrogen bond between complementary bases provides thermodynamic stability of the helix and the specificity of base pairing. Sometimes, individual base can protrude from the double helix. This phenomenon is known as base flipping. Flipping has some biological functions. The angle at which the two sugars protrude from the base pairs is about 120° and 240° for narrow and wide angle. By stacking of base pair one above the other is narrow angle between the sugars on one edge of the base pairs generates a minor groove and the large angle on the other edge generates a major groove. Thus, the double helix has minor and major grooves. The latter is rich in chemical information.

Z-DNA

Alexander Rich examined Z-DNA also known as d(c-G)3. It is different from the native DNA in following respects. Z-DNA contains about 12 bases/turn. In Z-DNA regions, the glycoside bond connects the base to the 1' position of 2'-deoxyribose. This bond can be in one of two conformations alternately, syn- and anti-. The alternating anti-syn conformation gives the backbone of left-handed DNA a zigzag look. In the left handed helix, the fundamental repeating unit usually is a purine-pyrimidine dinucleotide, with the glycosidic bond in the anti conformation at pyrimidine residues and in the syn conformation at purine residues. The syn-conformation at the purine nucleotides gives left-handedness of the helix. G-C pairs are distinctly displaced peripherally. Hence, there are deep groves in the Z-DNA forms. Z-DNA is left handed and the phosphodiester bond has a highly zig-zag orientation. If Z-DNA is present in plasmid, it modifies supercoiling. Z-DNA is less sensitive to nucleases. One complete helix is 45 Å in Z-zone.

Comparison of helix of normal B-DNA and Z-DNA

	B-DNA	Z-DNA
1. Helix sense	right handed	left handed
2. Residues/turn	10	12 (6 dimers)
3. Diameter	20	-18 Å
4. Rise per residue	3.4 Å	3.75 Å
5. Helix pitch	3.4 Å	45 Å
6. Base pair tilt	6°	7°
7. Rotation per residue	36°	60°
8. Glycoside torsion angle		
deoxyguanosine	anti	syn
deoxycytidine	anti	anti

9. Sugar pucker		
deoxyguanosine	C_2 endo	C_3 endo
deoxycytidine	C_3 endo	C_2 endo
10. Distance of P axis		
dep G	9 Å	8 Å
dep G	9 Å	69 Å

REPLICATION OF DNA

DNA is a biomolecule with the property to replicate and produce daughter DNA molecule. The base sequence of daughter molecule is as that of parent DNA molecule. The dispersive, conservative and semiconservative methods of DNA replication have been proposed. In dispersive mode of replication the parental DNA is broken down into smaller fragments. These are the precursors, which complete the remaining portion of the DNA molecule. The Conservative method proposes that the two strands of DNA molecule remains joined to each other while, they give programme for the formation to daughter DNA molecule. The Semiconservative method of DNA replication is now well accepted, the two strands of DNA molecule separate from each other and each strand gives template to the complementary strand.

M.S. Meselson and F.W. Stahl (1958) proved this method of DNA replication using *E.coli*, and labeled nitrogen (^{15}N). The culture grown in ^{15}N nitrogen source was later transferred to a medium containing ^{14}N. The cells developed in these culture media contained isotopic nitrogen in the nitrogenous base of DNA. Cells grown in ^{15}N will have a greater density (weight per unit volume) than the DNA of cells grown on medium containing ^{14}N. These molecules can be separated by equilibrium density-gradient centrifugation. When a heavy salt solution such as 6 M CsCl is centrifuged at vary high speeds (30,000-50,000 revolutions per minute) for 48-72 hours, an equilibrium density gradient is formed. At the equilibrium stage between sedimentation and diffusion, a linear gradient with increasing density, exist from the top to bottom of the tube. DNA molecule will settle where the density of the salt solution is equal to its own density. A DNA molecule with mixed chain i.e., one heavy chain (^{15}N) and one light chain (^{14}N) is present in the solution; two bands will appear in CsCl tube. This confirms the fact that DNA replicates in a semiconservative method.

DNA replication is a complex process and requires the four deoxynucleoside triphosphates (dGTP, dATP, dCTP, dTTP) as precursors and some enzymes, protein factors to complete this process. DNA polymerase binds two divalent metal ions (typically Mg^{2+} or Zn^{2+}). These metallic ions alter the chemical environment around the correctly base-paired dNTP and the 3'OH of the primer. The metal ion reduces the affinity of the 3'OH for its hydrogen.

DNA Polymerase

In 1957, Arthur Kornberg and his coworkers isolated an enzyme from *E.coli* and named it as DNA polymerase or "Kornberg enzyme" (Fig. 9.2). In *E.coli* five types of DNA polymerase enzymes have been identified and studied. These enzymes are distinguished by their enzymatic properties subunit composition, and abundance. These enzymes are Pol I made of a single subunit and functions as RNA primer removal, DNA repair. Pol II (Din A) is also made up of one subunit and acts as DNA repair. Pol III is the core enzyme with three subunit and takes part in chromosome replication. Pol III holoenzyme made of nine subunits also acts as chromosome replication. Pol IV (Din B) made up of only one subunit, acts as DNA repair, Trans Lesion Synthesis (TLS). Pol V (UmuC, UmuD'$_2$C) has three subunits and functions as TLS. Previously, DNA polymerase I was thought to be the major enzyme that joins together deoxyribonucleotides. Now it is established that it is actually DNA polymerase III, the polymerizing enzyme. DNA Pol III holoenzyme confers very high processivity. Pol I has a 5' exonuclease that allows DNA Pol I to remove RNA or DNA immediately upstream of the site of DNA synthesis. The 5' exonuclease activity us capable to remove the RNA-DNA linkage that is resistant to RNAse H. In Eukaryotes, there is a multiple DNA polymerase. These are named as Polα; Polβ; Polγ; Polδ; Polε; Polθ; Polξ; Polλ; Polμ; Polκ; Polη; Polτ, and Rev 1. Pol α (Polα/primase) is made up of four subunits, out of which two-subunit are DNA Polα and two-subunit are primase. Polγ is made up of three subunits, Polδ has 2-3 subunits, and Polθ is of four subunits, rest of the polymerases are made up of only one subunit. Polα; Polδ; Polε are essential for DNA duplication. Polα/primase; with DNA Polδ, and Polε is called polymerase switching.

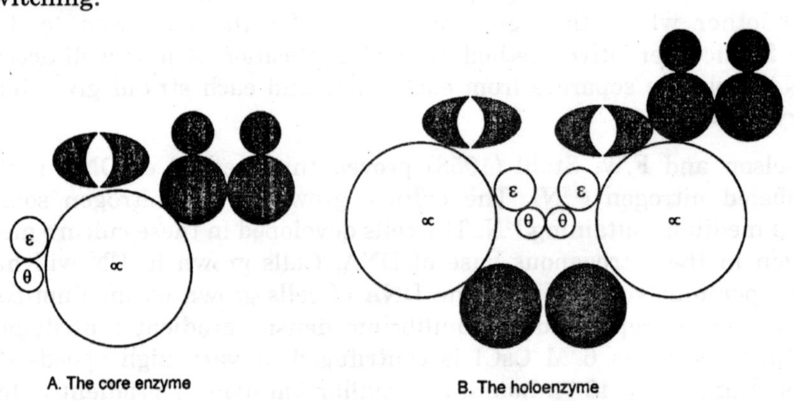

Fig. 9.2. Structure of DNA polymerase III.

The functions of these enzymes are as following :

Polα	Primer synthesis during DNA replication
Polβ	Base excision repair
Polγ	Mitochorndria DNA replication and repair
Polδ	DNA replication; nucleotide and base excision repair
Polε	DNA replication; nucleotide and base excision repair
Polθ	DNA repair of crosslinks
Polξ	Translesion synthesis (TLS)
Polλ	Meiosis-associated DNA repair
Polμ	Somatic hypermutation

Polκ	TLS
Polη	Relatively accurate TLS past cis-syn cyclobutane dimers
Polτ	TLS, somatic hypermutation
Rev1	TLS.

DNA polymerases are processive enzymes, i.e., they are capable of adding as many as 1,000 nucleotides per second to a primer strand (Fig. 9.3). Processivity is the average number of nucleotides added each time the enzyme binds a primer:template junction. Structure of DNA polymerase bound to primer:template junctions reveals that the DNA substrate sites in a large cleft that resembles a partially closed right hand. The three domains of the polymerase are called the thumb, fingers, and palm. The catalytic site resides in palm domain, which is composed of a β sheet. The residues located within the fingers bind to the incoming dNTP. The finger domains act on template region, leading to a nearly 90° turn of the phosphodiester backbone of the template. This enables to expose only the first template base after the active site. Thumb has two responsibilities, it maintains the correct position of the primer and the active site; and it helps to maintain a strong association between the DNA polymerase and its substrate. *Sliding DNA clamps* is associated with DNA polymerase to yield high processivity. The former is composed of multiple identical subunits. These subunits are assembled with a hole in the center of the clamp which can encircle DNA duplex and leave room for a layer of one or two water molecules between the DNA and the protein. *Sliding clamp loaders* are a special class of protein complexes, which catalyzes the opening and placement of sliding clamp on the DNA. They also remove the slide clamps after the replication is completed. Sliding clamp loaders need couple ATP binding and hydrolysis to the placement of the sliding clamp around primer:template junction. The sliding clamp proteins slide along the DNA without dissociating from it. A complex is formed consisting of primer:template junction, polymerase and newly synthesized double-stranded DNA. This complex moves efficiently along the DNA template during DNA synthesis. Sliding clamp prevents the polymerase from diffusing away from the DNA. Once an ssDNA template is completely copied, the DNA polymerase must be released from this DNA and the sliding clamp to act at a new primer:template junction. *DNA primase* is a specific RNA polymerase. It synthesizes 5-10 nucleotide long RNA primer on an ssDNA template. Primase does not require specific DNA sequence to initiate synthesis of a new RNA primer. These primers are needed for the initiation of replication of DNA, and are subsequently extended by DNA polymerase. Primase activity requires at least six other proteins and forms a complex primosome. Primosome, made up of some protein factors. The six proteins of primosomes are PriA, PriB, PriC, DnaT, DnaB and DnaC. The key event in localizing the primosome is the ability of PriA to displace SSB protein from single-stranded DNA. *DNA helicases* moves along ssDNA in a defined direction and separate the two strands of DNA, and form single stranded DNA at the beginning of the replication fork. This process needs energy. The movement of DNA helicase (either in 5'→3' or 3'→5' direction) is known as polarity. This direction is always defined according to the strand of DNA bound rather than the strand that is displaced. The movement of enzyme needs energy, which is derived by the breakdown of ATP. Two different helicases are often involved. One (helicase II or III) attaches to the template for the lagging strand and moves in the 5'→3' direction. The other helicase (Rep protein) binds to the strand directing synthesis of the leading strand and moves in the 3'→5' direction. *The exonuclease* enzyme acts in 3'→5' direction and carries out a critical proof reading. Proofreading of DNA synthesis is mediated by nucleases that remove incorrectly base-paired nucleotides. It takes a second opportunity to add the correct nucleotide. *DNA gyrase or topoiomerases* catalyze the formation of negative supercoils in DNA. During replication, enzyme helicase eliminates the base parts between the two strands. Thus, positively supercoils are formed in front of the fork. One DNA link must be removed approximately every ten base pairs of DNA unwound. DNA molecule is supercoiled under normal cellular conditions. At the time of replication, unwinding is essential so that the two strands separate. *The Replisome and Replicon* all the proteins (the multiple interacting machines) that function at the replication fork are combinedly known as replisome. These enzymes function in a coordinated interaction between them. A replicon is a unit of DNA in which undergoes replication. It is defined by its possession of the control elements needed for replication. A replicon has origin and terminus site, respectively where the replication initiates and terminates. Replicons may be linear or circular. The bacteria has a single chromosome, hence it has single copy replicons, while eukaryotic chromosome contains many replicons.

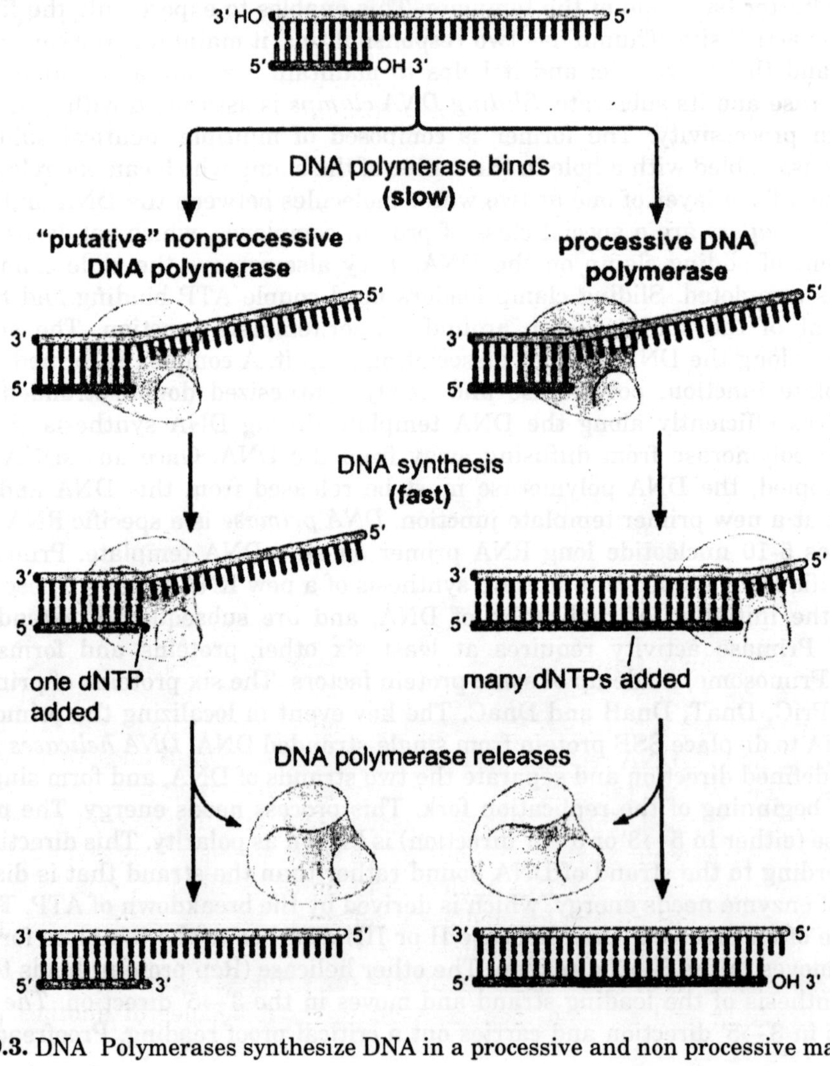

Fig. 9.3. DNA Polymerases synthesize DNA in a processive and non processive manner.

The Okazaki fragments In 1968 Reiji Okazaki and his colleges worked on *E. coli*, and explained the synthesis of complementary chain of 3'→5' strand. The cells were grown at 20°C with ^{3}H-thymidine. The small fragments remain attached to lagging strand of DNA. He found small fragments of DNA strands (1-2kd), each attached with an RNA primer. These fragments were named as Okazaki fragments, which represent transient, nascent DNA molecules. The length of Okazaki fragments vary from 1,000 to 2,000 nucleotides in prokaryotes and 100 to 400 nucleotides in eukaryotes.

The Mechanism

DNA polymerase uses a single active site to catalyze the addition of any of the four deoxynucleoside triphosphate. It monitors the ability of the incoming nucleotide to form an A:T or G:C base pairs rather than detecting the exact nucleotide that enters the active site. The unique property of DNA polymerase is to distinguish between ribo- and deoxyribonucleoside triphosphates. The nucleotide binding pocket is too small to allow the presence of a 2'OH on the incoming nucleotide. This space is occupied by two amino acids that make van der Waals contacts with the sugar ring. The specific site at which DNA unwinding and initiation of replication occurs is known as origins of replication. DNA replication occurs with only origin site as in prokaryotes or multiple origins as in eukaryotes. In 1963, Francois Jacob and Jacques Cuzin defined the site replicon. Jacob and Cuzin proposed replicon model, there are two components that controlled the initiation of replication: the replicator and initiator. The entire set of cis-acting DNA sequences that is sufficient to direct the initiation of DNA replication was named as replicator. The protein specifically recognizes a DNA element in the replicator and activates the initiation of replication was named as initiator. Origin of replication is a part of replicator. However, in eukaryotes, the origin of replication is only a fraction of the DNA sequence required to direct the initiation of replication. There are two common characteristics of replicator. First, they include a binding site for the initiator protein and the second they include a stretch of AT-rich DNA that unwinds readily but not spontaneously. Initiator protein carries three functions, it binds a specific DNA sequence within the replicator; after binding with DNA, it frequently distort or unwind a region of DNA adjacent to their binding site; and the third function is it interact with additional factors required for replication initiation, thus recruiting them to the replicator. In *E.coli*, there is only one origin site (present in 245 bp region) and is referred as `OriC' (Fig. 9.4). There are two repeated motifs that are critical for *OriC* function. This region has three repeats of a 13 bp long sequence and 4 repeats of 9 bp long sequences. The 9mer (bp) motif is the binding site for the initiator. The 13mer (bp) motif is repeated three times. It is the initial site of ssDNA formation. The replication initiates with the formation of a complex primosome. It initiates replication. It recognizes and binds at the origin site of replication. The initiator (DnaA) ssDNA recruits proteins DNA helicase, DnaB and other protein helicase loader DnaC. The helicase takes an active form, when it gets associated around the ssDNA. A replication fork is formed. Single strand specific DNA binding protein (SSB protein) is a tetramer of a 74KD protein. It rapidly binds to the separated strands and the binding at this site avoids the two strands of DNA molecule to unite. Binding of one SSB promotes the binding of another SSB to the immediately adjacent ssDNA. This is called cooperative binding and occurs because SSB molecules bound to immediately adjacent regions of ssDNA can also bind to each other. Now DnaB activates the primase, which forms one RNA primer. DnaB is a hexamer and to each monomer chain, six molecules of another protein DnaC become attached. At this stage, primer is initiated. RNA primer is complementary to 3'-end of template. Once the primer is synthesized enzyme DNA polymerase becomes functional and attaches new nucleotides by phosphodiester bonds. Yeast is an ideal experimental model to understand mechanism in eukaryotic cells. Six proteins form the initiator complex, which is known as origin recognition complex (ORC). It recognizes a conserved sequence the A-element (of replicator), as well as a second less conserved B1-element. ORC binds and hydrolyzes ATP. The remaining steps are governed by protein-protein interactions and protein-DNA interactions that are sequence independent. In eukaryotic cells replication is initiated in two steps to occur at distinct times in the cell cycles. These steps are replicator selection and origin activation, taking place in G1 and S phase of cell cycle, respectively. During replicator selection, the sequences are identified that will direct the initiation of replication. This process leads to the assembly of a multiprotein complex at each replicator in the genome. Origin activation, triggers the replicator-associated protein complex to initiate DNA unwinding and DNA polymerase recruitment. A pre-replicative complex (pre-RCs) mediates the replicator selection. The former is composed of four proteins that assemble in an ordered fashion at each replicator. In the first step, ORC recognizes replicator. Once ORC is bound, it recruits two helicase loading proteins (Cdc5 and Cdt1). Now these proteins recruit a protein that is thought to be the eukaryotic replication fork helicase, which is the Mcm 2-7 complex. The pre-RCs formed during G1 are only activated to initiate replication after cells pass from the G1 to the S phase of the cycle. CdK and DdK (the protein kinases) activate Pre-RCs to initiate replication. They are activated only in S phase of cell cycle. The role of these proteins is to covalently attach phosphate groups to target proteins. Phosphorylation of these proteins results in the assembly of additional replication proteins (Polα/primase; DNA Polε, Polδ, and some other proteins required for recruitment) at the origin and the initiation of replication. The DNA Polδ, and Polε associate first, followed by DNA Polα/primase. The Mcm complex and many of the factors required for DNA polymerase recruitment become part of the replication fork machinery. Cdc6 and Cdt1 are released or destroyed after their role is complete.

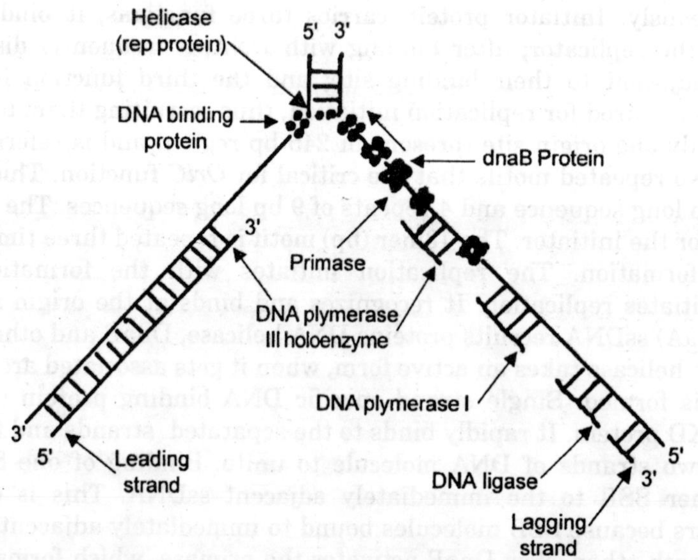

Fig. 9.4. Replication of DNA in prokaryotes

At the time of replication both the strands of DNA duplex separate and replicate at the same time. The junction between the newly separated template strands and the unreplicated duplex DNA is known as the replication fork (Fig. 9.5). DNA polymerase is not capable of separating the two strands of duplex. The strands are separated by enzyme DNA helicase. DNA helicases that act at replication forks are hexameric proteins that assume the shape of a ring. It encircles one of the two strands at the replication fork near the single-stranded: double-stranded. DNA is only synthesized by elongation a 3' end, and thus only one strand can be replicated continuously as the replication fork moves. This 3'→5' strand acts as a template strand, and directs the synthesis of new DNA strand. This strand is referred as leading strand. The second strand of DNA molecule directs the DNA polymerase to move in the opposite direction of the replication fork. It is known as lagging strand. The lagging strand takes sometime for synthesis of new polynucleotide chain. This is due to the fact that the lagging strand must wait for movement of the replication fork to expose a substantial length of nucleotide to act as template. The DNA polymerase is able to synthesize the polynucleotide chain complementary to this template chain. The function of primase differs in both leading and lagging strands. The leading strand needs only a single RNA primer, while the lagging strand needs a RNA primer for each okazaki fragment. Enzyme RNAse H replaces the RNA primers with DNA (Fig. 9.6). This enzyme removes all of the RNA primers except the ribonucleotide directly linked to the DNA end. The gap formed in the double-stranded DNA after the removal of RNA primer behaves as a primer:template junction. It is the site of binding of DNA polymerase. This enzyme fills this gap and all the nucleotides form base pairs. A nick is left remained in the backbone between the 3'OH and 5'P of the repaired strand. The enzyme DNA ligase fill ups the nicks, joins these fragments to form a DNA strand. DNA ligase uses a high-energy co-factor to create a phosphodiester bond between an adjacent 5' phosphate and 3'OH. The DNA synthesis is completed when all RNA primers are replaced, the nicks are sealed. When a mismatched base pair is detected by the polymerase, the primer:template junction slides away from the DNA polymerase active site. The mismatched DNA has a reduced affinity of the palm region, so the primer:template junction moves into the exonuclease site. When the incorrect base pair is removed, the correctly paired primer:template junction slides back into the DNA polymerase active site and DNA synthesis can continue.

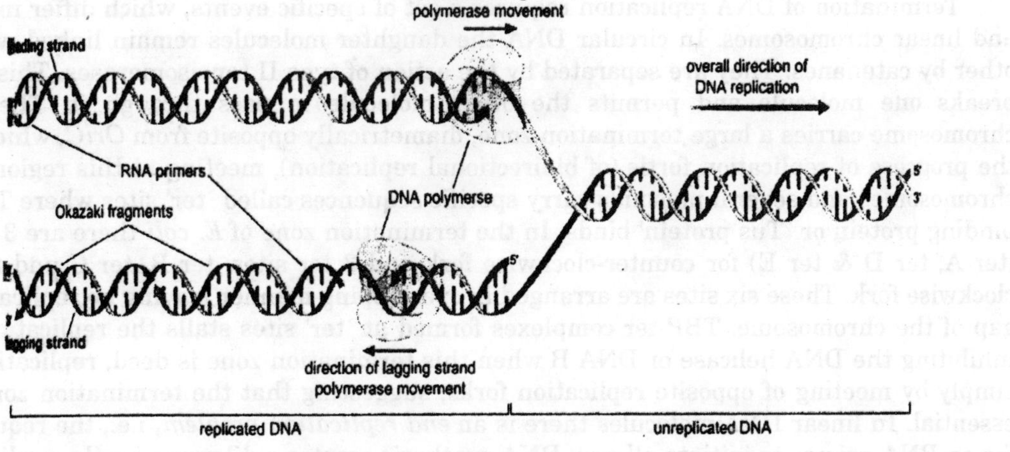

Fig. 9.5. The replication fork

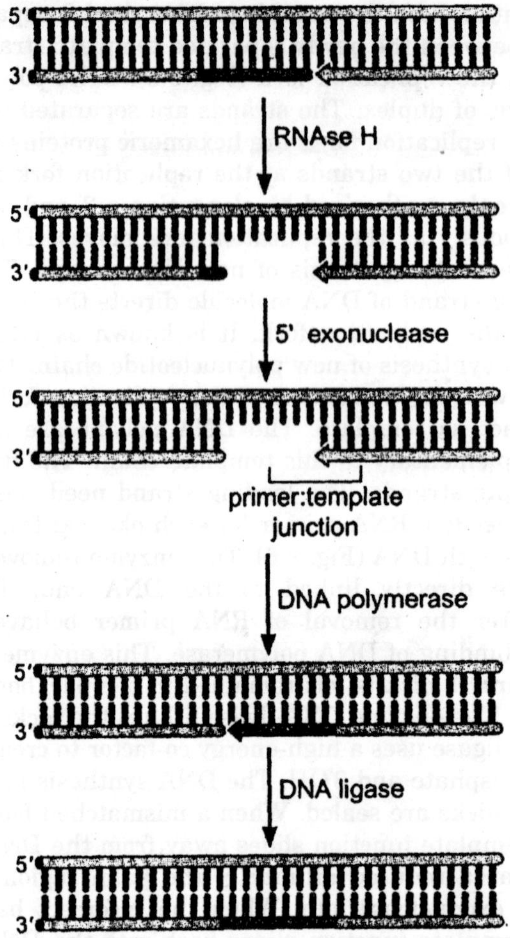

Fig. 9.6. Removal of RNA primers from newly synthesized DNA

Termination of DNA replication requires a set of specific events, which differ in circular and linear chromosomes. In circular DNA the daughter molecules remain linked with each other by catenanes. They are separated by the action of type II topoisomerases. This enzyme breaks one molecule and permits the other molecule to pass through it. The *E. coli* chromosome carries a large termination zone, diametrically opposite from *OriC*, which blocks the progress of replication fortis (of bidirectional replication), meeting at this region. *E. coli* chromosomes and several plasmids carry specific sequences called `ter' sites where TBPC ter binding protein or `Tus protein' binds. In the termination zone of *E. coli* there are 3 ter sites (ter A, ter D & ter E) for counter-clockwise fork and 3 ter sites, ter B, ter C and ter F for clockwise fork. These six sites are arranged in overlapping manner leaving no replication free gap of the chromosome. TBP-ter complexes formed at `ter' sites stalls the replication for by inhibiting the DNA helicase or DNA B when this termination zone is deed, replication stops simply by meeting of opposite replication forks, suggesting that the termination zone is not essential. In linear DNA molecules there is an *end replication problem*, i.e., the requirement for an RNA primer to initiate all new DNA synthesis creates a dilemma for the replication of the ends of linear chromosomes. In the leading strand, this problem is not observed, because it needs only a single RNA primer. In lagging strand, replication takes place in the form of Okazaki fragments, and each fragment requires an RNA primer. Even if the end of the last RNA primer for Okazaki fragment synthesis anneals to the final base pairs of the lagging strand template, once this RNA molecule is removed, there will remain a short region of unreplicated ssDNA at the end of the chromosome. Thus, after each round of DNA replication there is shortening of one of the two daughter DNA molecules. The bacterial cells adopt some ways to solve such type of shortening of base at the end of DNA molecule. It uses a protein (priming protein) instead of RNA as primer for the last Okazaki fragment at each end of the chromosome. The priming protein binds to the lagging strand template and uses an amino acid to provide an OH that replaces the 3'OH normally provided by an RNA primer. In eukaryotic cells, the ends of the chromosomes (the telomeres) are generally composed to head-to-tail repeats of a TG-rich DNA sequence. This sequence extends beyond 5' end as ssDNA. This unique structures act as a novel origin of replication that compensates for the end replication problem. A specialized DNA polymerase known as telomerase enzyme includes both protein and RNA components. It acts to extend the 3' end of its DNA substrate. It does not require an exogenous DNA template to direct the addition of new dNTPs. The template is served by the RNA component of the telomerase. The newly synthesized DNA is single-stranded.

RNA STRUCTURE AND TYPES

RNA is the primary product of gene expression, and plays a central role in gene expression. It is about ten times more than that of DNA. In general, RNA is single stranded with some exceptions of phage RNA which, are double stranded. The basic difference between RNA and DNA is that RNA has uracil pyrimidine instead of thymine and the pentose sugar is ribose in place of deoxyribose as found in DNA. On centrifugation, three main fractions separate, 23S to 25S, 16S to 17S, and 4S RNA. There are three types of RNA in a cell:

- ribsomal RNA (rRNA) are structural component of ribosome.
- transfer RNA (tRNA) are small RNAs. They carry amino acids to the site of polypeptide synthesis.
- messenger RNA (mRNA) carries the genetic information from DNA to the site of polypeptide synthesis.

In *E.coli* (all prokaryotic cells), there are three types of RNA, viz., mRNA, tRNA and rRNA. While, all eukaryotic cells have four major classes of RNA. Besides the three types as present in prokaryotes, eukayotes have a fourth class known as nuclear RNA (sRNA). It is involved in splicing of mRNA.

RNA molecules resemble single-stranded DNA with reference to the sequence of nitrogenous base. The RNA chains frequently fold back on themselves to form base-paired segments between short stretches of complementary sequences. A stem-loop structure may be formed when two stretches of complementary sequence are near each other. The interventing RNA is looped out from the end of the double-helical segment as in a hairpin. A complex structure (pseudonots) is formed between sequences that are not contiguous. RNA has enormous rotational freedom in the backbone of its non-base-paired regions, due to this RNA can form tertiary structures.

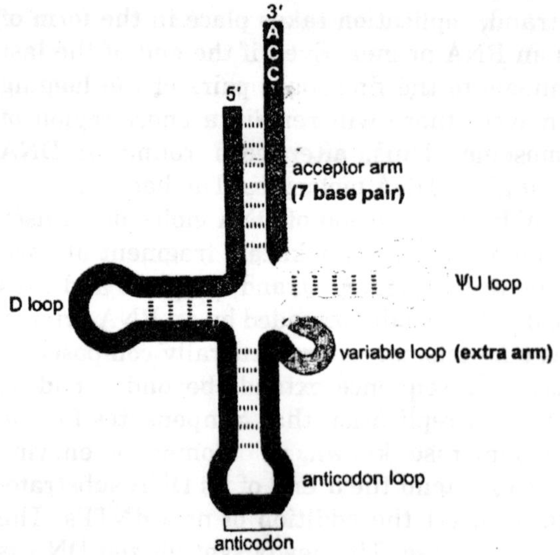

Fig. 9.7A. Cloverleaf representation of the secondary structure of tRNA

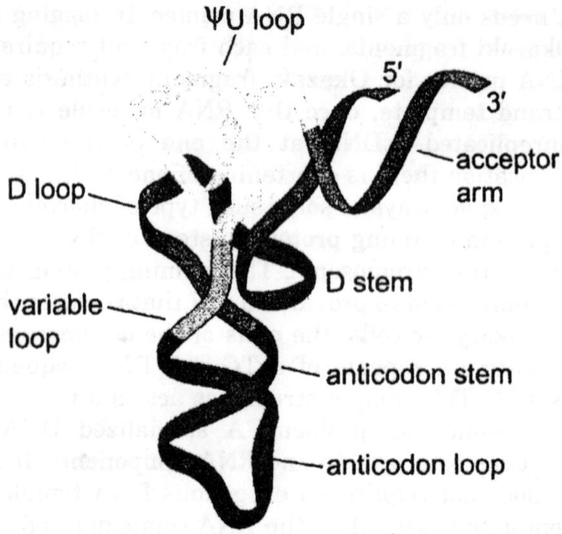

Fig. 9.7B. Conversion between the cloverleaf and the actual three dimensional structure

RNA has some biochemical properties and functions as a catalyst. It catalyzes the breakage and formation of covalent bonds between nucleotides during RNA splicing. The *ribozymes* are specific structures that cause particular chemical groups in one or more of its constituent nucleotides to become highly reactive. They catalyze the cleavage and rejoining of polynucleotide chains.

Messenger RNA

mRNA carries genetic information contained in DNA and makes a small fraction (5%-10%) of total RNA present in the cell. Messenger RNA (mRNA) is very heterogenous, varying less than 5S to more than 30S. It is synthesized by DNA template by the process of transcription. It contains information for amino acid sequence of several polypeptide chains. This is known as polycistronic mRNA. The genes contained in a polycistronic mRNA molecule often encode the different proteins of a metabolic pathway. All base of mRNA, are not translated into polypeptide chain. The polypeptide synthesis initiates from initiation codon. These codons may be present at a distance in 5'-P terminus of the RNA. This section of RNA present before the initiation codon is known as "leader". In certain cases, it contains regulatory sequences that influence the rate of protein synthesis. Polycistronic mRNA molecules contain spacer sequences ten bases long, which separate the coding sequences. This RNA is rather short lived and therefore has a rapid turnover. The molecular weight of this RNA varies, usually from 500,000-2,000,000.

Transfer RNA

Transfer RNA (tRNA) are small, single-stranded molecules, having 4s and 73 to 93 nucleotides. Amino acid attaches at 2' or 3' terminus of the tRNA molecule, which is always adenine. Amino acid is linked by an ester bond from its carboxyl group to the ribose sugar.

Internal complementary base sequences form short double stranded regions. These double stranded stems separate some folded structures known as loops. In two dimensions a tRNA molecule is drawn as a planar cloverleaf. Its three dimensional structure is more complex. In cloverleaf model, the complementary base paired region is called as stems and the single stranded regions as loops. The base pairing is virtually invariant or conserved. The base pairing is conventional type (A/U and C/G). However, unusual base pair are occasionally found. All tRNA have unusual bases. These bases are formed by modification of the four standard bases. The acceptor arm consists of a base-paired stem. The 3'OH group of unpaired region of this stem is the sites of amino acid linkage. When a tRNA is charged with the amino acid corresponding to its anticodon, it is called aminoacyl~tRNA. The anticodon arm has a loop containing anticodon. It contains base sequence complimentary to codon. There are no anticodons for nonsense codons. The TψC arm is named for the presence of pseudouridine, a modified base. The D arm has another unusual base, dihydrouridine. In D loop, there is variation of up to four residues. The extra nucleotides relative to the most common structure is denoted as 17 : 1 and 20 : 1 and 20 : 2. It shows that the extra nucleotide is lying between 17 and 18 positions and between 20 and 21 positions of nucleotide sequences. There is an extra arm lying between TψC and anticodon arm. Depending on the nature of the extra arm, tRNA can be divided into two classes; Class 1 tRNA and class 2 tRNA. Class 1 represents about 75% of the total tRNA. It consists of 3-5 bases. Class 2 has 13-21 bases and a stem of up to 5 base pair stem.

The secondary structure of each tRNA folds into a compact L-shaped tertiary structure in which the 3' end that binds amino acid is distant from the anticodon that binds mRNA. Double stranded stems are present in tRNA. Two double helices are present at right angle to each other. The acceptor stem and T ψ C stem form another continuous double helix with a single gap; the D stem and anticodon stem form another continuous double helix, also with the gap. The region between the double helices, where the turn in the L-shape is made, contains the Tψ C loop and the D loop.

GENETIC CODE

In 1961, Nirenberg worked on synthetic polynucleotides. He demonstrated that polyuridylic acid [poly(U)] directs the assembly of phenylalanine. The polypeptide chain synthesized was polyphenylalanine. In another system of studies, a trinucleotide was used to mimic a codon. It caused the corresponding aminoacyl-tRNA to bind to a ribosome. The meaning of codon may be presumed by identifying the amino acid component of the aminoacyl-tRNA. The translation of genetic information into polypeptide chain takes place on ribosomes. This information is carried by tRNA. There are four nitrogenous bases in DNA and their different combination code for twenty amino acid taking part in polypeptide synthesis. An RNA sequence corresponding to a particular amino acid is called a codon. Combination of three bases (triplets; $4^n = 64$), forms a codon. The different combinations of four nitrogenous bases forms total sixty four codons, out of which sixty one codon code for twenty amino acids. Genetic codes are universal. The genetic code is nonoverlapping and there is no punctuation. Three codons do not code for any amino acid, they are known as nonsense codon. The nonsense codons are UAA, UAG, and UGA. While AUG is the initiation codon and it is responsible for the initiation of polypeptide chain.

General features of genetic code

The genetic *code is degenerate* i.e., multiple codon must decode the same amino acid. As exhibited in the following add table, shows that many amino acids are specified by more than one codons. Thus, the codon specifying the same amino acids are synonyms, e.g., UUU and UUC code for phenylalanine, hence they are synonyms. The degeneracy is generally based on equivalence of the first two nucleotides. The genetic *code is unabmiguous* i.e., given a specific codon, only a single amino acid is indicated. To discriminate between degeneracy and ambiguity is important. Each codon has anticodon. For a given codon in the mRNA, only a single species of tRNA molecule possesses the proper anticodon. Each tRNA is charged with only one amino acid, hence only one amino acid will be coded by the codon of mRNA. Francis Crick (1966) proposed the *wobble concept*. According to this concept, the base at the 5' end of the anticodon is not as spatially confined as the other two. Thus, it may form hydrogen bonds with any of several bases located at the 3' end of a codon. There are eight codon families in which all four codons sharing the same first two bases have the same meaning. The pairing is shown in the table. There are seven codon pairs in which the meaning is the same whichever pyrimidine is present at the third position; and there are five codon pairs in which either purine may be present without changing the amino acid that is coded. An example of wobble concept is that U at the wobble position can pair with either adenine or guanine. The wobble rules do not permit any single tRNA molecule to recognize four different codons. The *code is highly redundant*. One amino acid may be coded by more than one codon. Only tryptophan and methonine are specified by one codon. While, understanding codon-anticodon binding two concepts have developed. The identity of third nitrogenous base of codon is not important. The number of distinct tRNA molecules that have been isolated from a single organism is less than the number of codons. Thus, the anticodons of some tRNA molecules must be able to pair with more than one codon. These features are explained by wobble hypothesis. It refers to the less stringent requirement for base pairing at the third position of the codon than at the first two positions. The first two bases (between codon and anticodon) should pair in usual type while, the third base pair can be of a different type (like G can pair with U). This concept came into existence while studying anticodon of $tRNA^{Ala}$ of yeast. It contains the nucleoside inosine, I, in the position that pairs with the third base of the codon (the first position of the anticodon).

In some species the universal genetic code are changed. Such changes are more common in mitochondrial genome. The phylogenetic tree of mitochondrial changes explains that there was a universal code that was changed at various points in mitochondrial evolution. In nuclear genomes, they are sporadic and usually affect only termination codon. In microbes, the universal termination codon code for some amino acids, e.g., in *Mycoplasma capricolum*, UGA codes for tryptophan, in *Tetrahymena thermophila*, UAA and UAG code for glutamine. The other common type of change is loss of tRNA that responds to a codon, so that the codon no longer specifies any amino acid.

SUMMARY

DNA is the genetic material; it is composed of polynucleotide chains twisted around each other. Alexander Rich examined Z-DNA contains about 12 bases/turn. DNA is a biomolecule with the property to replicate and produce daughter DNA molecule. The base sequence of daughter molecule is as that of parent DNA molecule. In 1957, Arthur Kornberg and his

coworkers isolated an enzyme from *E.coli* and named it as DNA polymerase or "Kornberg enzyme". In *E. coli* five types of DNA polymerase enzymes have been identified and studied. DNA polymerase uses a single active site to catalyze the addition of any of the four deoxynucleoside triphosphate. RNA is the primary product of gene expression, and plays a central role in gene expression. An RNA sequence corresponding to a particular amino acid is called a codon. Combination of three bases forms a codon. Genetic codes are universal. In some species the universal genetic code are changed.

Chapter - 10

AMINO ACID METABOLISM

Amino Acids are the building blocks of protein; the biochemistry of amino acid is discussed in chapter-3 of this section. In this chapter the metabolic pathways is only discussed herewith.

BIOSYNTHESIS OF AMINO ACIDS

In the synthesis of amino acids, inorganic ammonium nitrogen is incorporated to organic acid via two pathways. In both the pathways ammonium ions are assimilated to form the amino acid L-glutamate. The ammonium moiety can be transferred to other organic acids to form other types of amino acids.

Reductive deamination

In reductive deamination ammonium ion combines directly with α-keto carboxylic acids to form amino acids. An intermediate of Krebs cycle, α-ketoglutarate accepts an ammonium ion to form L-glutamic acid. This reaction is catalyzed by enzyme glutamate dehydrogenase and $NADPH+H^+$ acts as hydrogen donor.

$$\alpha\text{-ketoglutaric acid} + NH_4^+ \xrightarrow[\text{NADPH+H}^+ \ \text{NADP}^+]{\text{Glutamate dehydrogenase}} \text{L-glutamic acid}$$

The different pathways of Amino acid synthesis are presented in Fig. 10.1, 10.2, 10.3, 10.4 and 10.5.

Glutamate synthetase pathway

This pathway is also known as glutamine synthetase pathway. It takes place when concentration of ammonium is low. In the first step of this pathway, L-glutamate reacts with ammonium ions to form the amino amide L-glutamine. In this reaction energy is supplied by the hydrolysis of ATP. L-glutamine reacts with α-ketoglutrate to form two molecules of L-glutamate. These reactions are catalyzed by enzyme glutamine synthetase and glutamate synthetase, respectively. NADPH+H$^+$ acts as reducing power.

$$\text{glutamte} + NH_4^+ \xrightarrow[\text{ATP} \quad \text{ADP}]{} \text{Glutamine} + H_2O$$

$$\text{Glutamine} + \alpha\text{-ketoglutarate} \longrightarrow 2 \text{ Glutamate.}$$

After the synthesis of Glutamate, by the process of transamination six essential amino acids are synthesized. These amino acids act as precursors of six families of amino acids which form all the twenty amino acids required for the protein synthesis.

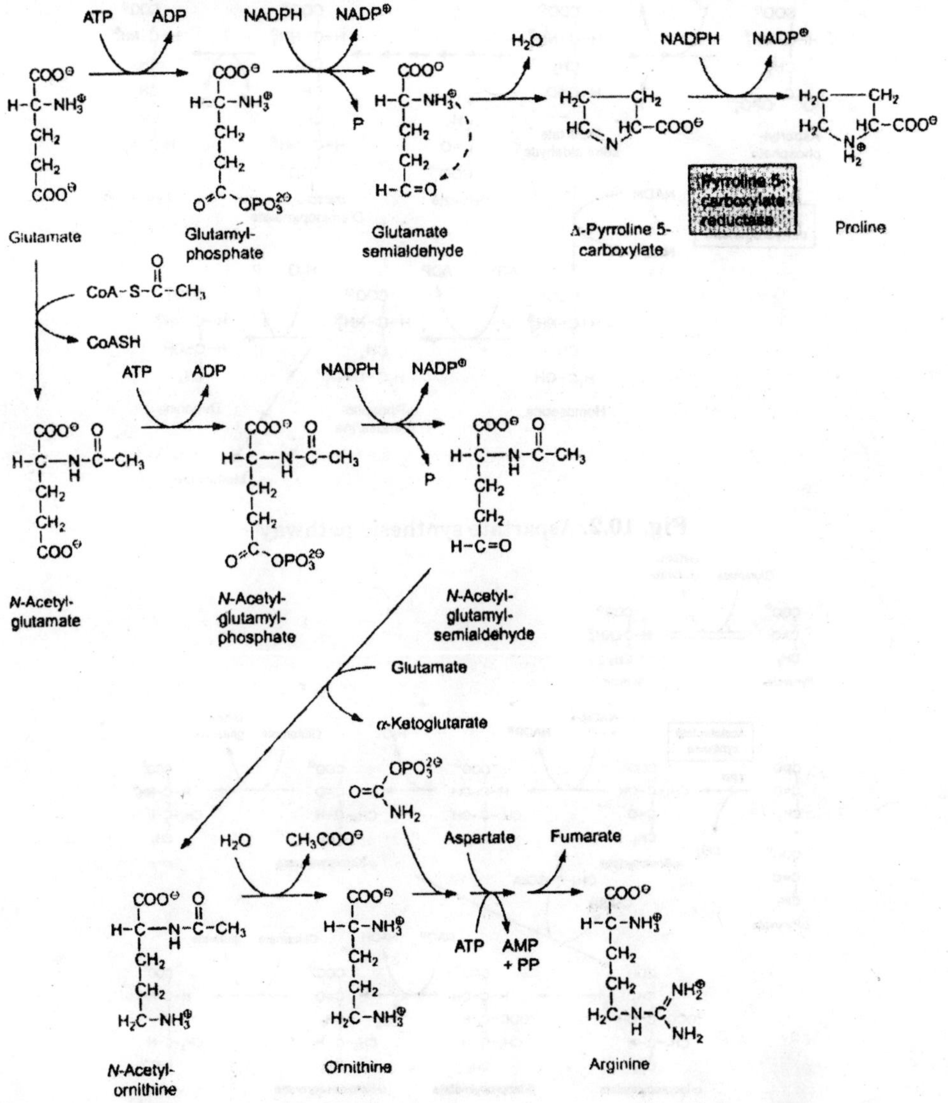

Fig. 10.1. Pathway for the formation of amino acid from glutamate.

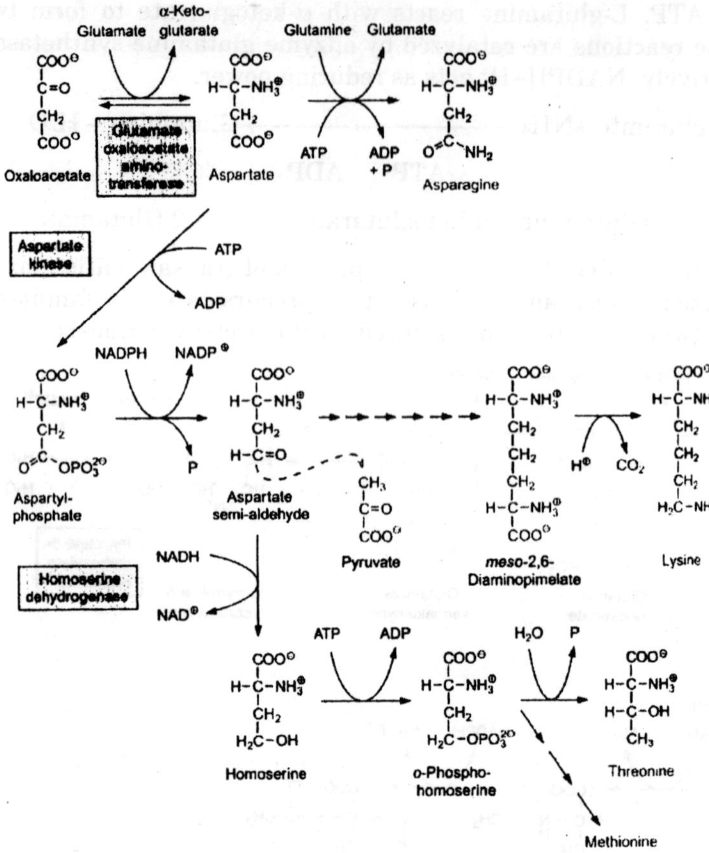

Fig. 10.2. Aspartate synthesis pathway

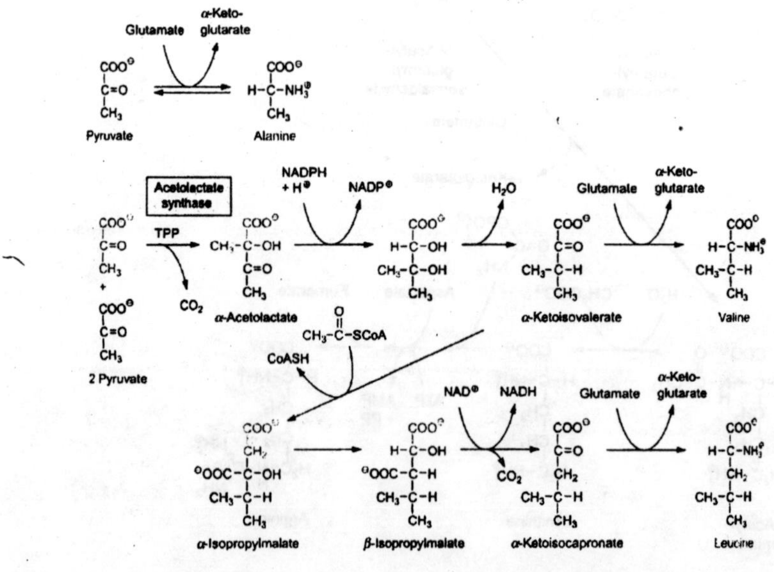

Fig. 10.3. Synthesis of amino acids from pyruvate

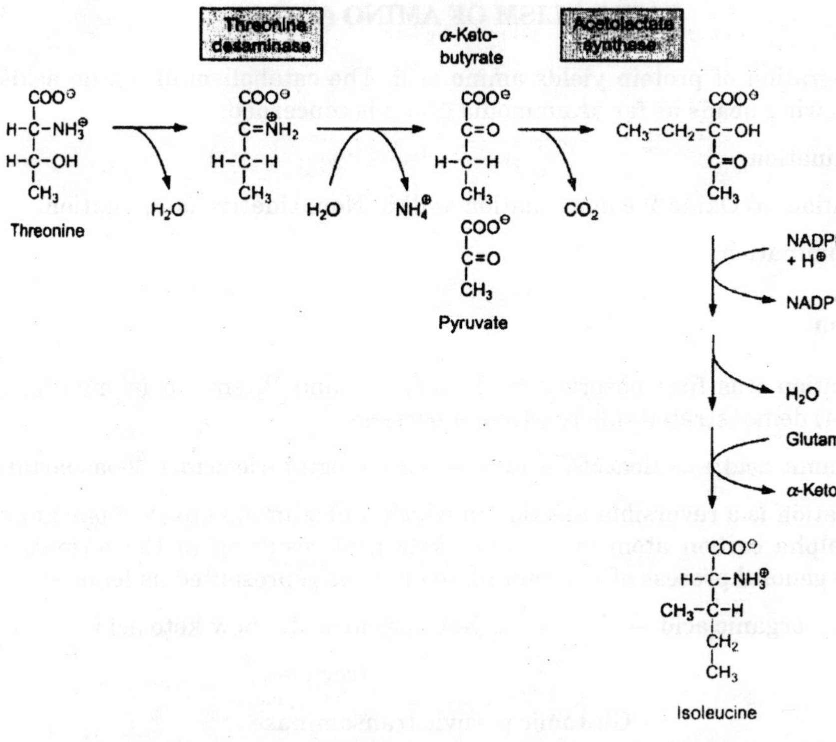

Fig. 10.4. Pathway for the synthesis from theronine and pyruvate

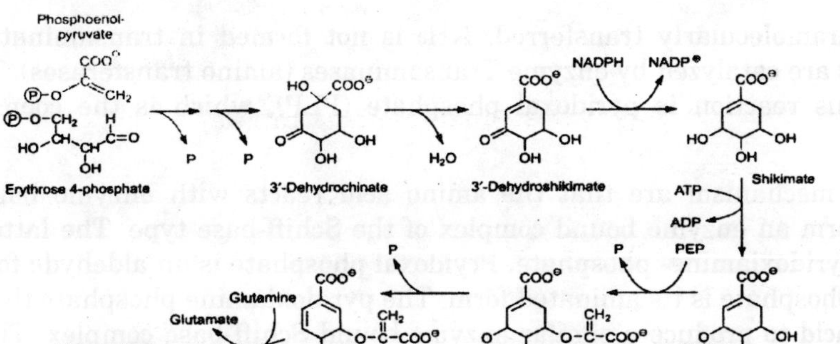

Fig. 10.5. Shikimate pathway (Synthesis of Aromatic Amino Acid)

CATABOLISM OF AMINO ACIDS

The disintegration of protein yields amino acid. The catabolism of amino acids may be discussed in following heads as far as ammonia group is concerned:

1. Transamination,
2. Deamination (a) Oxidative deamination and (b) Nonoxidative deamination.
3. Transdeamination.

Transamination

Transamination was first observed by Braunstein and Ritzmann in muscle. Lichstein and Cohen (1944) demonstrated this reaction in bacteria.

l(+) glutamic acid + oxaloacetic acid ⟶ α -ketoglutaric acid + l(-)aspartate

Transamination is a reversible reaction in which alpha amino group of one amino acid is transferred to alpha carbon atom of another keto acid resulting in the formation of new amino acid. The general process of transamination may be represented as follows:-

Amino acid + organic acid ⟶ New amino acid + new keto acid.

(donor) (recipient)

Glutamic pyruvic transaminase

Glutamic acid + pyruvic acid ⟶ ketoglutaric acid + alanine

NH_2 is intramolecularly transferred. NH_3 is not formed in transamination reactions. These reactions are catalyzed by enzyme Transaminases (amino transferases). The coenzyme required for this reaction is pyridoxal phosphate (PLP), which is the coenzyme from of vitamin B6.

Details of mechanism are that the amino acid reacts with enzyme bound pyridoxal phosphate to form an enzyme bound complex of the Schiff-base type. The latter then yields keto acid and pyridoxlamine- phosphate. Pryidoxal phosphate is an aldehyde form, while the pyridoxamine phosphate is its aminated form. The pyridoxlamine phosphate then reacts with a second keto acid to produce a similar enzyme bound Schiff-base complex. The later break down forming a second (new) amino acid and regenerates the pyridoxal phosphate.

There are some limitations to the reaction of transamination. Such as (I) Group transminases and specific transaminases and (II) Alanine transaminase (or Alanine amino-transferase).

Group Transaminases and Specific transaminases

Group Transaminases and Specific reactions Most of the amino acids may act as donor I and Keto acids acts as recipient. Recipient keto acids are the intermediates of TCA. They may be α-keto-(oxo) glutarate or oxalo-acetate or pyruvate. The new amino acids formed by these keto acids are glutamic acid, aspartic acid and alanine. The enzyme transaminases catalyzing the conversion of these keto acids into their respective amino acids are known as group transaminases.

There are few amino acids, which do not take part in transamination e.g., lysine, Threonine and proline. Transaminaton is not restricted to α -H_2 groups. The -NH_2 groups of orthine is readily transmitted forming glutamate-γ-semialdehyde.

Deamination

As the name suggests, deamination means loss of amino groups from amino acids. It is of two types: nonoxidative deamination and oxidative amination.

OXIDATIVE DEAMINATION

Enzyme amino acid oxidase deaminase catalyzes the deamination of amino acid. Depending upon the type of amino acid (D or L amino acid) the enzyme is named as D-amino acid oxidase and L-amino acid oxidase. Both the types of enzymes are flavoproteins. D-amino acid oxidase contains FAD, while L-amino acid oxidase contain FMN as coenzymes.

The mechanism of reaction completes in two steps. In the first step, amino acid is dehydrogenated under the influence of the enzyme, amino acid oxidase to form a α-imino acid. Flavin accepts the released hydrogen. In the second step, the α-imino acid is spontaneously cleaved in the presence of water to form a-keto acid and ammonia. The reduced flavoproteins are re-oxidized at substrate level directly by molecular oxygen forming H_2O_2 without participation of cytochrome or other electron carriers. The H_2O_2 formed is toxic and hence it is immediately converted to O_2 and H_2O. This is catalyzed by enzyme catalase.

NON-OXIDATIVE DEAMINATION

Some amino acids are not oxidized while deamination. Ammonia is formed in this reaction. The hydroxy amino acids (serine, threionine) are deaminated by specific enzymes called amino acid dehydrases which requires pyridoxal-P (B6-P) as coenzyme.

Serine + H_2O ⟶ Imino acid

Imino acid + H_2O ⟶ Pyruvic acid + NH_3

Deamination of Histidine Histidine is non-oxidatively deaminated. Ammonia and uronic acid are the products. This reaction is catalyzed by enzyme histidase.

$$\text{Histidine} \xrightarrow{\text{Histidase}} NH_3 + \text{Uronic acid}$$

Deamination of cysteine In the deamination of cysteine, first H_2S is removed and imino acid is formed. This is known as desulfhydration. This reaction is catalyzed by enzyme desulfhydrase. Imino acid thus formed is spontaneously hydrolyzed releasing ammonia and pyruvic acid is formed.

$$\text{L-cysteine} \xrightarrow{\text{desulfhydrase}} \text{iminoacid} + H_2S$$

Imino acid + H_2O ⟶ Pyruvic acid + NH_3

Deamination of L-Glutamic acid (Trans deamination): L-glutamic acid is deaminated by a specific enzyme L-glutamate dehydrogenase (Fig. III A, B). This enzyme is made up of four polypeptide chains, each containing a Zn ion. This enzyme requires NAD+ or $NADP^+$ as coenzyme. In the reaction first α-imino glutaric acid is formed. A molecule of water adds to this imino acid and α-ketoglutrate is formed, ammonia is liberated.

Deamination of Threonine Threonine is deaminated into α-oxobutyrate. Enzyme threonie dehydratase catalyzes this reaction.

Deamination of Glycine Ammonia, CO_2, and acetic acid are produced by anaerobic deamination of glycine

$$\text{Glycine} + H_2O \longrightarrow NH_3 + CO_2 + \text{acetic acid.}$$

Pathway of Amino acid degradation

There are 20 amino acids in proteins, with a variety of carbon skeletons. Correspondingly, there are 20 different catabolic pathways for amino acids degradation. The 20 catabolic pathways converge to form only five products, all of which enter the TCA. From here the carbons can be diverted to gluconeogenesis or ketogenesis, or they can be completely oxidized to CO_2 and H_2O. All or part of the carbon skeletons of ten of the amino acids are ultimately broken down to yield acetyl- CoA; five amino acids are converted into `ketoglutarate, four into succinyl CoA; two into fumarate and two into oxalate.

1. Ten amino acids are degraded to Acetyl CoA

The carbon skeletons of ten amino acids yield acetyl-CoA which enters the TCA cycle directly. Five of the ten are degraded to acetyl CoA via pyruvate. The other five are converted into acetyl CoA and/or acetoacetyl-CoA. The latter is then cleaved to form acetyl CoA. The five amino acids entering via pyruvate are alanine, glycine, serine, cysteine and tryptophan. Portions of the carbon skeleton of six amino acids- tryptophan, lysine, phenylalanine, tyrosine, leucine and isoleucine yield acetyl CoA and/or acetoacetyl CoA; the latter is then converted into acetyl CoA.

2. Five amino acids are converted in α-ketoglutarate

The carbon skeletons of five amino acids (argnie, histidine, glutamate, glutamine and proline) enter the TCA cycle via α-ketogluterate. Proline, glutamate and glutamine have five carbon skeletons. The cyclic structure of proline is opened by oxidation of the carbon most distant from the carboxyl group A to create a schiff's base and hydrolysis of schiff's base to a linear semialdehyde (glutamate semialdehyde). This is further oxidized at the same carbon to produce glutamate. The action of glutamine or any of several reactions in which glutamine donates its amide nitrogen to some acceptor, converts glutamine to glutamate. Trans-amination or deamination of glutamate produces the TCA cycle intermediate α-ketoglutarate.

Arginine and histidine contain five adjacent carbons and a sixth carbon attached through a nitrogen atom. The catabolic conversion of these two amino acids to glutamate is therefore slightly more complex than the path from proline or glutamine to glutamate. Arginine is converted to the five carbon skeleton of ornithine in the urea cycle and the ornithine is tranaminated to glutamate semialdehyde. The conversion of histidine to the five carbon glutamate occurs in 4 multistep pathways the extra carbon is removed in a step that employs tetrahydrofolate as a cofactor.

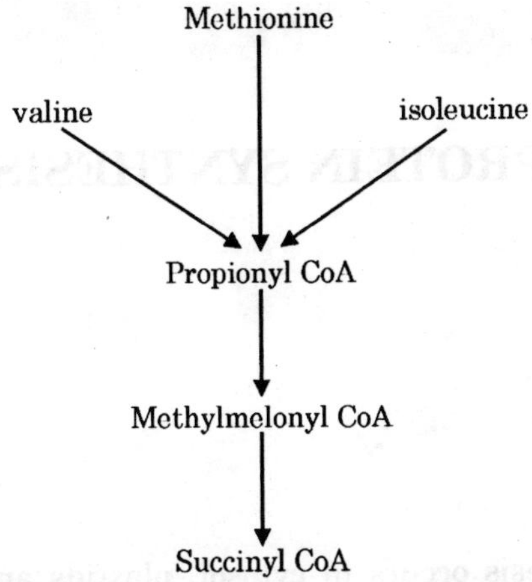

Pathway from Methionine, Valine and Isolecucine to Succinyl CoA

3. Four amino acids are converted into succinyl CoA

The carbon skeletons of methionine, isoleucine, threonine and valine are degraded by pathway, that yield succinyl CoA an intermediate of TCA cycle

4. Asparagine and Asparate are degraded to oxaloacetate

The carbon skeletons of asparagines and aspirate ultimately enter the TCA cycle via oxaloacetate. The enzyme asparginase catalyzes the hydrolysis of aspargine to yield aspirate, which undergoes a transmission reaction with ketoglutarate to yield glutamate and oxaloacetate. The latter enters the TCA cycle.

5. Some amino acids can be converted to glucose; others to ketone

Some carbon atoms from six of the amino acids (those that are degraded to acetoacetyl CoA /or acetyl CoA, tryptophan, lysine, phenylalanine, tyrosine, leucine and isoleucine) can yield ketones, β.

The amino acid that can be converted into pyruvate, α-ketoglutarate, succinyl- CoA, fumarate and oxaloacetate. These amino acids are converted to glucose by gluconeogenesis pathway.

SUMMARY

Amino acids are building blocks of proteins. The basic mechanism of biosynthese pathway of amino acid is the transfer of ammonium moiety to other organic acids to form other types of amino acids. The catbolism of amino acid takes place by transamination, deamination and transdeamination mechanism.

Chapter - 11

PROTEIN SYNTHESIS

In plants, protein synthesis occurs in cytosol, plastids and mitochondria. Thus, plant contains three different types of ribosomes, three groups of tRNAs, and three sets of auxiliary factors for protein synthesis. The ribosomes of cytosol, plastids, and mitochondria are different in size and composition. The cytosolic ribosomes with a sedimentation constant of 80S dissociates into a small subunit (40S) and large subunit (60S). The size of mitochondrial ribosome is approximately 78S. However, the size may vary from species to species. The plastidial ribosome is of 70S in size. About 20,000 different types of proteins are synthesized in cytoplasm, while mitochondria and chloroplast synthesizes about 50-100 and 30-40 proteins, respectively. The mechanism of protein synthesis in chloroplast and mitochondria resembles the prokaryotic pattern.

In plants, protein synthesis in cytosol does not occur at a constant rate, and it is regulated in response to numerous physiological and environmental variables. It is regulated during seed germination, during embryogensis, by light, and by hormones. Cytosolic protein synthesis is diminished in stress such as anaerobiosis, heat shock, or viral infection.

Some Concepts and Definitions

The central dogma: In 1958 Francis Crick proposed clearly the genetic information is transferred to cytoplasm through mRNA. The latter is being translated in the language of protein. The information flows is unidirectional. This whole concept is said to be the central dogma of molecular biology. It states that once "information" (sequence of nitrogenous base or amino acid) has passed into protein, it cannot get out again. The transfer of information from nucleic acid to nucleic acid or from nucleic acid to protein may be possible, but transfer from protein to protein, or from protein to nucleic acid is impossible.

Upstream and Downstream: The base pair sequences prior to the start-point are described as upstream while those after the start-point are known as downstream.

Transcription bubble: Transcription occurs at the site of DNA which is transiently separated into its single strands, and one strand is used as a template for synthesis of the RNA strand. The length of the transcription bubble varies with the phase of the elongation reaction. It may from 12-20 bp. DNA is unwound at the leading edge of the bubble and rewound into a helix at the trailing edge.

Pyrophosphorolytic and Hydrolytic editing these are the proofreading functions of RNA polymerase. In pyrophosphorolytic editing, the enzyme uses its active site, in a simple back-reaction. It catalyzes the removal of an incorrectly inserted ribonucleotide, by reincorporation of pyrophosphate. The enzyme can then incorporate another ribonucleotide in its place in the growing RNA chain. In hydrolytic editing, the polymerase backtracks by one or more nucleotides and cleaves the RNA product, removing the error-containing sequence.

Exons and Introns primary transcript of eukaryotes contains exons interrupted by introns. The nucleotide sequences of mRNA that code for amino acids are known as exons, while noncoding sequences are called introns. The latter are removed by splicing.

Rho factor Roberts (1969-70) first recognized this factor. It is a protein factor which enables RNA polymerase to complete transcription.

Primary Transcript: The immediate product of transcription is called the primary transcript. It is generally unstable. In prokaryotes, it is rapidly degraded (mRNA) or cleaved to give mature products (rRNA and tRNA). In eukaryotes, this is processed. Three types of mRNA processing are capping, polyadenylation and splicing.

Aminoacyl-tRNA synthetase enzyme is highly specific for one amino acid and its corresponding tRNA. For some amino acids that have two or more corresponding tRNAs the same aminoacyl tRNA synthetase usually aminoacylates all of them. The enzyme aminoacyl-tRNA synthetase has two different combining sites: one that recognizes the side group of an amino acid and another that recognizes the tRNA specific for that amino acid. Aminoacyl-tRNA synthetase attaches amino acid to the 3'OH, while others attach it to the 2'OH of the terminal ribose sugar of the tRNA.

Initiation Factors: In prokaryotes there are at least three Initation Factors: IF1, IF2, and IF3. IF-1 is required for 30S subunits to bind specifically to initiation sites in mRNA. It prevents tRNAs from binding to the portion of the small subunit that will become part of the A site. IF-2 is a GTPase. It binds a special initiator tRNA and controls its entry into the ribosome. IF3 binds to the small subunit and blocks it from reassociating with a large subunit, or from binding charged tRNAs. In eukaryotes, there are different classes of initiation factors (eukaryotic Initiation Factors eIF). Different classes of eukaryotic Initiation Factors, their members and functions are: eIF1 (eIF1, eIF1A) has multiple effects in enhancing initiation of complex formation. eIF2 (eIF2, eIF2B, eIF2C), the general role of this factor is GTP-dependent recognition of Met-tRNA and nucleotide exchange. eIF3 (eIF3) takes part in ribosomal subunit dissociation and promotes Met-tRNA binding to 40S subunits. eIF4 (eIF4A, eIF4B, eIF4E, eIF4F, eIF(iso)4F, eIFG, eIFH) takes the responsibility of recognizing 5'cap mRNA, and bind the 40S subunit to mRNA and unwinds the secondary structure of mRNA. eIF5 (eIF5) promotes eIF2 GTPase activity and release of factors from ribosome. eIF6 (eIF6), the role of this factor is unknown; however, it interacts with 60S subunit. Plants have two forms of eIF4: eIF4F and eIF(iso)4F. The latter may be able to interact with the cytoskeleton. As a result, it may help to localize the translation of certain mRNAs to specific regions of the cell.

Peptidyl Transferase Center and Decoding Center ribosome has two subunits. The large subunit contains the peptidyl transferase center. The latter catalyzes peptide bond formation. Decoding center is present in small subunit, in which charged tRNAs read or `decode' the codon units of the mRNA.

Transcription

Transcription means the process by which nucleotide sequence information is transferred from DNA to RNA (Fig. 11.1). Transcription is chemically and enzymatically very similar to DNA replication. Both need polymerizing enzymes that synthesize a new strand of nucleic acid complementary to a DNA strand. The identical strand of DNA is called coding strand while the other strand is known as template strand. The prokaryotes and eukaryotes show some major differences in the relation between the transcript and the mRNA used for polypeptide synthesis. In prokaryotes, primary transcript directly takes part in polypeptide synthesis. While in eukaryotes, the primary transcript must be converted to mRNA by a process known as RNA processing.

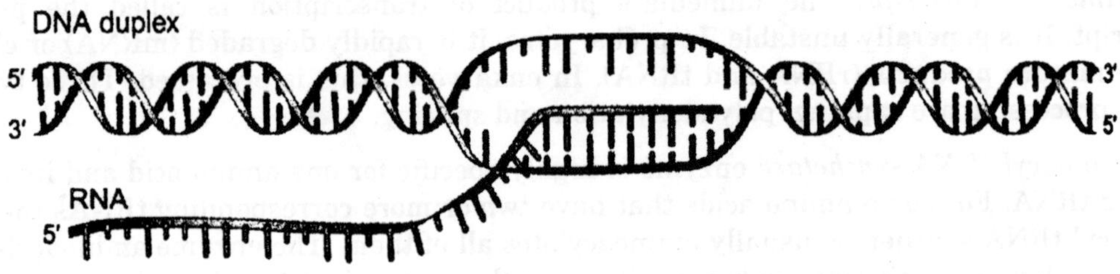

Fig. 11.1. Transcription of DNA into RNA

RNA polymerase

The enzyme RNA polymerase performs essentially the same reaction in all cells. It is implicated in all three stages of transcription (Initiation, elongation and termination). It does not need a primer and it can initiate transcription de novo. Transcription commences when RNA polymerase binds to a special region, the promoter, at the start of the gene. During elongation the enzyme moves along the template, synthesizing RNA, and termination of RNA takes place when this enzyme reaches a terminator sequence (Fig. 11.2). This action is known as a transcription unit. A transcription unit may include more than one gene. Multiple RNA polymerase molecules can transcribe the same gene at same time, each polymerase following closely along behind another. As a result cell can synthesize large number of transcripts from a single gene in a short time.

In bacteria there is only a single RNA polymerase, while in eukaryotic cells there are three polymerase Pol I, II and III. Pol II transcribes RNA involved in all proteins. Pol I transcribes the large ribosomal RNA (rRNA) precursor gene, whereas Pol III transcribes tRNA genes, some small nuclear gene, and the 5S rRNA gene.

Bacterial RNA polymerase About 7000 RNA polymerase molecules are present in an *E. coli* cell. The complete enzyme or holoenzyme in *E. coli* has a molecular weight of 465 kD. The ionic strength of the incubation medium is a critical parameter of enzyme activity. For *E. coli* RNA polymerase exist as a dimer but dissociate into monomer when the ionic concentration is raised. Travers and Burgess (1969) found that monomer is active and comprise of four types of polypeptide chains. It consists of five polypeptide chains including two polypeptide chains of alpha, one chain each of beta and sigma. The RNA polymerase molecule thus can be represented as α2ββ'σ. It comprises of 13-14S sedimentation in which (α-2 enzyme, mol. wt -40,000; β'-1 enzyme, mol. wt.- 1,55,000; β-1 enzyme, mol. wt. - 1,65,000; σ -1 enzyme, mol. wt. -1,95,000). The overall shape of each enzyme resembles a crab claw. The two pincers of the crab claw are made up predominantly of two largest subunits of each enzyme β and β'. The active site is made up of regions from both of these subunits. It is found at the base of the pincers. The holoenzyme can be separated into two components, the core enzyme (α2ββ') and the sigma factor (σ). The sigma factor binds at promoter site of DNA. After initiation sigma factor is released when the RNA chain reaches 8 to 9 bases, leaving the core enzyme to undertake elongation. Sigma factor is involved only in initiation. Free sigma factor becomes immediately available for use by another core enzyme.

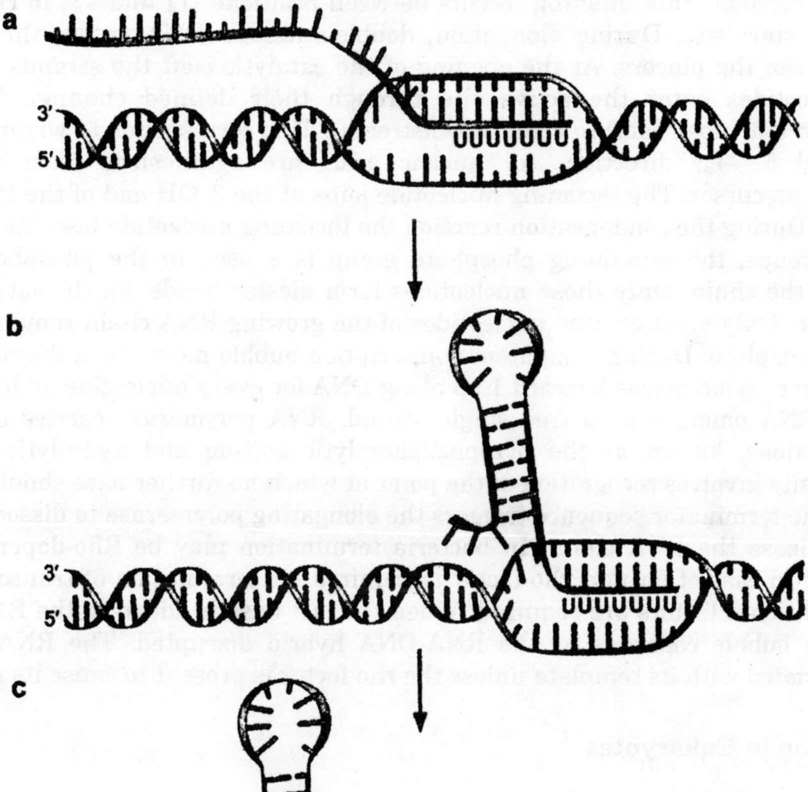

Fig. 11.2. Steps of Transcription A-B Elongation C-Termination

Transcription Process in Prokaryotes

RNA polymerase binds to the promoter sequence of DNA. This promoter-polymerase complex undergoes structural changes so that initiation proceeds. In *E.coli*, predominant σ factor is known as σ^{70}. The promoter and σ^{70} containing polymerase have two conserved sequences. These sequences contain each of six nucleotides, are separated by a nonspecific stretch of 17-19 nucleotides. These sequences are present at about 10 base pair, and 35 base pair σ^{70} upstream from start-point of RNA synthesis. These sequences are called -35 (minus 35) and -10 (minus 10) regions or elements according to the numbering scheme. By comparing many different promoters, a consensus sequence can be derived. Promoters with sequence closer to consensus are generally `stronger'. A transcription bubble is formed in DNA, which is a transition of closed system to open system of DNA. During this transition, enzyme undergoes structural changes, the double strand of DNA reveal the template and nontemplate strands. This "melting" occurs between positions -11 and +3, in relation to the transcription start site. During elongation, double-stranded DNA enters the front of the enzyme between the pincers. At the opening of the catalytic cleft the strands separate and the ribonucleotides enter the active site through their defined channel. Transcription proceeds from left (upstream) to right (downstream). This corresponds to writing the mRNA in the usual $5' \rightarrow 3'$ direction. All nucleic acids are synthesized from nucleoside 5' triphosphate precursor. The incoming nucleotide joins at the 3'-OH end of the last nucleotide of the chain. During the condensation reaction the incoming nucleotide loses its terminal two phosphate groups, the remaining phosphate group is a used in the phosphodiester bond linking it to the chain. Only those nucleotides form diester bonds which matches with the template base. Only eight or nine nucleotides of the growing RNA chain remain base paired to the DNA template. During elongation transcription bubble moves by a disruption of DNA molecule. The enzyme moves forward 1 bp along DNA for every nucleotide added to the RNA chain. The RNA emerges as a free single strand. RNA polymerase carries out two proof reading functions, known as the pyrophosphorolytic editing and hydrolytic editing. The termination site involves recognition of the point at which no further base should be added to the chain. The terminator sequence triggers the elongating polymerase to dissociate from the DNA and release the RNA chain. In bacteria termination may be Rho-depentent or Rho-independent. In case of former, rho-factor is required for termination of transcript, while in latter no additional factors are required. When the last base is added to the RNA chain, the transcription bubble collapses as the RNA-DNA hybrid disrupted. The RNA transcribed, remain associated with its template unless the rho factor is present to cause its release.

Transcription in Eukaryotes

The transcription process and polymerase in eukaryotes is much more similar like that of prokaryotes. However, there are some differences. In eukaryotes, there are three types of polymerase, Pol I, II, and III. Much more, several initiation factors are required for efficient and promoter-specific initiation. These initiation factors are called the general transcription factors (GTFs). GTFs help polymerase bind to the promoter and melt the DNA.

The Polymerase and GTFs

In eukaryotes, pre-initiation complex is formed at the TATA sequence of promoter. This sequence is recognized by general transcription factor known as TFIID. TFII stands for,

Transcription Factor Pol II, D stands for individual factor, which may be A, B, and so on. TFIID is a multi-subunit complex. TATA binding component of TFIID is called TBP (TATA binding protein), the other component of TFIID is called TAFs (TBP associated factor). There are ten TAFs associated with TBP. Out of which two TAFs bind DNA elements at the promoter. TFIIB factor is made up of single polypeptide. It forms a bridge between TATA-bound TBP and polymerase. TFIIE and TFIIF have two subunits. TFIIF binds with Pol II and stabilizes the DNA-TBP-TFIIB complex. TFIIH has nine subunits. Its two subunits functions as ATPase and another subunit acts as protein kinase. The large subunit of Pol II has a C-terminal domain (CTD), which has repeats of heptapeptides. Mediator is associated with the CTD tail. Mediator aids initiation by regulating the CTD kinase in TFIIH. There are activator proteins, which help recruit polymerase to the promoter and stabilizes the binding.

Pol I transcribe only one gene, coding for rRNA. There are many copies of this gene and it is expressed at much higher level than any other gene. The promoter of this gene has two components: the core element and the UCE (upstream control element). SL1 and UBF factors are also required for the initiation. SL1 encompasses TBP and three TAFs specific for Pol I. Some Pol III promoters contain two regions, called Box A and Box B, separated by a short element. Other Pol III consists of Box A and Box C; and still others contain a TATA element like those of Pol II. Genes for tRNA and 5S rRNA gene consists of Box A plus Box B, and Box A plus Box C, respectively.

The Transcription Process in Eukaryotes

Pol II is the major polymerase, involved in transcription (Fig. 11.3). For accurate transcription, Pol II requires a minimal set of sequence elements known as core promoter. It is about 40 nucleotides long, extending either upstream or downstream of the transcription start site. Four elements are found in Pol II core promoter, these are TFIIB recognition element (BRE), the TATA element (or box), the initiator (Inr) and the downstream promoter element (DPE). Typically, a promoter includes only two or three of these four elements. The complex formed by the binding of promoter, GTFs and polymerase, is known as *pre-initiation complex*. TBP binds to DNA at TATA sequence and distorts it. The distorted sequence provides a platform for other GTFs. The β-sheet of TBP recognizes the minor groove of the TATA element. TFIIB enters the pre-initiation complex after TBP. The crystal structure of the ternary complex of TFIIB-TBP-DNA shows specific TFIIB-TBP and TFIIB-DNA contacts. These include base-specific interactions with the major groove upstream and the minor groove downstream, of the TATA element. After initiation of transcription, elongation proceeds. The GTFs are replaced by elongation factors. These factors favour the phosphorylated form of the CTD. It results in the release of initiation factors and incorporation of factors required for elongation. Kinase P-TEFb is one of the elongation factors. It is recruited to polymerase by transcriptional activators. It plays three roles in elongation. The first role is that after binding with Pol II, it phosphorylates the serine residue at position 2 of the CTD repeats. Its second role is in activation of another elongation factor called hSPT5. The third role is that it recruits TAT-SF1. TFIIS protein has no role in initiation but this factor stimulates the overall rate of elongation by limiting the length of time polymerase pauses. Polymerase does not transcribe continuously but it takes periodic pauses. TFIIS reduces this pause time. This factor also contributes in proofreading.

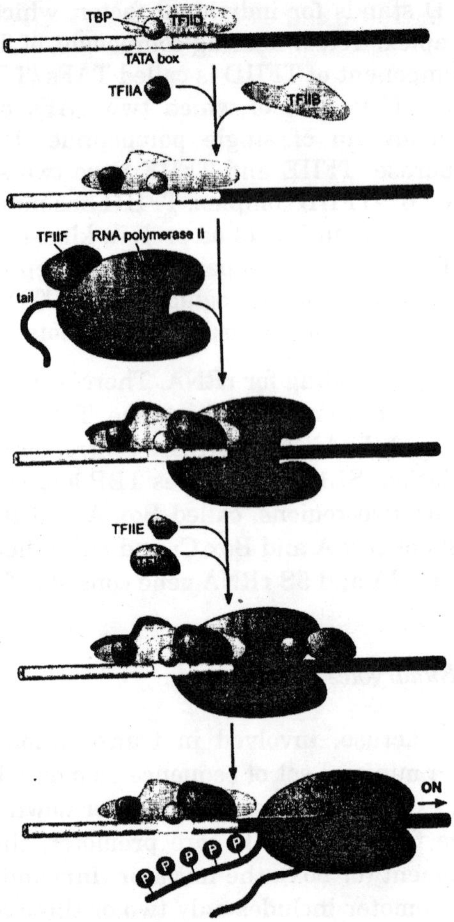

Fig. 11.3. Transcription initiation by RNA polymerase II.

RNA Processing in Eukaryotes

In eukaryotes, RNA is processed before it leaves nucleus and takes part in polypeptide synthesis. The processing events include the following: capping of the 5' end, splicing and polyadenylation of the 3' end. In capping a modified guanine base it added at 5' end of RNA in three enzymatic actions (Fig. 11.4). The bond is unusual 5' $\rightarrow$ 5' involving three phosphatase. In the first step, a phosphate group is removed from the 5' end of the transcript. This step is catalyzed by enzyme RNA triphosphatase. Second step is catalyzed by guanyl-transferase, and GTP is added at 5' of transcript. In the final step, nucleotide is modified by the addition of a methyl group. This step is catalyzed by enzyme methyl transferase. The RNA is capped when it is still only some 20-40 nucleotides long. This is the length of RNA, when trancription cycle is in the transition of initiation and elongation. After capping, dephosphorylation of Ser5 within the tail repeats leads to dissociation of the capping machinery. RNA splicing requires further phosphorylation. *RNA splicing* is carried out by two successive transesterification reactions, mediated by a large molecular `machine' known as spliceosome. In primary transcript, specific sequences mark the boundaries between exons and introns.

These sequences delineate where splicing will take place. The boundaries of 3' and 5' ends of introns are known as 3' splice site and 5' splice site, respectively. In transesterification reactions phosphodiester linkages within the pre-mRNA are broken and new ones are formed. *Polyadenylation* is intimately linked with the termination of transcription (Fig. 11.5). In this process, the polymerase CTD tail is involved in recruiting the enzymes necessary for polyadenylation. When polymerase reaches at the end of a gene, it transcribes some specific sequences. This event triggers the transfer of the polyadenylation enzymes to that RNA. As a result three processes are initiated: cleavage of the message; addition of many adenine residues to its 5' end; and subsequently, termination of transcription by polymerase. CPSF (cleavage and polyadenylation specificity factor) and CstF (cleavage stimulation factor), are two proteins of CTD. When polymerase reaches at the end of a gene, these factors activate poly-A signals. Polyadenylation is mediated by enzyme poly-A polymerase. It adds about 200 adenines to the 3' end of RNA. ATP is the precursor of polyadenylation and it is added to RNA as other nucleotides are added to the RNA during elongation. During polyadenylation addition of adenine does not requires templates. The mature RNA is transported from the nucleus to the cytoplasm.

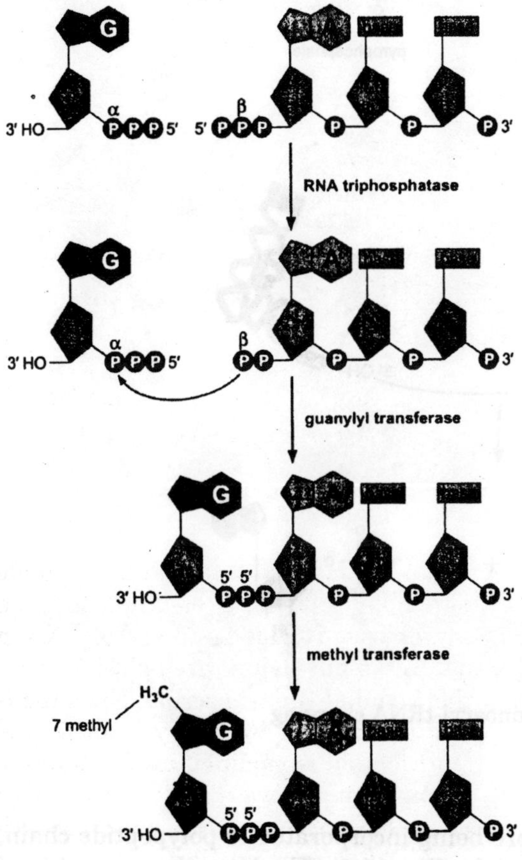

Fig. 11.4. The structure and formation of the 5' RNA cap

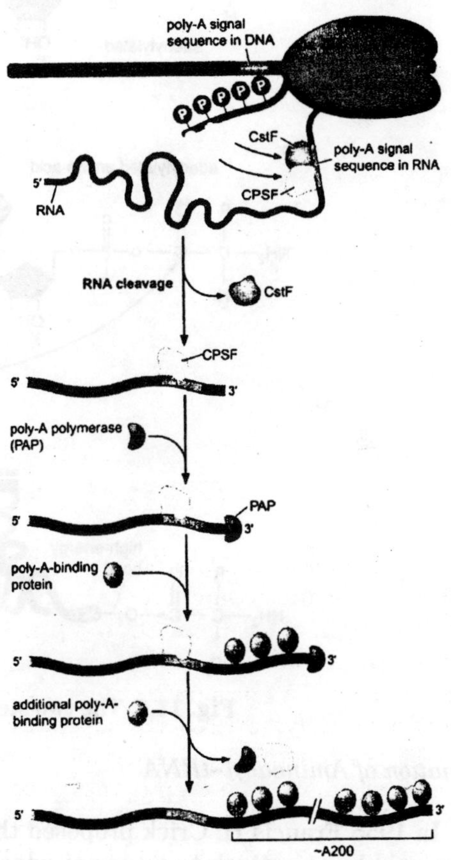

Fig. 11.5. Polyadenylation and termination

Translation

The messages transferred from DNA in the form of mRNA, is translated in the language of polypeptide chain. This is a linear sequence of amino acids. This process is called translation.

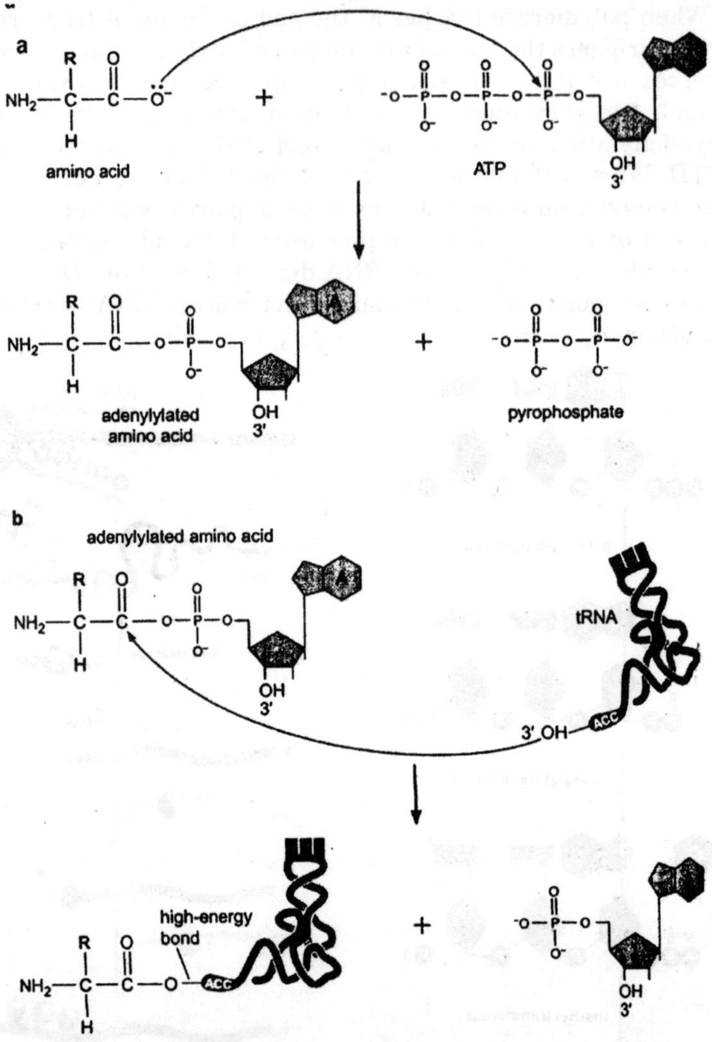

Fig. 11.6. The two steps of aminoacyl-tRNA charging

Formation of Aminoacyl~tRNA

In 1955 Francis H. Crick proposed that before being incorporated to polypeptide chain, amino acid has to attach to a special adaptor molecule (Fig. 11.6). The latter has a quality of directly interacting with and recognizing the three-nucleotide-long coding units of the mRNA. Paul C. Zamecnik and Mahlon B. Hoagland (1957) demonstrated that prior to their incorporation into proteins, the twenty different amino acids are esterified to their corresponding tRNA. The amino acid links with 3' terminal adenosine of tRNA by covalent bond. It is a two step process catalyzed by enzyme aminoacyl tRNA synthetase. Step one is adenylylation (transfer of AMP), in which the amino acid reacts with ATP to become adenylylated coupled with the release of pyrophosphate. Adenylylation results in attachment of amino acid to adenylic acid via a high-energy ester bond. The ester bond is formed between carbonyl group of the amino acid and phosphoryl group of AMP. In the second step, the adenylylated amino acid is transferred to the 3' end of the tRNA via the 2' or 3'-hydroxyl group and AMP is released. This step is known as tRNA charging.

(1) E + ATP + amino acid → E(aa AMP) + iPP

(2) E(aa AMP) + tRNA → aa tRNA + AMP + E

There are two classes of tRNA synthetases. Class I catalyze the attachment of amino acid to the 2'OH of the tRNA and it is generally monomeric. The, attachment of amino acid at 3'OH of tRNA is catalyzed by Class II synthetase. It is typically dimeric or tetrameric. There are usually several types of tRNA for the same amino acid. They are known as isoacceptors. Such isoacceptors tRNA molecules that bind to different codons are activated by the same aminoacyl-tRNA synthetase.

An individual aminoacyl~tRNA synthetase is specific for a single amino acid and certain tRNA as well. The interaction between aminoacyl~tRNA synthetases and tRNAs has been referred to as the second genetic code. Genetic, biochemical, and X-ray crystallographic evidence indicates that the specificity determinants are clustered at two distant sites on the molecule: the anticodon loop. The tRNA conformation is also important for such recognition.

Initiation of Polypeptide Chain

Initiation is most specific and complicated step of polypeptide synthesis.

Initiation of Polypeptide Chain in Prokayrotes

In bacteria, the initiation codon is AUG and initiator aminoacylated-tRNA is N-formylmethionyl-tRNA (Fig. 11.7). The formyl group is enzymatically added to methionine after the methionine has become attached to its tRNA. All methionine tRNA molecules can be formulated, rather only one known as tRNA formylmethionine permits the formulation reaction. Some molecules of tRNA are designated as tRNA meti permits formylation reaction. Both Met-tRNA types have same anticodon sequence. There are two differences between the initiating and elongating met-tRNA: the tRNA moieties themselves are different and the second difference is that the amino acids differ in the state of the amino group. The enzyme deformylase removes the formyl group from the growing chain very soon after synthesis commences. As there is no formylated methionine group in pure protein, it shows that in addition to deformylase (an aminopeptidase enzyme) subsequently removes the terminal methionine from many proteins. In prokaryotes, first mRNA binds to 30S subunit of ribosome. It is interesting that only specific mRNA remains binded to the ribosome subunit and other mRNA are trimed by ribonuclease. A group of 3 to 9 purine nucleotides appearing in the 5' end of the mRNA are critical for ribosome binding. These nucleotides which are upstream from the initiator codon, the polypurine stretch is known as the Shine-Dalgarno sequence. These nucleotides form base pairs with complementary residues found in a pyrimidine-rich sequence at the 3' end of 16S rRNA. Hence, it is clear that initiation starts with binding of mRNA-rRNA pairing. When sequences of mRNA roughly 30-nucleotide fragments were analyzed, they were found to contain AUGs roughly near the middle. Following this initiating AUG the other codons code for first three or four amino acids of their respective polypeptide chains. During elongation of polypeptide chain the 16S rRNA-mRNA base pair dissociates. These factors are found only on 30S three separate proteins, which are only loosely bound by ribosomes, called subunits. They are released when the 30S subunit associates with 50S subunits to generate 70S ribosome. In initiation cycle, first the 30S subunit forms a complex with IF3. The mRNA binds to complex I, placing initiation codon in a proper place. To this complex IF1 binds. Following this, f-met-tRNA is incorporated to initiation complex. Initiation codon, AUG recognizes this species of tRNA. GUG or occasionally UUG may also take part in initiation. These codons are not recognized equally well. The extent of initiation declines about half when AUG is replaced by GUG and declines by about half yet again when UUG is employed, the initiating tRNA and GTP binds to IF2 and the resulting complex then combines with the 50S subunit to form a complete 70S initiation complex. GTP is hydrolyzed to GDP + iP. IF-1, IF2 and IF3 are released after initiation is completed.

Initiation in Eukaryotes

Initiation of polypeptide synthesis in eukaryotes is almost the same as in bacteria. However, there are some differences in order of events and the number of accessory factors. The small subunit of ribosome is 40S. These subunits first recognize the 5' end of the mRNA, and then move to the initiation site.

In eukaryotes, in contrast to prokaryotes, binding of the initiator tRNA to the small subunit always precedes after association with the mRNA. Initiation begins with the formation of so-called ternary complex. The latter is formed when eIF2 binds with an initiator aminoacylated tRNA (Met-tRNA), GTP supplies energy to this reaction. Two GTP-binding factors eIF2 and eIF5B, mediate the recruitment of the charged initiator tRNA. The complex formed as a result of two GTP-binding proteins position the Met-tRNAiMet in the future P site of the small subunit, which is known as 43S pre-initiation complex. Other eIF incorporate in initiation process. eIF4 family of initiation factors, which recognize the cap structure and it facilitate the interaction between the 40S subunit and mRNA. Recognition is mediated by a three-subunit protein, known as eIF4F. One of the three subunits binds directly to the 5' cap and the other two subunits binds non-specifically to the associated RNA. The subunit must then identify the correct AUG codon at which reading the mRNA begins. The start codon is recognized through base-pairing between the anticodon, of the initiator tRNA and the start codon. Correct base-pairing triggers the release of eIF2 and eIF3. Typically, the Met-tRNA-bound 40S subunit migrates away from the cap along the mRNA in a 5' to 3' direction. This process is known as scanning. In most examples, only the first AUG can be used as the start site of translation. In selection process, hydrogen bond is established between AUG and the Met-tRNA bound to the ternary complex on the 40S subunit. Finally, the large ribosomal subunit binds to the small subunit while holding the mRNA and Met-tRNA in the correct orientation. At this stage, GTP is hydrolyzed and eIF2 is released.

In eukaryotes, the nontranslated 5' leader is relatively short. It is generally less than 100 bases. The nontranslated 3' trailer is often rather long, sometimes -1000 bases. Initiation is by met-tRNA. It is not formylated. It is known as $tRNA^{met}$.

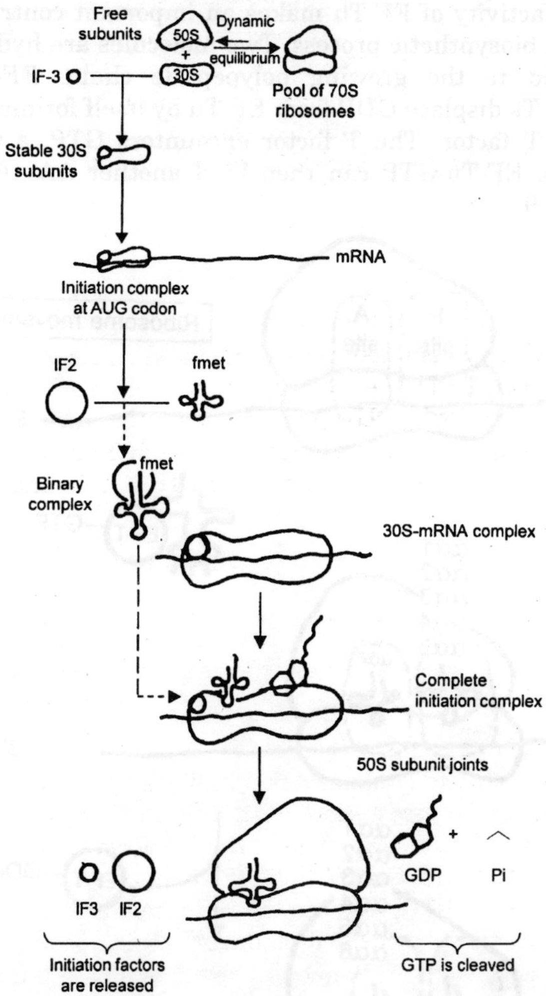

Fig. 11.7. Stages of inition of polypeptide chain

Elongation of Polypeptide Chain

Elongation of polypeptide chain is sequential addition of amino acids (Fig. 11.8).

Elongation of Polypeptide Chain in Prokaryotes

The initiation complex with initiator aminoacyl-tRNA on P site is ready for elongation. During elongation of polypeptide chain the incoming aminoacyl~tRNA enters to the A site and form the peptide bond with amino acid of P site. The incoming aminoacyl-tRNAs are delivered to A site by elongation factor (EF-Tu). The latter binds to the tRNA's 3' end, masking the coupled amino acid. This interaction prevents the bound aminoacyl-tRNA from participating in peptide bond formation until it is released from EF-Tu. GTP acts as energy source. EF-Tu binds with GTP and AA~tRNA to form an AA~tRNA-Tu-GTP complex. This complex is transferred to the ribosomal A site with release of a free EF-Tu-GDP complex and phosphate. The GTPase activity of EF-Tu makes an important contribution to the rate and conformity of the overall biosynthetic process. Two molecules are hydrolyzed for each amino acid subsequently added to the growing polypeptide chain. EF-Ts is responsible for regeneration of GTP. EF-Ts displace GDP from EF-Tu by itself forming a complex, EF-Tu.Ts. The latter is known as T factor. The T factor encounters GTP, a GTP-EF-Tu complex is formed, releasing EF-Ts. EF-Tu-GTP can then bind another AA~tRNA complex and thus repeats the cycle (Fig. 11.9).

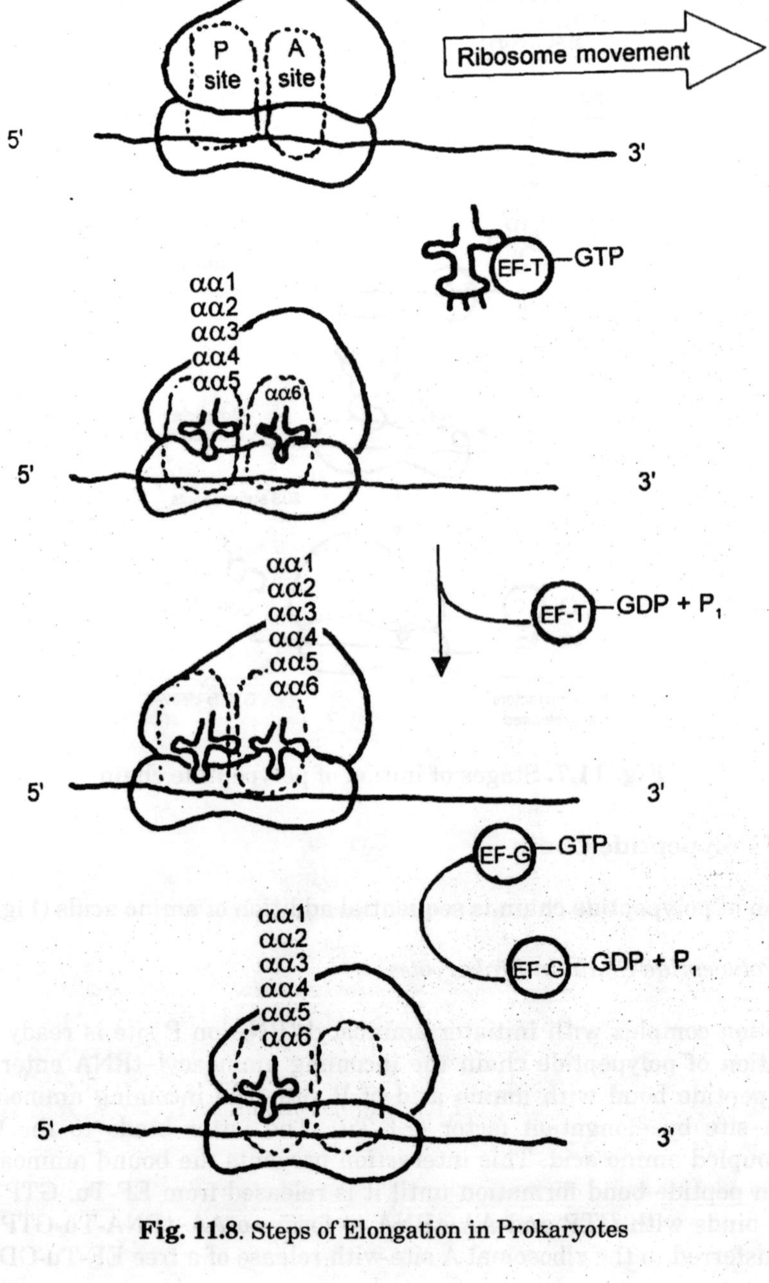

Fig. 11.8. Steps of Elongation in Prokaryotes

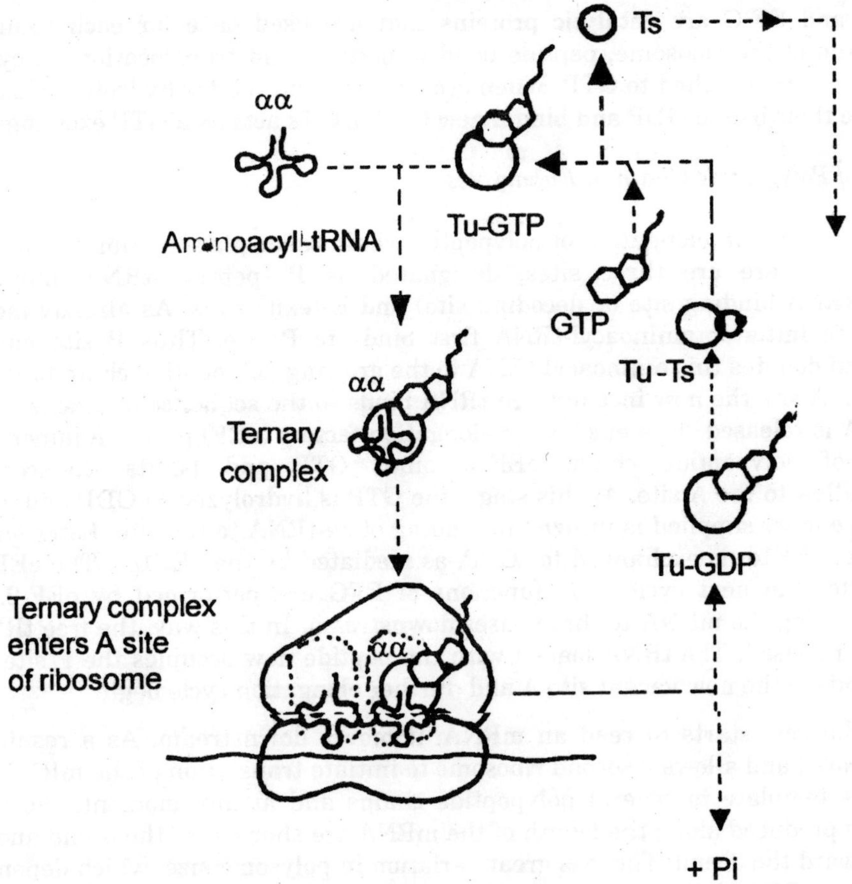

Fig. 11.9. Elongation factor cycle in Parokaryotes

The incoming aminoacyl~tRNA if suits with mRNA, then codon-anticodon hydrogen bond is established. Ribosome also performs selection of correct aminoacyl-tRNA. Two adjacent adenine residues in the 16S rRNA form a tight interaction with the minor groove of each correct pair formed between the anticodon and the first two bases of the codon. This gives fidelity in selection of correct aminoacyl-tRNA to be placed at A site. The peptide bond formation is catalyzed by enzyme peptidyl transferase. The bond formation is due to nucleophilic attack of the amino group of aminoacyl~tRNA to carboxyl group of fmet~tRNA. Peptidyl bond formation does not require energy source (GTP or ATP).

Peptide bond formation is followed by translocation i.e., movement of ribosome towards the 3' end of mRNA. As a result the aminoacyl~tRNA of A site is shifted towards the P site. This is a complex process and requires elongation factor G (EF-G) also known as translocase. EF-G can only bind to the ribosome when associated with GTP. An EF-G-GTP-ribosome complex first forms translocation then occurs coupled ejection of the free tRNA from the P site. GTP is hydrolyzed to GDP and inorganic phosphate and EF-G is gained in reusable form. Changes in the small subunit of the ribosome also contribute to translocation. After this process the A site of 70S ribosome is open and is ready to receive another aminocyl~tRNA.

EF-Tu and EF-G are catalytic proteins that are used once for each round of tRNA loading on top of the ribosome, peptide bond formation, and translocation. They are active only after they are attached to GTP. When cycle is over and GTP is hydrolyzed, both proteins must release their bound GDP and bind a new GTP. EF-Ts acts as a GTP exchange factor.

Elongation of Polypeptide Chain in Eukaryotes

In eukaryotic cell, elongation of polypeptide chain takes place in similar mechanism. In 80S ribosome there are three sites, designated as P (peptidyl-tRNA-binding site), A (aminoacyl-tRNA binding site or decoding site) and E (exit) sites. As already mentioned in initiation, the initiator aminoacyl-tRNA first binds to P site. Thus P site participate in initiation and donates this aminoacyl-tRNA to the growing polypeptide chain to the incoming aa-tRNA. On A site the new incoming aa-tRNA binds to the sequence of codons, while E-site empty tRNA is released. Two eukaryotic elongation factors (eEF) plays an important part in elongation of polypeptide chain. eEF1α binds GTP and points the corresponding aminoacyl-tRNA to the A site. At this stage, the GTP is hydrolyzed to GDP and a phosphate moiety. The energy supplied is utilized in binding of aa-tRNA to the site. Later on, the GDP, still bound to eEF1α, is exchanged for GTP, as mediated by the eEF1βγ. The eEF1α-GTP is now ready for the next cycle. The functions of EFG are performed by eEF-2. Ribosome translocates along the mRNA to three bases downstream. In this way, the free tRNA arrived at site E, is released. The tRNA loaded with the peptide now occupies the P site. The third aa-tRNA binds to the now vacant site A and further elongation cycle begin.

As a ribosome starts to read an mRNA, it moves downstream. As a result, the start codon is exposed and allows a second ribosome to initiate translation of the mRNA. The same mRNA gives template to several polypeptide chains and at any moment, the polypeptide chains being produced along the length of the mRNA are shortest at the 5' end and gradually lengthen toward the 3' end. There is great variation in polysome size, which depends both on the size of the mRNA chain and how closely the ribosomes are packed.

Termination of Polypeptide Chain

The polypeptide biosynthesis continues until one of the three stop codon enters the A site. The first one is presence of nonsense codon and the second is the release factor (RF). There are three nonsense codons UAA, UAG and UGA, which do not code for any amino acid. These codons are named by three names. These casual names are from the history of their discovery. The UAG is called the amber codon; UAA is ochre codon; and UGA is opal codon. The release factors are proteins and recognize the nonsense codons. It is believed that recognition is mediated by a protein-RNA interaction. There are two classes of release factors, Class I release factors. In prokaryotes, RF1 and RF2 are included in Class I release factor. RF1 recognizes the stop codons UAG and RF2 recognizes the stop codons UAG. UAA is recognized by both RF1 and RF2. In eukaryotes, there is a single class I release factor called eRF1 that recognizes all three stop codons. Class II release factors stimulate the dissociation of the class I factors from the ribosome after release of the polypeptide chain. The third release factor is RF3 in prokaryotes and eRF3 in eukaryotes belongs to Class II. This factor is regulated by GTP. However, it is not clear whether GTP hydrolysis is required for termination. All Class I release factors share a conserved, three-amino acid sequence GGQ (glycine glycine glutamine) essential for polypeptide release. A complex of release factors including eRF1 and GTP-bound eRF3 binds to the stop codon exposed in the A-site. During termination final tRNA splits off from the polypeptide chain. The peptidal transferase hydrolyzes the ester bond that links the completed polypeptide to the final tRNA. The 30s ribosomal subunit is released from mRNA. Another protein factor may be involved in ribosomal release.

Post transitional modification of proteins

The polypeptide released from the ribosome reaches to their appropriate subcellular location and faces some changes. The proteolytic action removes few amino acids or even a segment of polypeptide chain. Newly synthesized polypeptide chain should be folded and takes a three dimensional structure so that it is in a position to undergo biological functions. Polypepetides first folds into a structure in which, most of the elements of secondary structure are present. The partially folded molecule, referred to as a molten globule. The latter is a starting material from which the final three-dimensional structure of the protein will emerge. The folding of proteins are regulated by a group of proteins known as molecular chaperones. It promotes proper folding of polypeptides and suppresses folding patterns that would result to protein aggregation or nonfunctional protein products.

SUMMARY

In plants protein synthesis occurs in cytosol, plastids and mitochondria, and protein synthesis in cytosol does not occur at a constant rate, and it is regulated in response to numerous physiological and environmental variables. Transcription means the process by which nucleotide sequence information is transferred from DNA to RNA. The prokaryotes and eukaryotes show some major differences in the relation between the transcript and the mRNA used for polypeptide synthesis. In prokaryotes, primary transcript directly takes part in polypeptide synthesis. While in eukaryotes, the primary transcript must be converted to mRNA by a process known as RNA processing. In eukaryotes, RNA is processed before it leaves nucleus and takes part in polypeptide synthesis. The processing events include the following: capping of the 5' end, splicing and polyadenylation of the 3' end. The messages transferred from DNA in the form of mRNA, is translated in the language of polypeptide chain. The twenty different amino acids are esterified to their corresponding tRNA. Initiation is most specific and complicated step of polypeptide synthesis. Elongation of polypeptide chain is sequential addition of amino acids. The polypeptide biosynthesis continues until one of the three stop codon enters the A site. The first one is presence of nonsense codon and the second is the release factor (RF).

Chapter - 12

FAT METABOLISM

Fats are produced from glycerol and fatty acids by condensation reaction. Fatty acids are not synthesized by a reversal of the breakdown process. For synthesis of fatty acids acetyl CoA in the cytoplasm is carboxylated to malonyl CoA in presence of ATP. The malonyl part is then transferred to the SH group of acetyl carrier protein (ACP) also present in the cytoplasm. After a series of reactions a long chain of fatty acid is produced which is elongated by C2 fragments coming from acetyl coenzyme A and the process is repeated until a long chain fatty acetyl CoA is produced which is then released from the enzyme synthetase. In mitochondria acetyl CoA condenses repeatedly with a short chain fatty acid molecule until the long chain results.

During fatty acid biosynthesis, a repeated series of reactions incorporates acetyl moieties of acetyl-CoA into an acyl group of 16 or 18 carbons long chain. The enzymes involved in this synthesis are aectyl-CoA carboxylase (ACCase) and fatty acid synthase (FAS). The name fatty acid synthase refers to a complex of several individual enzymes that catalyze the conversion of acetyl-CoA and malonyl-CoA to 16:0 and 18:0 fatty acids. Acyl-carrier protein (ACP), an essential protein cofactor, is generally considered a component of FAS (Fig. 12.1).

Fatty acid biosynthesis is initiated by the ATP-dependent carboxylation of acetyl-CoA to form malonyl-CoA. The malonyl group is transferred next to ACP. Subsequent decarboxylation of the malonyl moiety acts to drive a condensation reaction, in which a carbon-carbon bond is formed between C-1 of acetate "primer" and C-2 of the malonyl group on ACP. This two-carbon chain length extension results initially in the formation of acetoacetyl-ACP. Subsequently, a sequence of three reaction-reduction, dehydration, and reduction take place. This again leads to the formation of the fully reduced acyl-ACP. This sequence, progressing in three steps from a 3-ketoacyl group to a saturated acyl group, is a common reaction series found in biochemical pathways. For example, both β-oxidation and the citric acid cycle use the same series of reactions, but in reverse order.

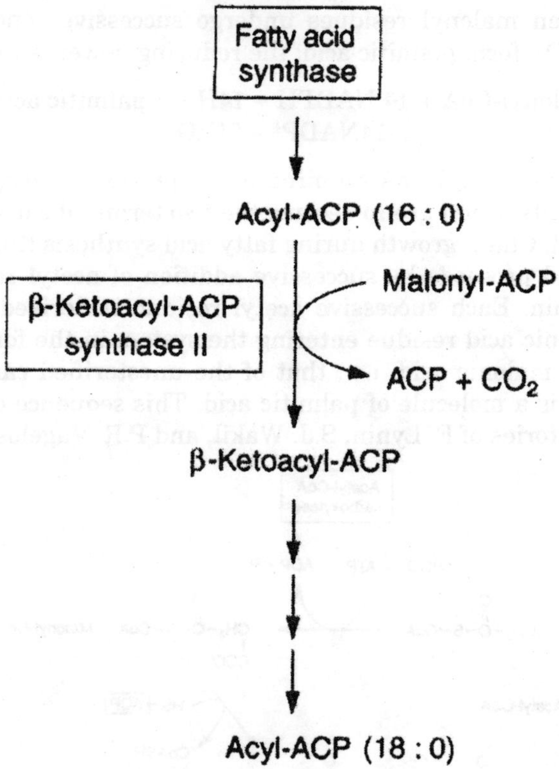

Fig. 12.1. Elongation and denaturation of fatty acid in plaslids.

Biosynthesis of saturated fatty acids

Plants are not capable of long-distance fatty acid transport. The synthesis of fatty acids occurs only for the cell's own requirements, except in seeds and fruits. Since fatty acids are present as constituents of membrane lipids in every cell, each cell must contain the enzymes for the synthesis of membrane lipids and thus also for the synthesis of fatty acids. In plants the *de novo* synthesis of fatty acids always occurs in the plastids: in the chloroplasts of green cells and the leucoplasts and chromoplasts of non-green cells. Although in plants cells enzymes of fatty acid synthesis are also found in membrane of the ER, these enzymes appear to be involved only in the modification of fatty acids, which have been synthesized earlier in the plastids. These modifications include a chain elongation of fatty acids, as catalyzed by elongases, and the introduction of further double bonds by desaturases.

Acetyl CoA is the precursor of biosynthesis of saturated fatty acid. In the first place total biosynthesis of fatty acids takes place in cytosol (Fig. 12.2). Citrate is necessary for maximal rates of synthesis of fatty acids. In the overall reaction of fatty acid synthesis, which is catalyzed by a cluster of seven proteins in the cytosol, the fatty-acid synthetase complex, acetyl-CoA derived from carbohydrate or amino-acid sources is the ultimate precursor of all the carbon atoms of the fatty acid chain. However, of the eight acetyl units required for biosynthesis of palmitic acid, only one is provided by acetyl-CoA; the other seven arrive in the form of malonyl CoA, formed from acetyl-CoA and HCO_3^- in a carboxylation reaction. One acetyl residue and seven malonyl residues undergo successive condensation steps. Seven released molecules of CO_2, form plamitic acid; the reducing power is furnished by NADPH

$$\text{Acetyl-CoA} + 7\ \text{malonyl-CoA} + 14\ \text{NADPH} + 14H^+ \rightarrow \text{palmitic acid} + 7CO_2 + 8\text{CoA} + 14\text{NADP}^+ + 6H_2O$$

The single molecule of acetyl CoA required in the process serves as a primer, or starter; the two carbon atoms of its acetyl group become the two terminal carbon atoms (15 and 16) of the palmitic acid formed. Chain growth during fatty acid synthesis thus starts at the carboxyl group of acetyl-CoA and proceeds by successive addition of acetyl residues at the carboxyl end of the growing chain. Each successive acetyl residue is derived from two of the three carbon atoms of a malonic acid residue entering the system in the form of malonyl-CoA; the third carbon atom of a malonic acid, i.e., that of the unesterified carboxyl group, is lost as CO_2. The final product is a molecule of palmitic acid. This sequence of reactions was largely elucidated in the laboratories of F. Lynin, S.J. Wakil, and P.R. Vagelos.

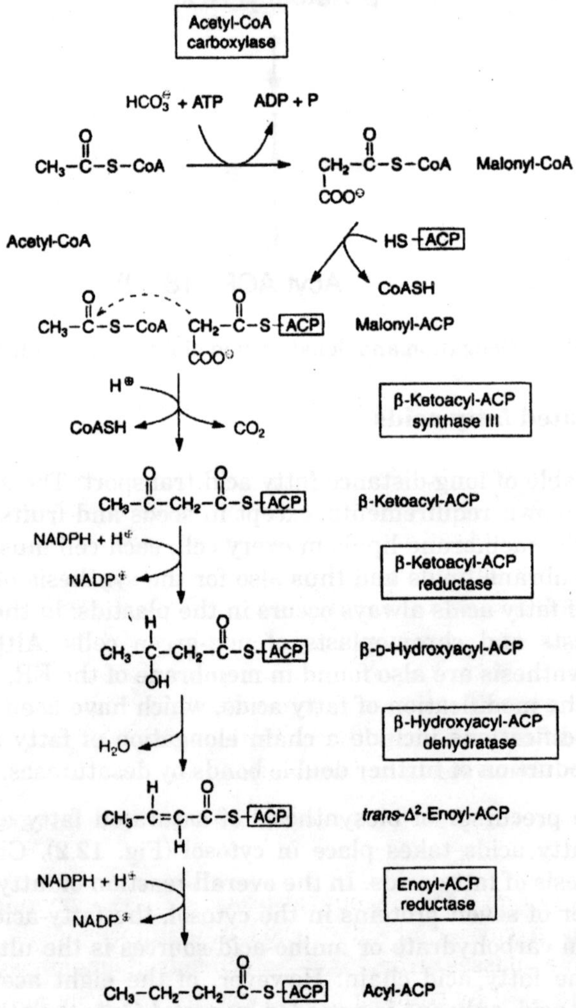

Fig. 12.2. Biosynthesis of fatty acids

A distinctive feature of the mechanism of fatty acid biosynthesis is that the acyl intermediates in the process of chain lengthening are thio esters. These are not from CoA, as in fatty acid oxidation, but of a low-molecular-weight conjugated protein called acyl carrier protein (ACP). This protein can form a complex or complexes with the six other enzyme proteins required for the complete synthesis of palmitic acid.

The fatty acid synthetase (FAS) refers to all enzyme activities in fatty acid biosynthesis except ACCase. FAS are essentially the same for all organisms. Two distinctly different types of FAS are found in nature. Animals and yeast use a Type I FAS, a single multifunctional enzyme complex characterized by large subunits By contrast, plants and most bacteria have a Type II FAS, in which each enzyme activity resides on an individual protein that can be readily separated from the other activities participating in fatty acid synthesis. Type II FAS also includes ACP. FAS complex of the plastids consisting of several proteins is similar to those of many bacteria, it is called prokaryotic fatty acid synthase complex. The assembly of a ^{18}C fatty acid from acetyl-CoA in Type II fatty acid synthesis requires 48 reactions involving at least 12 different proteins.

The carbon source for fatty acid synthesis

Acetyl-CoA is the initial substrate for synthesis of the carbon backbone of all fatty acids. It is also a central intermediate in many aspects of cellular metabolism, it is produced and consumed by dozens of reactions in the cell. Until recently, there has been some uncertainty as to which of these reactions produce the acetyl-CoA used in fatty acid biosynthesis. Most likely, acetyl-CoA from more than one reaction finds its way into fatty acid synthesis.

The ultimate source of all the carbon atom of fatty acids is acetyl-CoA, which may be formed in the mitochondria by the oxidative decarboxylation of pyruvate, or by the oxidative degradation of some of the amino acids, or by the β-oxidation of long-chain fatty acids. Acetyl-CoA itself cannot pass out of the mitochondria into the cytosol; however, its acetyl group is transferred through the membrane in other chemical forms. Citrate formed in the first step of TCA may pass through the mitochondrial membrane to the cytoplasm via the triacarboxylic transport system. In the cytosol acetyl-CoA is regenerated from citrate by ATP-citrate lyase, also called as citrate clevage enzyme, which catalyzes the reaction:

$$\text{Citrate + ATP + CoA} \rightarrow \text{acetyl-CoA + ADP +Pi + oxaloacetate.}$$

Like mitochondria, plastids contain a pyruvate dehydrogenase by which pyruvate is oxidized to acetyl CoA, accompanied by the reduction of NAD, but it is still not clear how acetylCoA is formed from the products of CO_2 fixation.

When chloroplasts are supplied with radioactively labeled acetate, the radioactivity is very rapidly incorporated into fatty acids.

In a second pathway the acetyl group of acetyl-CoA is enzymatically transferred to carnitine, which acts as a carrier of fatty acids into mitochondria preparatory to their oxidation. Acetylcarnitine passes from the mitochondrial membrane into the cytosol; acetyl-CoA is then regenerated by transfer of the acetyl group from acetylcarnitine to cytosol CoA.

Formation of Malonyl-CoA: Acetyl-CoA Carboxylase

Although malonyl-CoA is used in several pathways in the cell, fatty acid biosynthesis is the only known fate for malonyl-CoA within the plastid. Therefore, the ACCase reaction in plastids is the first committed step for fatty acid synthesis. Several lines of evidence indicate that plastid ACCase activity is tightly regulated and that this regulation determines, in large part, the overall rate of fatty acid synthesis. The concentration of malonyl-CoA in chloroplasts changes quickly during light-dark transition and remains proportional to the rate of fatty acid synthesis. Analysis of the substrates and products of plastid fatty acid metabolism indicates this regulation occurs at the ACCase reaction. Thus, plastid-localized ACCase appears to be a highly regulated enzyme subject to feedback and other biochemical controls.

Malonyl-CoA is the immediate precursor of 14 of the 16 carbon atoms of palmitic acid. Malonyl-CoA is formed from acetyl-CoA and bicarbonate in the cytosol by the action of acetyl-CoA carboxylase, a complex enzyme that catalyzes the reaction:

$$\text{Acetyl-CoA} + HCO_3^- + H^+ + \text{ATP} \longleftrightarrow \text{malonyl-CoA} + \text{ADP} + \text{Pi}$$

The carbon atom of the HCO_3^- is the active species rather than CO_2. It becomes the distal or free carboxyl carbon of malonyl-CoA. The above reaction has overall at least three intermediate reactions.

Acetyl-CoA carboxylase contains biotin as its prosthetic group. The carboxyl group of biotin is bound in amide linkage to the n-amino group of specific lysine residue of the biotin carboxyl carrier protein. It contains an -NH-group that forms a carbamate with HCO_3^- this reaction is driven by the hydrolysis of ATP. Acetyl CoA carboxylation requires two steps:

1. Biotin is carboxylated at the expense of ATP by biotin carboxylase.
2. Bicarbonate is transferred to acetyl CoA by carboxyl transferase.

All three enzymes-biotin carboxyl carrier protein, biotin carboxylase, and carboxyl transferase-form a single multienzyme complex. Since the biotin is attached to the carrier protein by a long flexible hydrocarbon chain it reacts alternately with the carboxylase and carboxyl transferase in the multienzyme complex.

The covalently bound biotin serves as an intermediate carrier of a molecule of CO_2 in a two-step reaction cycle. The reaction catalyzed by acetyl-CoA carobxylase, an allosteric enzyme, is the primary regulatory, or rate-limiting step in the biosynthesis of fatty acids. Acetyl-CoA carboxylase is virtually inactive in the absence of its positive modulators citrate or isocitrate.

Acetyl-CoA carboxylase occurs in both an inactive monomeric form and an active polymeric form. The inactive enyme monomer has a molecular weight of 410,000 and contains one binding site for HCO_3- (one biotin prosthetic group), one binding site for acetyl-CoA, and one for citrate. Citrate shifts the equilibrium between the inactive monomer and the active polymer, to favour the latter;

$$\underset{\text{(inactive)}}{\text{Monomer}} \underset{-\text{citrate}}{\overset{+\text{citrate}}{\rightleftarrows}} \underset{\text{(active)}}{\text{polymer}}$$

The acetyl-CoA caroboxylase reaction is actually more complex. In fact, the monomeric unit of the enzyme contains four different subunits. The sequence of reaction in the formation of malonyl-CoA has been deduced from studey of the four subunits of the monomer. One of these subunits, biotin carboxylase (BC), catalyzes the first step of the overall reaction, namely, the carboxylation of the biotin residue covalently bound to the second subunit, which is called biotin carboxylase-carrier protein (BCCP):

$$\text{Biotin-BCCP} + HCO_3^- + H^+ + \text{ATP} \xrightarrow{\text{BC subunit}} \text{carboxybiotin-BCCP} + \text{ADP} + \text{Pi}$$

The second step in the overall reaction is catalyzed by the third type of subunit, called carboxyl transferase (CT)

$$\text{Carboxybiotin-BCCP} + \text{acetyl-CoA} \xrightarrow{\text{CT subunit}} \text{biotin-BCCP} + \text{malonyl-CoA}$$

In these reactions the biotin residue of the carboxyl carrier protein serves as a swinging arm. It transfers the bicarbonate ion from the biotin carboxylase subunit to the acetyl-CoA. The latter is bound to the active site of the carboxyltransferase subunit. The change from the inactive monomeric, form of the enzyme occurs when citrate is bound to the fourth subunit of each monomeric unit.

Plants contain both homomeric and heteromeric forms of ACCase

ACCase catalyzes the initial step in fatty acid synthesis. The plastid form of ACCase has four subunits namely biotin caroboxyl carrier protein (BCCP), biotin carboxylase (BC), and α-and β-subunits of carboxyltransferase (CT). These four subunits form a heteromeric complex of more than 650 kDa, which may be membrane-associated. Three of the subunits are encoded by distinct nuclear genes, whereas the fourth (β-CT) is encoded in the chloroplast genome.

The cytosolic form of ACCase is a single large (greater than 500 kDa) homodimeric protein. The four subunits described above are integrated into domains of a single polypeptide, two of which associate to make up the homodimer. This structure is similar to the ACCase found in animals and fungi. Almost all monocot and dicot plants so far examined have the two types of ACCase described above-with the heteromeric, multisubunit form in the plastid and the homodimeric form in the cytosol.

The Reactions of the Fatty-Acid Synthetase System

Once malonyl-CoA has been generated from acetyl-CoA by the action of the rather complex acetyl-CoA carboxylase reaction, the ensuing reactions of fatty acid synthesis occur in a sequence of six successive steps catalyzed by the six enzymes of the fatty-acid synthetase system (Fig. 12.3). The seventh protein of this system, which has no enzymatic activity itself, is acyl carrier protein, to which the growing fatty acid chain is covalently attached.

Acyl Carrier Protein (ACP)

Acyl carrier protein, universally symbolized as ACP. It was first isolated in pure form from *E.coli* by Vagelos and his colleagues and has since been studied from many other sources. This small protein is about 80 amino acids long. It contains a phosphopantetheine prosthetic group covalently linked to a serine residue near the middle of the polypeptide chain. The single sulfhydryl group of ACP, to which the acyl intermediates are esterified, is contributed by its prosthetic group, a molecule of 4'-phosphopentetheine. It is covalently linked to the hydroxyl group of serine residue 36 of the protein. The 4'-phosphopantetheine moiety is identical with that of coenzyme A, from which it is derived. ACP can be regarded as a CoA, which is covalently bound to a protein. The function of ACP in fatty acid synthesis is analogous to that of CoA in fatty acid oxidation; it serves as an anchor to which the acyl intermediates are esterified.

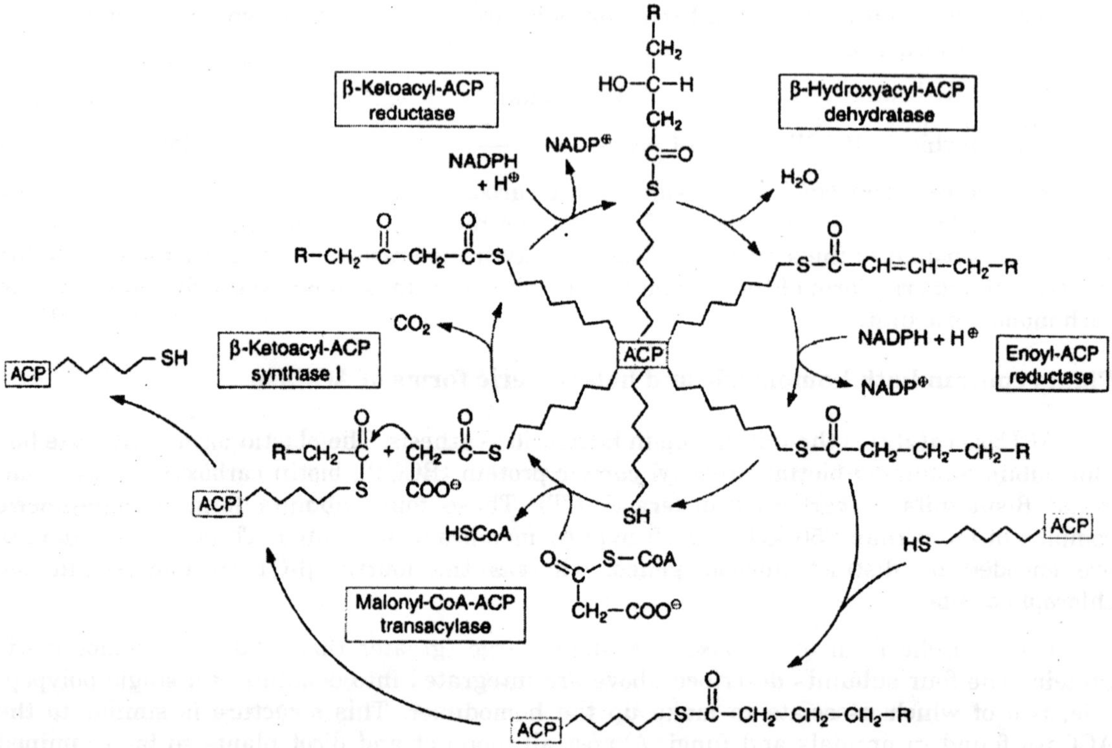

Fig. 12.3. The interplay of the various enzymes during fatty acid synthesis

The Priming Reaction

To prime the fatty-acid synthetase system, acetyl-CoA first reacts with the sulfhydryl group of ACP by the action of one of the six enzymes of the synthetase system, ACP-acyltransferase, which catalyzes the reaction:

$$\text{Acetyl-S-CoA} + \text{ACP-SH} \rightarrow \text{acetyl-S-ACP} + \text{CoA-SH}$$

This acetyl group does not, remain on the ACP but is transferred to a specific cysteine residue of another enzyme of the fatty-acid synthase complex, β-ketoacyl-ACP synthase, in the reaction

$$\text{Acetyl-S-ACP} + \text{synthase-SH} \rightarrow \text{ACP-SH} + \text{acetyl-S-synthase}$$

In which β-ketoacyl-ACP synthetetase is simply designated as synthase. With this reaction the fatty-acid synthase complex is now primed and ready to carry out the sequence of reactions required to add a two-carbon unit in the chain-lengthening process.

The Malonyl Transfer Step

The next reaction is catalyzed by ACP malonyltransferase and malonyl-CoA formed. The acetyl-CoA carboxylase reaction reacts with the -SH group of the 4'-phosphopantetheine arm of ACP, with loss of free CoA, to form malonyl-S-ACP :

$$\text{Malonyl-S-CoA} + \text{ACP-SH} \rightarrow \text{malonyl-S-ACP} + \text{CoA-SH}$$

As a result of this step and of the preceding priming reaction, a malonyl group is now esterified to ACP and an acetyl group is esterified to an -SH group on the β-ketoacyl-ACP synthase molecule.

The Condensation Reaction

In the next reaction of the sequence, catalyzed by β-ketoacyl-ACP synthase, the acetyl group esterified to the cysteine residue of this enzyme is transferred to carbon atom 2 of the malonyl group on ACP, with release of the free carboxyl group of the malonyl residue as CO_2.

$$\text{Acetyl-S-synthase} + \text{malonyl-S-ACP} \rightarrow \text{acetoacetyl-S-ACP} + CO_2 + \text{synthase-SH}$$

The molecule of CO_2 so released contains the same carbon atom that was introduced as HCO_3^- in the acetyl-CoA carboxylase reaction. The liberation of CO_2 makes this reaction irreversible. The acetoacetate thus formed, remains bound as a thioester to ACP. It is reduced by NADPH to β-D-hydroxyacyl-ACP. Following the release of water, the carbon-carbon double bond formed is reduced by NADPH to acyl-ACP. In essence, HCO_3^- plays a catalytic role in fatty acid synthesis since it is regenerated as CO_2 as each two-carbon residues for fatty acid synthesis. If precursor of two-carbon residues for fatty acid synthesis. If acetoacetyl-CoA was to be formed from two molecules of acetyl-CoA by the action of acetyl-CoA acetyltransferase

$$\text{Acetyl-S-CoA} + \text{acetyl-S-CoA} \rightarrow \text{acetoacetyl-S-CoA} + \text{CoA-SH}$$

The reaction would be endergonic, with its equilibrium lying to the left. On the other hand, the formation of acetoactyl-S-ACP by the reaction is strongly exergonic. Its equilibrium lies far to the right, in the direction of synthesis, because the decarboxylation of the malonyl residue, which is strongly exergonic, it provides a strong thermodynamic pull toward completion.

$$\text{Acetyl\_S\_synthase} + \text{malonyl-S-ACP} \rightarrow \text{acetoacetyl-S-ACP} + \text{synthase-SH} + CO_2$$

The First Reduction Reaction

The initial reductive step of fatty acid biosynthesis is the conversion of 3-ketoacyl-ACP to a 3-hydroxyacyl-ACP. The acetoacetyl-S-ACP now undergoes reduction by NADPH to form β-hydroxybutyryl-S-ACP. This reaction is catalyzed β-ketoacyl-ACP reductase:

$$\text{Acetoacetyl-S-ACP} + \text{NADPH} + H^+ \rightarrow \text{D-}\beta\text{-hydroxybutyryl-S-ACP} + \text{NADP}^+$$

3-ketoacyl-ACP reductase is a tetramer with atleast two isomer. It is noteworthy that the product is the D stereoisomer of the β-hydroxy acid, whereas the isomer formed during fatty acid oxidation has the L-configuration

The Dehydration Step

D-β-hydroxybutyryl-S-ACP is next dehydrated to the corresponding trans-α,β- or Δ^2-unsaturated acyl-S-ACP, namely crotonyl-S-ACP. It is catalyzed by enoyl-ACP hydratase:

$$CH3CHOHCH_2CO\text{-}S\text{-}ACP \underset{\longleftarrow}{\longrightarrow} CH_3CH{=}CH\text{-}CO{=}S{+}ACP + H_2O$$

Crotonyl-S-ACP

The Second Reduction Step

Crotonyl-S-ACP is now reduced to butyryl-S=ACP by enoyl-ACP reductase (NADPH): the electron donor is NADPH in *E. coli*

$$\text{crotonyl-S-ACP} + NADPH{+}H_{+} \leftarrow\rightarrow \text{Butyryl-S-ACP} + NADP^{+}$$

This reaction also differs from the corresponding reaction of fatty acid oxidation in mitochondria. It is because a pyridine nucleotide rather than a flavoprotein is involved. Since the $NADP{+}NAD^{+}$ couple has a more negative standard potential than the fatty acid oxidizing flavoprotein, NADPH favours reductive formation of the saturated fatty acid.

The formation of butyryl-ACP completes the first of seven cycles in route to palmitoyl-S-ACP. To start the next cycle the butyryl group is transferred from ACP to the -SH group of the β-ketoacyl-ACP synthase molecule. Thus allowing ACP to accept a malonyl group from another molecule of malonyl-CoA. Then the cycle repeats, the next step being the condensation of malonyl-S-ACP with butyryl-S-β-ketoacyl-ACP synthase to yield β-ketohexànoyl-S-ACP and CO_2.

After seven complete cycles, palmitoyl-ACP is the end product. The palmitoyl group may be removed to yield free palmitic acid by the action of a thioesterase, or it may be transferred from ACP to CoA, or it may be incorporated directly into phospholipids and triacylglycerols.

It is remarkable that in most organisms the fatty-acid synthetase system stops with the production of palmitic acid and does not yield stearic acid. The latter has only two more carbon atoms than palmitic acid and thus does not differ greatly in physical properties. This chain-length specificity is conferred on the system by β-ketoacyl-ACP synthase. It is very active in accepting the tetradecanoyl group (14 carbons) from ACP but does not accept the hexadecanoyl group (16 carbons). It may be presumably because the binding site of the enzyme for the saturated acyl group can accommodate only a specific range of chain lengths. In addition, palmitoyl-CoA functions as a feedback inhibitor of the fatty-acid synthetase system.

Saturated fatty acids having an odd number of carbon atoms are also made by the fatty-acid synthetase complex. In this case the synthesis is primed by a starter molecule of propionyl-S-ACP (instead of acetyl-S-ACP), to which are added successive two-carbon units via condensations with malonyl-S-ACP.

Thioesterase reactions terminate the fatty acid biosynthesis cycle

Each cycle of fatty acid synthesis adds two carbons to the acyl chain. Typically, fatty acid synthesis ends at 16:0 or 18:0, when one of several reactions stops the process. The most common, reactions are hydrolysis of the acyl moiety from ACP by a thioesterase of by transfer of the actyl moiety from ACP directly onto a glycerolipid by an acyl transferase, or an acyl-ACP desaturase. The thioesterase reaction yields a sulfhydryl ACP.

Two principla types of acyl-ACP thioesterases occur in plants, FatA, and FatB. In both cases, however, the metabolic consequence of these reactions is the same. The cleavage of the acyl moiety from ACP prevents extension and targets the acyl groups for export out of the plastid by an unknown mechanism.

The prokaryotic and Eukaryotic pathways

In plastid lipid biosynthesis, the acyl residues are transferred directly from acyl-ACP to glycerol 3-phosphate. For the first acylation step, mostly an 18:1-, less frequently a 16:0-, and more rarely an 18:0-acyl residue is esterified to position 1. The C-2 position, however, is always esterified with a 16:0-acyl residue. Since this specificity is also observed in cyanobacteria, the glycerolipid biosynthesis pathway of the plastids is called the prokaryotic pathway.

For glycerolipid synthesis in the ER membrane, the acyl residues are transferred from acyl-CoA. Here again, the hydroxyl group in C-1 position is esterified with an 18:1-, 16:0-, or 18:0-acyl residue, but position C-2 is always esterified with a some what desaturated 18:n-acyl residue. The glycerolipid pathway of the ER membrane is called the eukaryotic pathway.

Specialized elongase systems produce long-chain fatty acids

Fatty acid synthesis generally results in C16 and C18 fatty acids, yet plants have numerous requirements for longer fatty acids. All plants produce waxes, usually derived form C26 to C32 fatty acids. Often, spingolipids contain C22 and C24 fatty acids. In some plants, triacylglycerols contain large amounts of C20 and C22 fatty acids.

Plants and most other eukaryotic organisms have a specialized elongase system for extension of fatty acids beyond C18. Although the enzymology of the elongase is not well understood, several important differences between the two systems (FAS and elongase) are known

(i) Fatty acid elongases are localized in the cyotosol and are membrane-bound.

(ii) ACP is not involved in this process.

(iii) The elongase 3-ketoacyl-CoA synthase (elongase KCS) catalyzes the condensation of malonyl-CoA with an acyl primer.

The ER membrane is the site of fatty acid elongation and desaturation

The plastids produce 16:0-, 18:1-, and to a lesser extent 18:0-acyl residues. However, some storage lipids contain fatty acids with a greater chain length. This also applies to waxes, which are esters of a long chain fatty acid (C20-C24) with very long chain acyl alcohols (C24-C32). The elongation of fatty acids greater than C18 is catalyzed by elongases, which are located in the membranes of the ER. Elongation proceeds in the same way as fatty acid synthesis, with the only differences being that different enzymes are involved and that the acyl- and malonyl residues are activated as acyl CoA thioesters.

The ER membrane is also the site for further desaturation of the acyl residue. For desaturation, the acyl groups are first incorporated into phospholipids, such as phosphatidylcholine. Desaturases bound to the membrane of the ER convert oleate (18:1) to linoleate (18:2) and then to linolenate (18:3). The 18:2- or 18:3-acyl residues in the C-2-position of glycerol can be exchanged for an 18:1 acyl residue, and the latter can then be further desaturated.

The first double bond in a newly formed fatty acid

The stearoyl-ACP formed in the plastid stroma is desaturated there to oleoyl-ACP (Fig. 12.4). This reaction can be regarded as a monooxygenation, it is because one O-atom from an O_2 molecule is reduced to water and the other is incorporated into the hydrocarbon chain of the fatty acid as hydroxyl group. A carbon-carbon double bond is formed by subsequent liberation H_2O analogous with the B-hydroxyacyl-ACP dehydratase reaction. It is in contrast to fatty acid synthesis and has a cis-configuration. The monooxygenation requires two electrons, which are provided by NADPH via reduced ferredoxin. Monooxygenase are widespread in bacteria, plants, and animals. In most cases, O_2 is activated by a special cytochrome, cytochromeP450. However, in the stearoyl-ACP desaturase, the O_2 molecule reacts with a di-iron-oxo-cluster. Iron-sulphur-cluster is a redox carrier in which the Fe-atoms are bound to the protein via cysteine residues. In the di-iron-oxo-cluster of the desaturase, two iron atoms are bound to the enzyme via the carboxyl groups of glutamate and aspartate. The two Fe-atoms alternate between oxidation state +IV +III and +II. An O_2 molecule is activated by the binding of the two Fe-atoms.

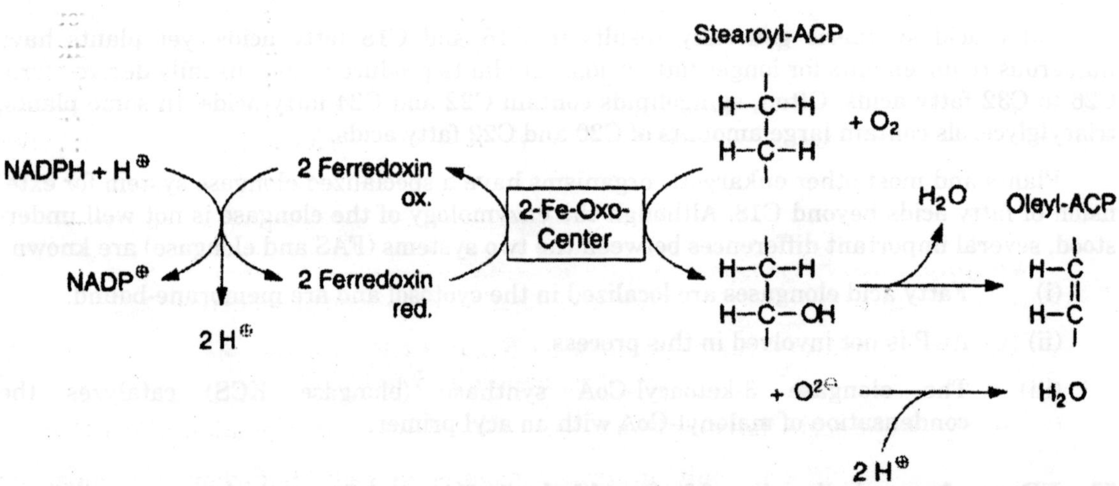

Fig. 12.4. Desaturation of Stearoyl-ACP to Oleyl-ACP as catalyzed by stearoyl-ACP-desaturase

Stearoyl-ACP desaturase is a soluble protein that is localized in chloroplasts and other plastids. The enzyme is so active that normally the newly formed stearoyl-ACP is almost complelelty converted to oleyl-ACP. This soluble desaturase is capable of introducing only one double bond into fatty acids. The introduction of further double bonds is catalyzed by other desaturases, which are integral membrane proteins of the ER. They react only with fatty acids that are constituents of membrane lipids. For this reason, they are termed acyl lipid desaturases. These membrane-bound desaturases also require O_2 and reduced ferredoxin, similar to the ACP-desaturases, but have a different electron transport chain. The required reducing equivalents are transferred from NADPH via an FAD containing NADPH-cytochrome b5-reductase to cytochrome b5. They are further transferred to the actual desaturase, which contains two Fe atoms probably bound to histidine residues of the protein. The family catalyzes the introduction of hydroxyl groups (hydroxylases), epoxy groups (epoxygenase), conjugated double bonds (conjugases), and carbon triple bonds (acetylenases) into fatty acids of acyl lipids.

Biosynthesis of Triacylglycerols

For glycerol synthesis dehydroxy acetone phosphate is reduced by a NADP-linked dehydrogenase to glycerol phosphate, from which the phosphate residue is removed by a phosphatase.

Two major precursors are required for the synthesis of triacylglycerols : L-glycerol 3-phosphate and fatty acyl-CoA. L-Glycerol-3-phoshphate is derived from two different sources. Its normal precursor is dihydroxyacetone phosphate, which is a product of the aldolase reaction of glycolysis. It may also be formed from free glycerol arising from degradation of triacylglycerols, through the action of glycerol kinase

$$\text{ATP + glycerol} \xrightarrow{Mg^{++}} \text{L-glycerol 3-phosphate + ADP}$$

The first stage in triacylglycerol formation is the acylation of the free hydroxyl groups of glycerol phosphate by two molecules of fatty acyl-CoA to yield first a lysophosphatidic acid and then a phosphatidic acid. Free glycerol is not acylated. These reactions occur preferentially with 16- and 18-carbon saturated and unsaturated acyl-CoA's. An alternative pathway for the formation of phosphatidic acid involves esterification of the fatty acyl group with dihydroxyacetone phosphate. This pathway consists of the reactions.

Dihydroxyacetonephsophate acyltransferase

Dihydroxyacetone phosphate + fatty acyl-CoA ⟶ Fatty acyl dihydrox- yacetone phosphate + CoA

Fatty acyl dihydroxyacetone phosphate + NADPH + H^+ ⟶ Lysophosphatidic acid + $NADP^+$

Lysophosphatidic acid + fatty acyl-CoA ⟶ Phosphatidic acid + CoA

Phosphatidic acids occur only in trace amounts in cells, but they are important intermediates in the biosynthesis of triacylglycerols and phsophoglycerides.

In the pathway to triacylglycerols, phosphatidic acid undergoes hydrolysis by phsophatidate phosphatase to form a diacylglycerol:

$$\text{L-Phosphatidic acid} + H_2O \rightarrow \text{diacylglycerol + Pi}$$

The diacylglycerol then reacts with a third molecule of a fatty acyl-CoA to yield a triacylglycerol by the action of diacylglycerol acyltransferase

Fatty acyl-CoA + diacylglycerol → triacylglycerol + CoA

The lipid content of mature seeds can amount to 45% of the dry weight, but in some cases e.g., Ricinus, it is up to 85%. The precursor for the synthesis of triacylglycerol is again glycerol 3-phosphate. There are at least four pathways for its synthesis, the two most important are

1. Phosphatidic acid is formed by acylation of the hydroxyl groups in the C-1 and C-2 positions of glycerol. After hydrolysis of the phosphate residue, the C-3 position is also acylated. The total cytosolic acyl CoA pool is available for these acylations, but because of the eukaryotic pathway, position-2 is mostly esterified by a C18 acyl residue.

2. Phsophatidylcholine is formed first, and its acyl residues are further desaturated. The choline phosphate residue is then cleaved off by transfer to CMP, and the corresponding diacylglycerol is esterified with a third acyl residue. This pathway operates frequently in the synthesis of highly unsaturated triacylglycerols.

LIPID CATABOLISM

During seed germination enzyme lipase is synthesized, which catalyze the hydrolysis of triaclglycerols to glycerol and fatty acids. Lipases bind to the oleosins of the oil bodies. The glycerol formed by the hydrolysis of triacylglycerol, after phosphorylation to glycerol 3-phosphate, can be fed into the gluconeogenesis pathway. The released free fatty acids are first activated as CoA-thioesters and then degraded to acetyl CoA by β-oxidation. This process occurs in specialized peroxisomes called glyoxysomes.

Although in principle β-oxidation represents the reversal of the fatty acid synthesis, there are distinct differences that enable high metabolic fluxes through these two metabolic pathways operating in opposite directions. The differences between β-oxidation and fatty acid synthesis are:

1. In acyl CoA dehydrogenation, hydrogen is transferred via an FAD-dependent oxidase to form H_2O_2. A catalase irreversibly eliminates the toxic H_2O_2 at the site of its production by conversion to water and oxygen.

2. β-L Hydroxyacyl CoA is formed during the hydration of enoyl CoA, in contrast to the corresponding D-enantiomer during synthesis.

3. Hydrogen is transferred to NAD system in the cell. It is highly oxidized driving the reaction in the direction of hydroxyacyl CoA oxidation. It is not known which reactions utilize the NADH formed in the peroxisomes.

4. In an irreversible reaction, CoASH-mediated thiolysis cleaves β-ketoacylCoA to form one molecule of acetyl CoA and one of acyl CoA and one of acyl CoA shortened by two C atoms.

During the degradation of unsaturated fatty acids, intermediate products are formed during the degradation of oleic acid, is converted by an isomerase shifting the double bond to Δ^2-trans-enoyl CoA, an intermediate of β-oxidation.

β-oxidation takes place in peroxisomes and glyoxysomes

The major site of fatty acid oxidation in animal cells is the mitochondrial matrix. In contrast, fatty acid oxidation in plants occurs primarily in the peroxisomes of leaf tissue and the glyoxysomes of germinating seeds. Unlike in animals, the β-oxidation pathway is not used solely for production of metabolic energy in plants. The biological role of β-oxidation in peroxisomes and glyoxysomes is to provide biosynthetic precursors from stored lipids. During germination, triacylglycerols stored in seeds are converted into glucose and a wide variety of essential metabolites. Fatty acids released from tiracylglycerols are activated to form their CoA derivatives and are oxidized in glyoxysomes by the same four-step process that occurs in peroxisomes. The acetyl-CoA produced is converted by way of the glyoxylate cycle to C4 precursors for gluconeogenesis. Glyoxysomes, like peroxisomes, contain high concentrations of catalase. The latter converts the H_2O_2 produced by β-oxidation to H_2O and O_2.

In animal mitochondria, each of the four enzymes of β-oxidation is a separate, soluble protein. These structure is similar to the analogous enzyme in Gram positive bacteria. In contrast, the enzymes of peroxisomes and glyoxysomes form a complex of proteins. As in the oxidation of fatty acids in mitochondria, the intermediates are CoA derivatives, and the process consists of four steps

(i) dehydration to aΔ^2-trans unsaturated structure.

(ii) Addition of water to the resulting double bond.

(iii) Oxidation of the β-hydroxyacyl-CoA to a ketone

(iv) Thiolytic clevage by CoASH.

The difference between the peroxisomal and mitochondrial pathways lies in the first step. In peroxisomes, the flavoproteins dehydrogenase that introduces the double bond passes electrons directly to O_2, producing H_2O_2. This strong and potentially damaging oxidant is immediately cleaved by catalase to H_2O and $1/2O_2$. In contrast, in mitochondria, the electrons removed in the first oxidation step pass through the respiratory chain upto O_2, leaving H_2O as the product, a process accompanied by ATP synthesis. In peroxiomes, the energy released in the first oxidative step of fatty acid breakdown is dissipated as heat.

The biochemical steps of β-oxidation

Mitochondrial oxidation of fatty acids takes place in three stages. In the first stage of β-oxidation-the fatty acids undergo oxidative removal of successive two-carbon units in the form of acetyl-CoA, starting from the carboxyl end of the fatty acyl chain. e.g., the 16 carbon fatty acid palmitic acid undergoes seven passes through this oxidative sequence, in each pass losing two carbons as acetyl-CoA (Fig. 12.5). At the end of seven cycles the last two carbons of palmitate are left as acetyl-CoA molecules. Formation of each molecule of acetyl-CoA requires removal of four hydrogen atoms (two pairs of electrons and four H^+) from the fatty acyl moiety by the action of dehydrogenases.

In the second stage of fatty acid oxidation the acetyl residues of acetyl-CoA is oxidized to form CO_2 via the citric acid cycle. Acetyl-CoA derived from fatty acid oxidation thus enters a final common pathway of oxidation along with acetyl-CoA derived from glucose via glycolysis and pyruvate oxidation.

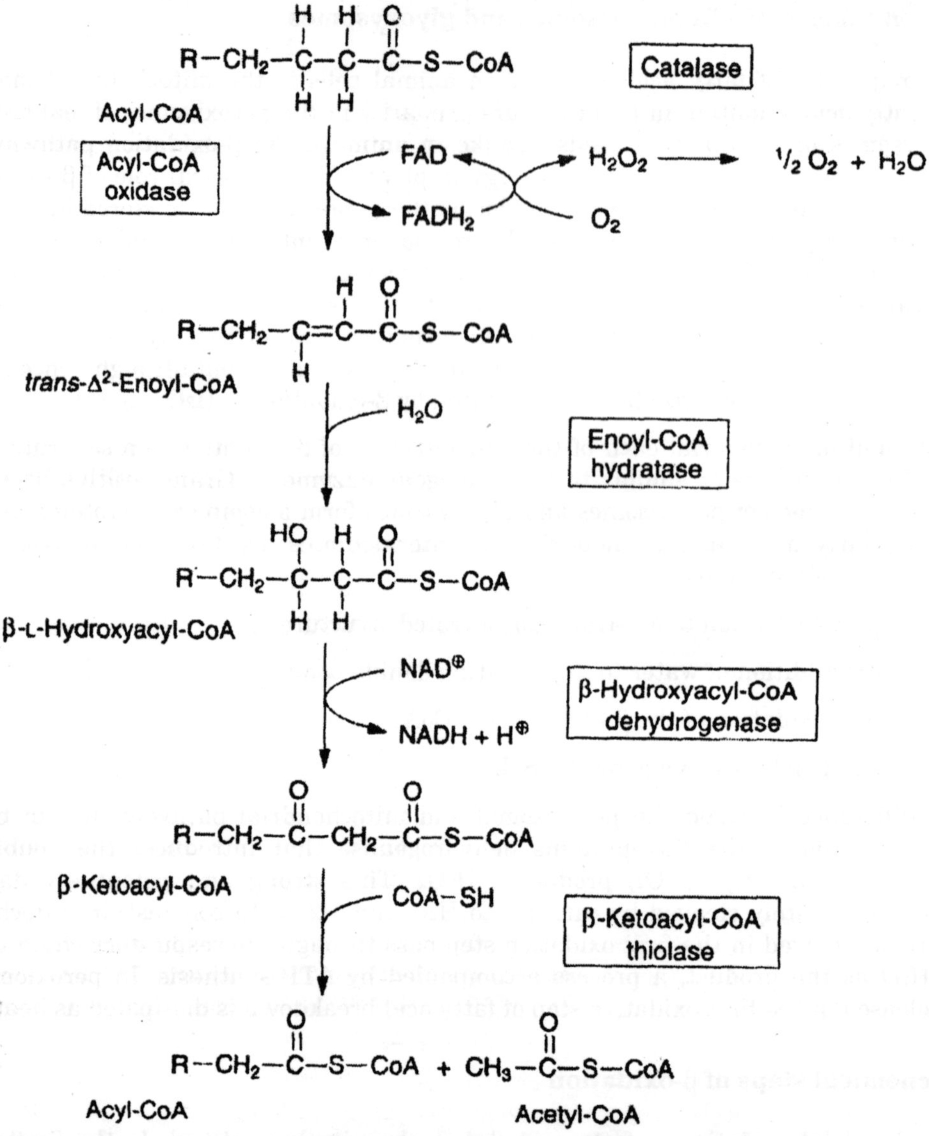

Fig. 12.5. β-oxidation of fatty acids

The first two stages of fatty acid oxidation produce the reduced electron carriers NADH and $FADH_2$, which in the third stage donate electrons to the mitochondrial respiratory chain. The electrons are carried to oxygen through this chain. Coupled to this flow of electrons is the phsophorylation of ADP to ATP. Thus, energy released by fatty acid oxidation is conserved as ATP.

β-oxidation of saturated fatty acids has four basic steps

Four enzyme-catalyzed reactions are involved in the first stage of fatty acid oxidation. First, dehydrogenation produces a double bond between α and β carbon atoms (C-2 and C-3), yielding a trans-Δ^2-enoyl-CoA. The symbol Δ^2 designates the position of the double bond. The new double bond has the trans configuration (naturally occurring unsaturated fatty acids normally have their double bonds in the cis configuration).

The enzyme responsible for this first step, acyl-CoA dehydrogenase, includes FAD as a prosthetic group. The electrons removed from the fatty acyl-CoA are transferred to the FAD. The reduced form of the dehydrogenase then immediately donates its electrons to an electron carrier, the electron-transferring flavoprotein (ETFP). ETFP, is an integral protein of the mitochondrial respiratory chain. The transfer of a pair of electrons from the $FADH_2$ of acyl-CoA dehydrogenase to O_2 via the respiratory chain provides the energy for the synthesis of two ATP molecules.

The oxidation catalyzed by acyl-CoA dehydrogenase is analogous to succinate dehydrogenation in the citric acid cycle. In both reactions the enzyme is bound to the inner membrane, and a double bond is introduced into a carboxylic acid between the α and β carbons. FAD is the electron acceptor, and electrons form the reaction ultimately enters the respiratory chain. These elections are carried to O_2 with the concomitant synthesis of two ATP molecules per electron pair.

In the second step of the fatty acid oxidation cycle, water is added to the double bond of the trans-Δ^2-enoyl-CoA to form the L stereoisomer of β-hydroxyacyl-CoA. This reaction, catalyzed by enoyl-CoA hydratase, is formally analogous to the fumarase reaction in the citric acid cycle.

In the third step, the L-β-hydroxyacyl-CoA is dehydrogenated to form β-ketoacyl-CoA by the action of β-hydroxyacyl-CoA dehydrogenase; NAD^+ is the electron acceptor. This enzyme is absolutely specific for the L-stereoisomer. The NADH formed in this reaction donates its electrons to NADH dehydrogenase (Complex I). The latter is an electron carrier of the respiratory chain. The reaction catalyzed by β-hydroxyacyl-CoA dehydrogenase is closely analogous to the malate dehydrogenase reaction of the citric acid cycle.

The fourth and last step of the fatty acid oxidation cycle is catalyzed by acetyl-CoA acetylatransferase (more commonly called thiolase). This enzame promotes reaction of β-ketoacyl CoA with a molecule of free coenzyme A. This splits off the carboxyl-terminal two-carbon fragment of the original fatty acid, (now shortened by two carbon fragment of the original fatty acid) as acetyl-CoA. The other product is the coenzyme A thioester of the original fatty acid, now shortened by two carbon atoms. This reaction is called thiolysis, by analogy with the process of hydrolysis, because the β-ketoacyl-CoA is cleaved by reaction with the thiol group of coenzyme A.

The carbon-carbon single bond that connects methylene (CH_2^-) groups in fatty acids is relatively stable. The b-oxidation sequence represents an elegant solution to the problem of breaking these bonds. The first three reactions of b-oxidation have the effect of creating a much less stable C-C bond, in which one of the carbon atoms (the carbon, C-2) is bonded to two carbonyl carbons. The ketone function on the b carbon (C3) makes it a good point for nucleophilic attack by SH of coenzyme A, catalyzed by thiolase. The acidity of the carbon makes the terminal $-CH_2CO$-S-CoA a good leaving group, facilitating breakage of a-b bond.

SUMMARY

Fats are produced from glycerol and fatty acids by condensation reaction. During fatty acid biosynthesis, a repeated series of reactions incorporates acetyl moieties of acetyl-CoA into an acyl group 16 or 18 carbon long chain. Accetyl CoA is the precursor of biosynthesis of saturated fatty acid. ACCase catalyzes the initial step in fatty acid synthesis. Each cycle of fatty acid synthesis adds two carbons to the acyl chain. Oxidation of fatty acid is known as β-oxidation, which takes place in mitochondria and there is successive removal of two carbon units at the end of the cycle acetyl-CoA molecules are left as the final product.

SECTION - II :

DEVELOPMENTAL PHYSIOLOGY

Chapter - 13

SIGNAL TRANSDUCTION

Signal transduction is a complex process, involving the passage of information. It is defined as a phenomenon in which information is relayed across the plasma membrane to the cell interior and often to the cell nucleus. Cell constructs responses to a signal, and signal transduction is the ability of cells to receive and act on signals from beyond the plasma membrane. The extracellular stimuli are known as first messenger and the stimuli formed as a result of interaction with G protein and released within the cell is called second messenger. The signals received by plant cells alter their physiology, morphology, and development. The cells respond to a stimulus from the environment by conveying information to the internal compartment of the cell. In general the stimulus is a molecule that has been secreted into the extracellular space by another cell. However, a stimulus may also originate due to contact by another cell or a noncellular substrate. Some signals are carried by xylem and phloem. The nature of the stimulus received by the cell-surface receptor is entirely different from the signal that is released to the cell interior. Plant cells receive signals from membrane receptors related to the surrounding like light, mineral nutrients, organic metabolites, gravity, water status, turgor, soil quality, mechanical tension, wind, heat, cold, hormone, pH, gases, woundings, and disease.

Some Basic Concepts

Signal transduction is an extraordinarily specific and delicately sensitive phenomenon. Specificity of signal transduction is achieved by precise molecular complementarity between the signal and receptor molecules, mediated by the same kinds of weak (noncovalent) forces. Three factors account for the extraordinary sensitivity of signal transducers: the high affinity of receptors for signal molecules, cooperativity in the ligand-receptor interaction, and amplification of the signal by enzyme cascades. A final noteworthy feature of signal-transducing systems in integration, the ability of the system to receive multiple signals and produce a unified response appropriate to the needs of the cell or organism.

The Receptors and Lignads The *receptors* are proteins, which binds with a signal. In other words, signals are first sensed by receptors. They are present in limited abundance, and have quality to select biologically active molecules. Most of the receptors are present in the plasma membrane. Some are located in the cytosol or other cellular compartments. In some transmembrane protein receptors, the C-terminal region is phosphorylated by protein kinases. The membrane potential can act as a receptor. Plants have another class of receptors known as receptor-like protein kinase, which consists of a large extracytoplasmic domain, a single membrane-spanning segment, and a cytoplasmic domain. Intracellular membranes also have receptors, which can act as Ca^{2+} channels. Ca^{2+} increase is thought to be central to signal transduction. The increase in the concentration of free Ca^{2+} activates downstream targets. This phenomenon transduces the initial signal into the end response. Voltage-gated Ca^{2+} permeable channels are located in the plasma membrane. The channels are activated by membrane depolarization. Modifications of membrane potential open a group of voltage-gated channels. This allows Ca^{2+} to enter and thus activate a transduction sequence. A *ligand* is the agent that binds to the receptor at the outer cell surface. The nature of ligand is not that important. When ligand binds to receptors the receptors change from inactive form to active state. The ligand on the extra cellular domain induces the activity of the cytosolic domain or intracellular domain.

Receptor-ligand binding is reversible, allowing the system to respond to changes in ligand concentration. Ligand binds to a specific site on their receptors. The binding is of relatively high affinity, and saturates at a certain ligand concentration. The binding of the ligand to its specific receptor causes a change in the conformation of the receptor that increases its affinity for the G protein. The ligand-bound receptor binds to the G protein. This is discussed in details under subsection G protein.

Genetic and Epigentic Information the genetic information means a usual flow of message in plant cell i.e., DNA→RNA→protein→phenotype. The epigenetic information concept means phenotypic characteristics strongly modified by the environment, in which the plant growing. In these cases the phenotypic character is constructed from epistatic genes and pleiotropic genes. The latter influences multiple traits. These characters vary with the environment. The variation may be on production of biomass, duration of growth, branching, partitioning of photosynthate between reproductive and vegetative structures, and responses to stress.

Glucagon and Epinephrine Glucagon and epinephrine are the first messengers. Rodbell and his colleagues worked on the activities of glucagons and epinephrine on stimulating adenylyl cyclase, and in 1971 they published as series of papers on this field of research. These compounds activate the same effector (adenylyl cyclase). Glucagon is a small protein made up of 29 amino acids. Epinephrine is a small hydrophilic amine. The studies were based on the aspect about the nature of glucagon molecule to bind to their corresponding receptors on the isolated plasma membrane. In these studies it was established that GTP is required for the activation of adenylyl cyclase. Presence of GTPase in plasma membrane was observed in mid 1970s. Cassel and Selinger proposed that the hydrolysis of GTP by the GTPase serves to turn off the activated adenylyl cyllase, returning the enzyme to the basal inactive state.

The Second Messenger it is a voluntarily diffusible molecule involved in conveying information from an extracellular source to the principal target enzymes within the cell. In plant cell $[Ca^{2+}]_I$ transduce many signals. It is a prominent second messenger. During signaling, $[Ca^{2+}]_I$ increase (Ca^{2+}transients) are often associated with initiation of responses. Changes in ($[Ca^{2+}]_I$)can initiate diverse responses that vary according to cellular structure and are sensitive to interaction between, specific components of signaling systems. *Protein kinases* are similarly ubiquitous, and cell uses hundreds of different protein kinases. In a cell about 3% to 4% of genome contains genes encoding for the members of superfamily protein kinases (Fig. 13.1). Members of this superfamily catalyze the reversible transfer of the γ-phosphate from ATP to serine, threonine, or tyrosine amino acid side chains on target proteins. Thus, protein kinases are grouped into two classes: serine/threonine residues or tyrosine residues. A few protein kinases phosphorylate all three amino acids. Tyrosine residue phosphorylating protein kinases are exclusively rare in plant cells.

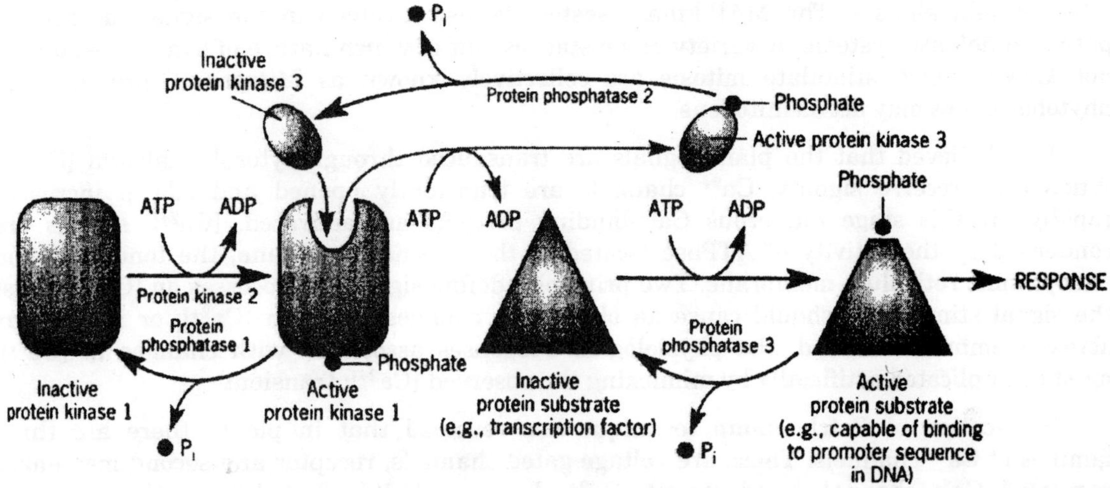

Fig. 13.1. Signal transuction pathways involving primarily of protein kinases and protein phosphatases

Signal Transduction Pathway Plant cell has two principal elements in the signal transduction pathways. They are intracellular Ca^{2+} ($[Ca^{2+}]_I$) and protein kinases, enzymes that carry out phosphorylation. The interesting part of signal transduction is that one protein of the pathway alters the activity of the next protein in the series. It may be by adding or removing phosphate groups. Amplification by enzyme cascades results when an enzyme associated with a signal receptor is activated and, in turn, catalyzes the activation of many molecules of a second enzyme, each of which activates many molecules of a third enzyme, and so on. The signal pathway typically contains protein kinase and phosphatase. Protein kinase transfers phosphate from ATP to amino acids of target proteins, this is counter-balanced by the action of specific protein phosphatase. The latter removes the phosphate from protein. They direct rapid, reversible changes in cellular activities. Some of these enzymes are soluble cytoplasmic proteins, others are integral membrane proteins. In general these enzymes can phosphorylate or dephosphorylate. These enzyme has other enzymes as substrate. Some of these enzymes act on numerous proteins, while other may act on a single protein. The protein substrate of these enzymes may include ion channels, transcription factors, and various types of regulatory subunits. The addition or removal of phosphate groups from the substrate results in conformational changes in protein. This change is supposed to be responsible for stimulation or inhibition of activity, thereby leading to the specific response. GTP-binding proteins are involved in switching the signal transduction pathway on or off. During signaling GTP transmits the message along `cellular information circuits'. The receptors have a binding site for signal substances at the outside and the binding site for G-protein at the cytoplasmic site of the plasma membrane. Thus, it behaves in an appropriate manner to pass external signals into the cell. Recently channels for another second messenger, cyclic ADP-ribose (c-ADPR) have been reported. In eukaryotic cells superfamily protein kinase includes the cyclin-dependent-protein kinases (CDK). CDKinases activate a number of proteins that are involved in mitosis. One of the proteins of this superfamily is known as *cyclin*, which has a significant function in the cell cycle. Another member of the superfamily is the nitrogen-activated-protein kinases (MAPK). It plays an important role in protein kinase cascades, where protein kinases are regulated through phosphorylation by other protein kinases. The MAP-kinase system is also involved in the signal cascades of pathogen defense systems. A variety of substances, out of which nature of some substance is not known which stimulate mitosis are collectively known as *Mitogen*. G-proteins and phytohormones may act as mitogens.

It is believed that the plant signals are transduced through cytosolic calcium $[Ca^{2+}]_I$. When cells receive signals, Ca^{2+} channels are transiently opened and $[Ca^{2+}]_I$ increases rapidly. At this stage numerous Ca^{2+}-binding proteins are activated. $[Ca^{2+}]_I$ signals are condensed by the activity of ATPase located in the plasma membrane, the tonoplast, and endoplasmic reticulum membrane. Two principles define signaling processes on $[Ca^{2+}]_I$. First the signal stimulation should cause an observable changes either in $[Ca^{2+}]_I$ or in Ca^{2+}flux across membrane. Second, the physiological responses associated with changes in $[Ca^{2+}]_I$ must be duplicated artificially by mimicking the observed $[Ca^{2+}]_I$ transient.

Application of Patch clamp technique has revealed that in plants there are three families of Ca^{2+} channels. These are voltage-gated channels, receptor and second messenger regulated Ca^{2+} channels, and stretch Ca^{2+} channels. Voltage-gated have their opening probability determined by a particular value of membrane potential. The second category of Ca^{2+} channels is present in vacuole and most probably, in ER and nuclear membrane. These types of Ca^{2+} channels of vacuolar membrane are opened by the cyclic nucleotide second messenger cyclic ADP-ribose. The stretch channels sense tension in the membrane and are opened when the tension is altered. Such channels get implemented for mechanical signals and turgor status.

Many signals interact cooperatively and in a synchronized manner to produce the final response. Different signaling pathways are often interconnected. The benefit of such action is that signals from a variety of unrelated ligands can converge to activate a common effector. Signaling pathways may end in apoptosis. The latter is a programmed death of the cell.

The Mechanism of Signal Transduction

Different models are proposed for signal transduction viz. (1) Cyclic AMP mediated signal transuction. (2) Cyclic GMP mediated signal transduction. (3) G-protein mediated signal transduction. (4) Inositol mediated signal transduction. (5) Ca^{++} mediated signal transduction. (6) Protein kinase mediated signal transduction and (7) Ras mediated signal transduction. Some of these models explain the mechanism in animal cells. In the present text, the mechanisms of signal transduction only in a plant cell are discussed.

G-Protein

Alfred Gilman (1980) purified GTP-binding protein. The G-proteins (or G-binding proteins) bind with GTP or GDP and can alternate between two conformational states (Fig.

13.2). GTPase are converted from an empty state to an active state by association with GTP. Hydrolysis of GTP to GDP inactivates the protein, and subsequent dissociation of GDP returns the protein to the empty state. These heteromeric proteins interact with receptor proteins located in the plasma membrane. The SDS-PAGE studies of purified GTP-binding protein established that this protein is made up of three distinct polypeptide with molecular weight of 52,45, and 35 kDa. This protein is composed of three different subunits namely G_α, G_β, and G_γ. G_α has binding site for GTP or GDP, and binding results in conformational changes in this subunit. After this conformational change, G_α dissociates from the other two units. The G_α-GTP functions as an activator of various enzymes forming messengers for signal transduction. The former can stimulate GMP-cyclase, phospholipase C (PLC). GMP-cyclase forms the messenger substance guanosine-3'-5'-monophosphate (cGMP) from GTP (Fig. 13.3). Stimulation of phospholipase C liberates Ca^{2+} as a messenger. G_α-GTP has a short half life, and after that bound GTP is hydrolyzed to GDP by an intrinsic GTPase activity. The confromational change in this complex (Fig. 13.4), $G\alpha$ is released and thus looses its activator function. At this stage G_α is in a position to join the remaining two subunits (dimer) to form trimer and proceed for the next cycle. The other class monomeric GTPase includes Ras, Ypt, Rab, Sec, Rho, and EFTu. The monomeric G-protein of Ras superfamily have various regulatory functions, such as the regulation of defense reactions, ABA responses, vesicle transport, cell polarity, and the growth of pollen tubers and root hairs. These proteins have only one subunit related to α-subunit of heterotrimeric G-protein. They are present in cytosol, have binding domains for GDP/GTP and an effector domain. On stimulation with signal, this protein interacts with an exchange factor, which converts the GDP-bound inactive form to a GTP-active form by GTP/GDP replacement. The active GTP-form interacts with effector proteins like G_α-GTP mechanism. The effector domain takes part in this interaction.

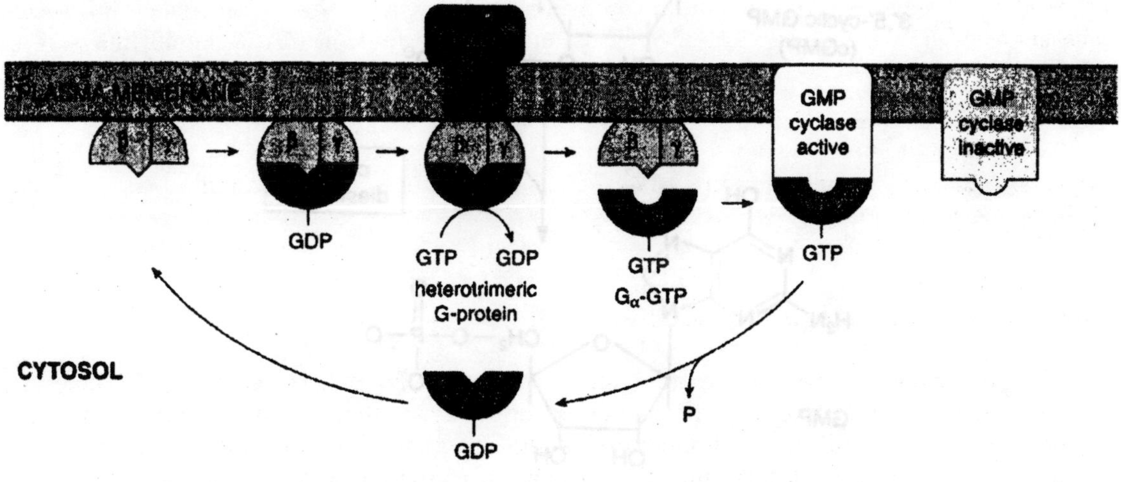

Fig. 13.2. Scheme of the function of a G-protein

Calcium

Ca^{2+} is mainly present in plasma membrane, endoplasmic reticulum, cytosoloic region and in mitrochrondrial inner membrane. In plant cells the concentration of Ca^{2+} is high in sacroplasmic reticulum. The concentration of Ca^{2+} in cytosolic region is as low as 10^{-7} M. If

Ca^{2+} gets increased beyond 1 x 10^{-5}M, the cell may get damaged. Cell is unable to tolerate high concentration of Ca^{2+}, as it competes with Mg^{2+} and ultimately makes adverse effects on metabolism. Ca^{2+}-ATPases domains are present in the plasma membrane of plant cells. It uses the free energy released by ATP hydrolysis to translocate Ca^{2+} into extracytosolic compartments against its electrochemical gradient. The active sites of both enzymes require a phosphorylated aspartyl residue as an intermediate in the use of ATP. It is controlled by phosphorylation/dephosphorylation reactions.

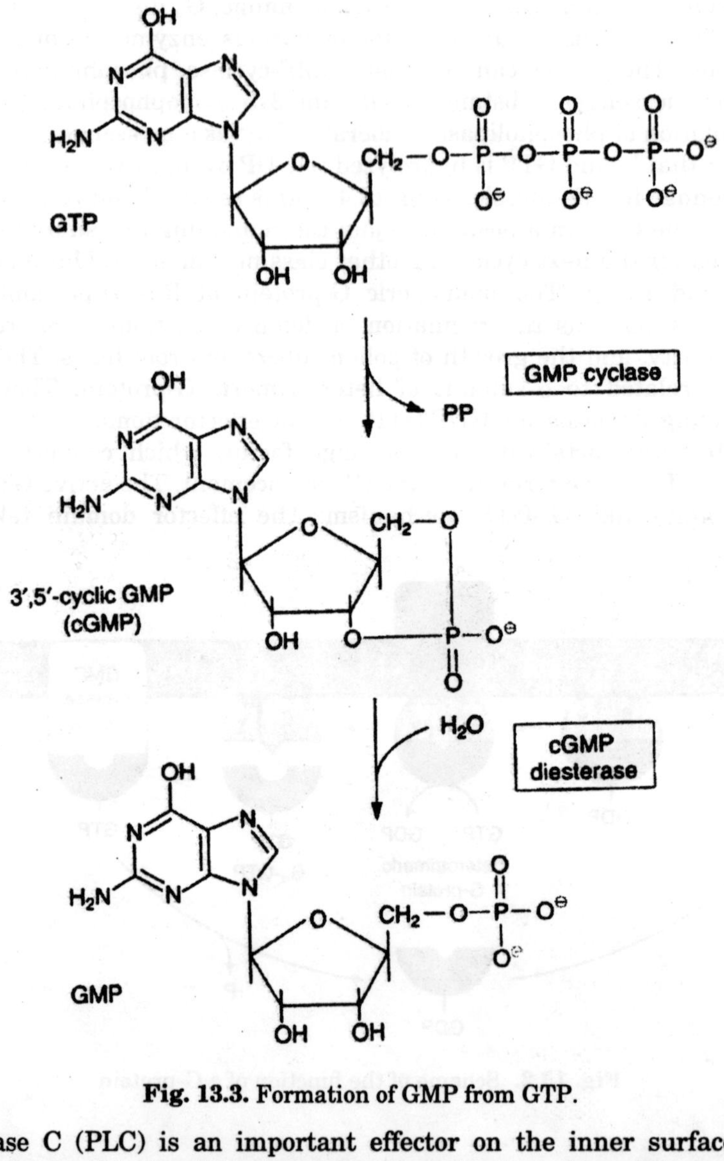

Fig. 13.3. Formation of GMP from GTP.

Phospholipase C (PLC) is an important effector on the inner surface of the plasma membrane. It is involved in inositol phospholipid synthesis and breakdown. Phosphoinositol signal transduction cascade controls Ca^{2+} channels. It was first reported in animal cells and was later discovered in plant cells. G-proteins stimulate PLC, which cleaves the lipid to inositol-1,4,5-triphosphate (IP_3) and diacylglycerol (DAG). Cytosolic Ca^{2+} concentration gets increased by IP_3. Many kinases and phsophatases also participate in this cycle. The inositol residue is phosphorylated at the hydroxyl groups in 4' and 5' positions by a kinase. The substrate for PLC-phosphatidylinositol 4,5-bisphosphate (PIP_2) is cleaved to yield two products. These products are inositol 1,4,5-triphosphate (IP_3) and diacylglycerol (DAG). The former is a soluble second messenger, while the latter remains membrane-bound. DAG functions within the plasma membrane, where it is likely quite mobile. DAG remains in plasma membrane and activates an enzyme called protein kinase *c* (PKC). PKC phosphorylates serine and threonine residues on a variety of target proteins. Two types of kinases catalyze the formation of PIP_2 from phosphatidylinositol (PI). PI is first converted into phatidylinositol 4-phosphate (PIP) and successively to PIP_2. The modified membrane lipid functions as a signal for vesicle transfer. Activation of receptors by red or blue light in plants results in exchange of bound GDP for GTP on the G-protein. The α-subunit bound to GTP is released and activates PLC. IP3 is a small, water soluble molecule. It diffuses freely in the cytoplasm and binds to specific Ca^{2+} channels in the vacuole and the rough ER. This ultimately results in release of Ca^{2+} in the cytoplasm. In plants, the phosphoinositol cascade plays important role in passing signals from the environment to cellular functions. Patch clamp technique developed by Erwin Neher and Bert Sakmann demonstrated that in plants, IP_3 causes Ca^{2+} open in vacuole and other Ca^{2+} stores.

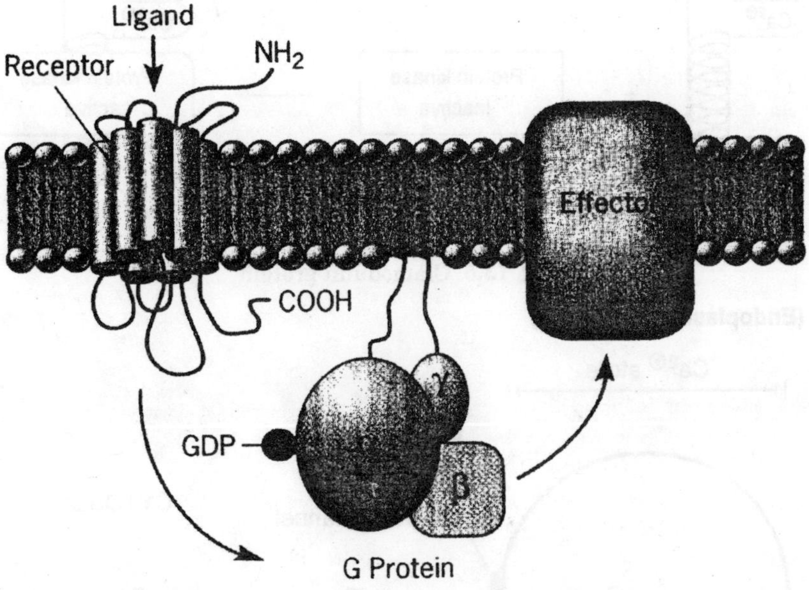

Fig. 13.4. The membrane-bound machinery for a heterotrimeric G protein.

The Calcium-Calmodulin mediated Signal Transduction

Calmodulin is a soluble protein mainly present in cytosol (Fig. 13.5). However, it is also present in nuclear compartments and can be attached to the plasma membrane. Its molecular mass is 15- 17 kDa. It is the primary calcium receptor in both plant and animal cells; however, there are some differences in the sequences of amino acids in the two cells. The characterized calmodulin-binding proteins in plants are myosin V, kinesin, NAD^+kinase, glutamate decarboxylase, protein kinases, and Ca^{2+}-ATPase. It is an acidic protein and consists of a flexible helix containing two loops at each end. The molecule has four Ca^{2+}-binding regions or loops. These loops contain 12 amino acids each, and are rich in aspartate and glutamate. It is termed as helix-loop-helix protein, due to the presence of two helices found on either side of the loops. Sometimes these loops are also known as EF hands, because these loops contain glutamate (E) and phenylalanine (F). When all of four EF hands are occupied by Ca^{2+}, conformational change takes place in this protein. When intracellular $[Ca^{2+}]$ rises to about 10^{-6}M (1μM), the binding of Ca^{2+} to calmodulin drives a conformational change in tertiary structure. Due to this conformational change, hydrophobic domain gets exposed. This patch is rich in metheonine, leucine, and phenylalanine. The target protein i.e., certain protein kinase (calmodulin-binding kinases, CBK) recognizes these patches and combine with the calmodulin. This results in activation of target proteins. Prominence of methionine in the calmodulin-binding region makes the interaction flexible. Ca^{2+}-calmodulin also binds to other proteins. Plant calmodulin is also important in the formation of actin filaments and cytoplasmic streaming, polarized growth, and the cell division cycle. Therefore, Ca^{2+}-calmodulin is an important member of signal transduction chain (Fig. 13.6).

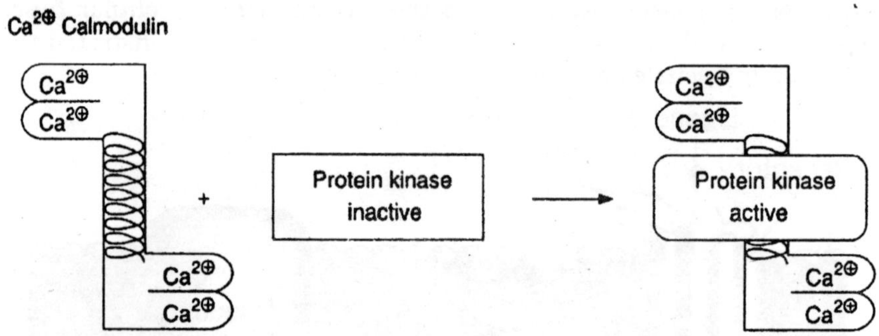

Fig. 13.5. Clamodulin protein

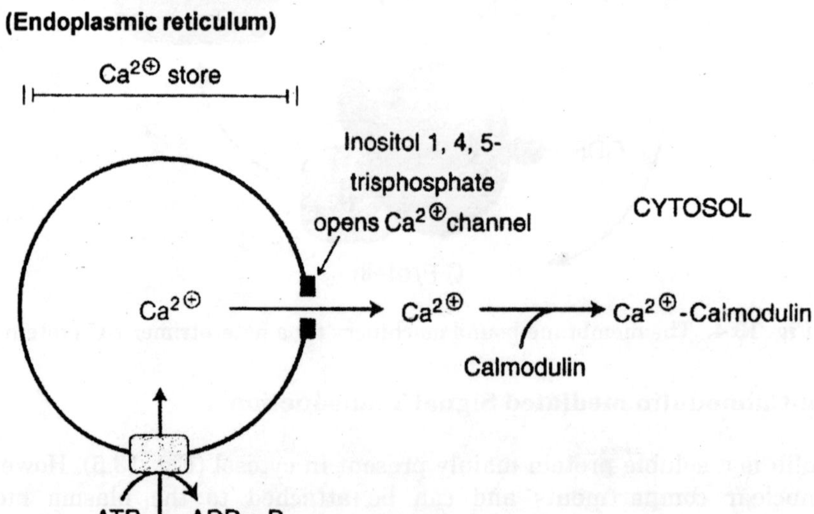

Fig. 13.6. Calcium-Calmodulin channel

Cyclic Nucleotides

Cyclic AMP was discovered by Sutherland in 1965. It is present in all animal tissues, fungi, and bacteria. In these organisms it acts commonly as a second messenger. However, its role in plant cell is still not clear. In animal cells, several different types of adenyl cyclase enzymes take part in synthesis of cAMP. Enzyme adenyl cyclase is a large protein, and is present in isoforms. Some of these isoforms are present in cytoplasm, while others are located in plasma membrane. Some forms are Ca^{2+} dependent. This enzyme is activated or inhibited by some signals, which are mediated by specific types of G-proteins. cAMP was recently reported in higher plants, with the sequence similar to adenyl cyclase. In plants cAMP is involved in some specific processes. Such processes includes, signals that initiate stomatal closure may be transduced through adenyl cyclase (Fig. 13.7). The activity of guard cell channels is modified by cAMP-dependent phosphorylation. The cloned plant potassium channel proteins also have a cAMP-binding region. cAMP may also regulate pollen tube growth. Its main function is to transmit primary hormonal message from the cell surface to the cell interior. Its role as secondary messenger in plants is still to be worked out.

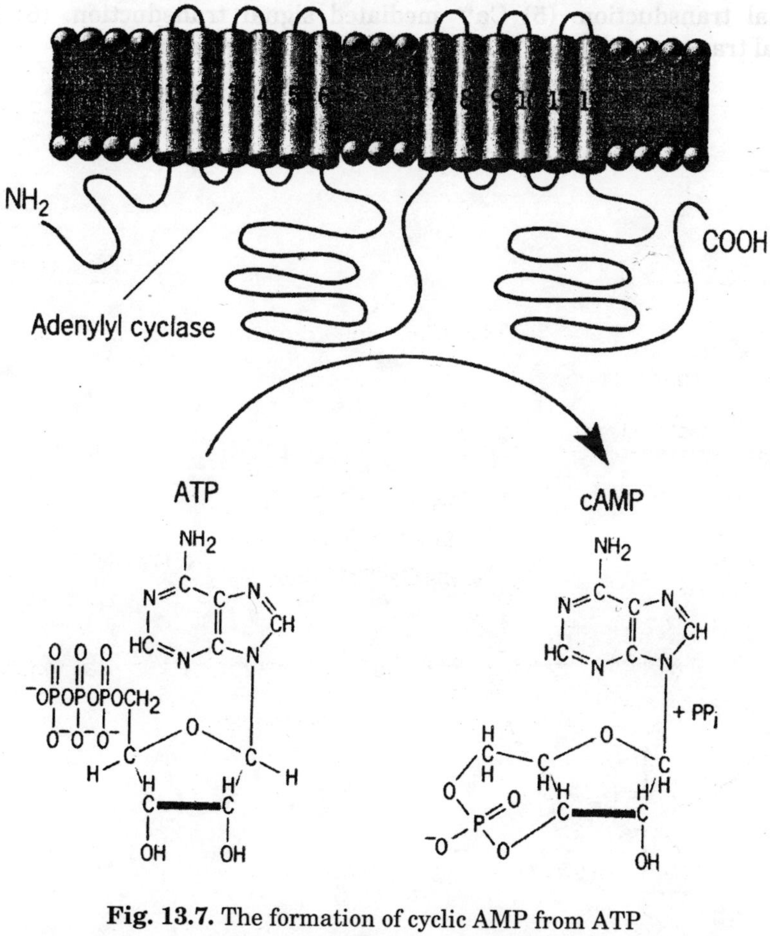

Fig. 13.7. The formation of cyclic AMP from ATP

Cyclic 3'-5' guanosine monophosphate. cGMP is commonly a second messengers found in all animal tissues and bacteria. The production of cGMP is increased by the activation of guanlyate cyclase, and GTP acts as a substrate. In plant cells, cGMP may participate in transducing signals of fungal invasion, red light signals mediated by phytochrome. It also takes part in gibberellin signals that regulate synthesis of aleurone amylase.

SUMMARY

Signal transduction is a complex process, involving the passage of information. It is defined as a phenomenon in which information is relayed across the plasma membrane to the cell interior and often to the cell nucleus. Specificity of signal transduction is achieved by precise molecular complementarity between the signal and receptor molecules, mediated by the same kinds of weak (noncovalent) forces. The receptors are proteins, which binds with a signal. A ligand is the agent that binds to the receptor at the outer cell surface. Plant cell has two principal elements in the signal transduction pathways. They are intracellular Ca^{2+} ($[Ca^{2+}]_I$) and protein kinases, enzymes that carry out phosphorylation. Different models are proposed for signal transduction viz. (1) Cyclic AMP mediated signal transuction. (2) Cyclic GMP mediated signal transduction. (3) G-protein mediated signal transduction. (4) Inositol mediated signal transduction. (5) Ca^{++} mediated signal transduction. (6) Protein kinase mediated signal transduction and (7) Ras mediated signal transduction.

Chapter - 14

PLANT GROWTH REGULATORS AND ELICITORS

A plant hormone is an organic compound (signal molecules) synthesized in one part of a plant and translocated to another part. At this site, in very low concentrations, it causes a physiological response. In 1930s a concept developed, which believed that growth substances were inductive agents mediating and controlling cell and tissue interactions and cell growth during development. Hormones are usually effective at internal concentrations of 1 mM or less. Plants have a variety of hormones with diverse structures and functions (Fig. 14.1). A brief account of types of phytohormones and their functions are as following.

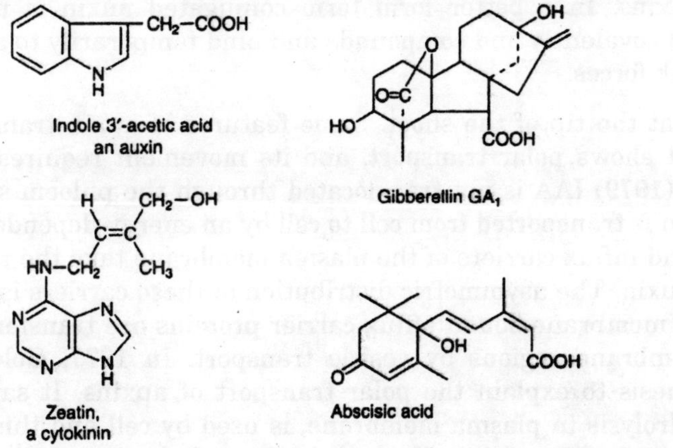

Fig. 14.1. The phytohormones.

Auxin (Indole Acetic Acid) generally found in apical meristems, derived from indole. It is associated with apical dominance and tropism. It generally influences cell division and differentiation, and stimulates shoot elongation, induction of cambial cell division, and root initiation. *Gibberellins* are derivatives of gibberellane, induce elongation growth of internodes. Cytokinin is a prenylated adenine and stimulates cell division. *Abscisic Acid* which is formed from carotenoid, regulates the water balance. The most universal response of cells to ABA is growth inhibition. *Ethylene* in an endogenous plant growth regulator and has major function in fruit ripening. *Jasmonic acid* a derivative of fatty acids and is involved in disease resistance. It is involved in reproductive development and enhances senescence. It is involved growth inhibition. *Brassinosteroids* is a steroid phytohormone with a key function in the regulation of cell development. Peptide hormones regulate plant development.

INDOLE-3-ACETIC ACID

Frits Went (1926) isolated a growth stimulating substance from the tip of oat seedlings and named it as `auxin'. Later this was identified as indoleacetic acid (IAA). Previously it was believed that IAA is the only auxin. In 1982, Wightman and Lighty, observed that two other compounds viz. 4-chloro-indoleacetic acid (4-chloroIAA) and phenylacetic acid (PAA) should be considered as auxin. Besides IAA, some other substances are known with auxin properties (e.g., phenylacetic acid). IAA concentrations are found in the main growth zones of the shoot. Another compound of auxin activity, indole-3-butyric acid has been found in a number of higher plants. It is used to induce root formation. Some compounds are synthesized artificially show auxin like properties. These are known as growth regulators. When the carboxyl group of IAA is combined covalently with other molecules to form derivatives it is known as bound auxins. In a better form term conjugated auxin is used. It is because conjugates are bound covalently and compounds and bind temporarily to auxins by hydrogen bonding or other weak forces.

Auxin is found at the tip of the shoot. Some features of auxin transport is that auxin movement is slow, it shows polar transport, and its movement requires metabolic energy. According to Jacobs (1979) IAA is not translocated through the phloem sieve tubes and nor through xylem. Auxin is transported from cell to cell by an energy-dependent polar transport. The Specific efflux and influx carriers of the plasma membrane take the responsibility of cell to cell transport of auxin. The asymmetric distribution of these carriers is the major cause of polar transport. The membrane-bound efflux carrier proteins are transferred in a reversible fashion between membrane regions by vesicle transport. In 1977, Goldsmith proposed a chemiosmotic hypothesis to explain the polar transport of auxins. It says that the energy released by ATP hydrolysis in plasma membrane, is used by cell and this energy pumps H^+ from cytosol into cell walls. Thus, the pH of cell walls is about 5.5, which keeps the carboxyl group of an auxin less-dissociated. The pH in the cytosol is near 7. Under such physiological conditions, the non-charged auxin move from the wall into the cytosol, where the higher pH causes their carboxyl group to dissociate and attain a negative charge. As the concentration of charged auxin builds up in the cytosol, its outward movement would be favoured more thermodynamically. It is assumed that some carrier in membrane transport charged auxins out toward the cell wall. The outward movement of auxin from the end of the cell opposite to that from where it enteres. At this site the low pH again causes more of them to become noncharged. It is interesting to understand that how charged auxins are transported out at the basal ends of cells. For this purpose, the cells should be polarized in such a way that it can absorb at one end and secrete at the opposite end. Naphthylthalamic acid (NPA) and 2,3,5-triodobenzoic acid (TIBA) are strong inhibitors of polar auxin transport. The blockage of transporter by NPA and blockage at the same site by TIBA explains the polar transport of auxin.

Different plants show different types of mode of synthesis of auxins. Tyrptophan acts as a precursor of IAA synthesis. Much more the precursors of tyrptophan can act as a precursor of IAA synthesis. The synthesis may take place by three different routes: the indole-3-pyruvic acid, the indole-3-acetaldoxmine, and the tryptamine pathways. In cabbage, IAA synthesis occurs by plasma membrane-bound enzymes, and L-tyrptophan is the precursor. The pathway proceeds via indole-3-acetaldoxime and indole-3-acetonitrile. Indole-3-acetonitrile is converted to IAA by the action of nitrilases. Major pathways of IAA synthesis proceed via indole-3-pyruvic acid and indole-3-acetaldehyde. Indole-3-ethanol and its conjugates are produced in a side shunt from indole-3-acetaldehyde. The intermediate of IAA synthesis from L-tryptophan via indole-3-pyruvic acid are: L-tryptophan → indole-3-pyruvic acid → indole-3-acetaldehyde → IAA. IAA biosynthesis independent of L-tryptophan has been obtained with the orange pericarp (*orp*), a tryptophan auxotrophic mutant of maize. Seeds of *orp* mutant contain increased concentrations of two L-tyrptophan precursors. These precursors are anthranilate and indole. The IAA precursor is probably an indole intermediate of tyrptophan biosynthesis. It is downstream from anthranilate and upstream from L-tryptophan. In *Arabidopsis*, *trp1, trp2,* and *trp3* mutants are reported to be tryptophan auxotrophs. The tryptophan dependent and tryptophan independent IAA synthesis varies among the plant species. It is interesting to note that such pathway may occur among different developmental stages in the same life cycle.

Philips (1971b), and Leshem (1973), Davis (1973) and Ray (1974) were the pioneers to explain the biochemical basis of auxin action. The evidence of role of auxin in cell growth can be noticed by two effects, when IAA is added to a cut tissue, after a few minutes, the rigid cell wall starts loosening. It was learned that growth promotion by auxins requires continued formation of both RNA and proteins. Rayle and Cleland (1979), Cleland (1980), Taiz (1984) reviewed a hypothesis, known as the acid-growth hypothesis. It states that auxins cause receptive cells to secrete H^+ into their surrounding primary walls and that these H^+ then lower the wall pH so that wall loosening and growth occur. By further studies, it was presumed that auxin activates an H^+-P-ATPase of the plasma membrane via a signal transduction chain and probably a protein kinase are involved. The cell wall region undergoes acidification due to the involvement of these proteins. As a result, those enzymes get activated which loosen the rigid cell wall and in this way enable a turgor-driven cell enlargement. 14.3.3 proteins act as regulatory proteins, and they have property to bind with a variety of proteins and change their activity. 14.3.3 proteins form a large family of multifunctional regulatory proteins, and its many isoforms are present in a single plant. These proteins bind to a specific binding site of the target protein with six amino acids, containing a serine phosphate in position 4. Thus, 14.3.3 proteins regulate in plants the activity of the H^+-P-ATPase of the plasma membrane. The second effect can be observed after one hour of treatment. At this stage, proteins required for growth are found to be synthesized. Thus, the effect of auxin on gene expression can be recognized.

IAA stimulates cell division in the cambium, enhances apical dominance by suppression of lateral bud growth, and controls embryo development. Low concentration of IAA promotes root elongation. However, a higher concentration inhibits it. There is ample evidence that auxins from stems strongly influence root initiation. Auxins also promote adventitious root development on stems. In most of the species, the apical bud exerts an inhibitory influence upon the lateral (axillary) buds, preventing or slowing their development. This phenomenon is known as apical dominance. Thimann and Skoog (1933, 1934) opined that auxin inhibited lateral buds. Jacobs et al., (1959) had controversial opinion. Libbert (1964) opined about the role of auxin in apical dominance. Hillman (1984) reviewed apical dominance. It is difficult to analyze hormone level and correlate it in bud growth inhibition. During early embryogenesis, auxin governs the formation of the main axis of polarity, with the shoot meristem at the top and the root meristem at the opposite pole.

IAA prevents the formation of an abscission layer on leaves and fruits and is thus an antagonistic to ethylene. Increased IAA concentration can induce the synthesis of ethylene.

Auxin induces fruit formation.

GIBBERELLINS

A rice plant disease caused by *Gibberella* was studied in Japan. Excessive elongation of shoot was observed in this disease. Eiichi Kurozawa (1926) isolated gibberellins from fungus *Gibberella fujikuroi*. He was investigating the causative agent of bakanae the foolish seedling disease of rice. This substance was found to induce unnatural growth. Brain and Hemming (1955) observed that when a few micrograms of GA_3 was added to dwarf pea plants gave greater elongation than GA_3 added to normal tall forms of pea. All gibberellins are based on the gibbane skeleton. Structural analysis revealed that gibberellin is a mixture of various substances with similar structures. Such structural compounds also occur in plants and act there as phytohormones. Radley (1956, 1958); Radley and Dear (1958), Phinney *et al.*, (1957) could perform bioassay. GA was found in extracts of higher plants. Gibberellins are a group of tetracyclic diterpenes. About 125 types of gibberellins have been identified in plants, and are numbered in the order of their identification. However, this numbering gives no information about structural relationships or functions. The difference is on the basis of the position of hydroxyl groups or saturation of double bonds in the ring. The GAs is divided into two groups, C_{20}-GAs and C_{19}-GAs, with 20 and 19 carbon atoms, respectively. The C_{20}-GAs have either δ-lactone ring or are characterized by a C-20 that exists as part of a methyl, hydroxymethyl, aldehyde, or carboxyl group. In C19-GAs one carbon atom is absent and they contain γ-lactone ring. The most important gibberellins are GA_1 and GA_4.

Gibberellins have more than one primary site of action. The simple role of gibberellins in shoot elongation has three contributing events. These are stimulation of cell division at shoot apex, hydrolysis of polysaccharides, and increase in cell wall plasticity. Gibberellins promote cell division because they stimulate cells in the G1 phase to enter the S phase and because they also shorten the S phase. The cell growth is promoted by gibberellins, as they increase hydrolysis of starch, fructans, and sucrose into glucose and fructose molecules. These monosaccharides are consumed in respiration and provide energy. This is also a cause of making the cell's water potential more negative. This results in inwards flow of water and cells get expanded.

The buds are the active site of gibberellin synthesis. In contrast, internode sections were not synthesizing GA. Immature seeds contain relatively high amounts of gibberellins compared with other plant parts. Jones and Phillips (1964) experimented with diffusion technique have showed that mature leaves have little ability to synthesize gibberellins. Roots also synthesize gibberellins in significant quantities. Gibberellins are derived from the hydrocarbon *ent*-gibeberellane. Many of these gibberellins are intermediates or by-products of the biosynthetic pathway, but all of them do not act as phytohormones. Isoprenoid geranylgeranyl pyrophosphate is the precursor, and *ent*-kaurene is an intermediate of GA_1 synthesis, the pathway completes in 13 steps. Isoprenoids are produced by a nonmevolonate, pyruvate/glyceraldehydes 3-phosphate pathway. The initial steps are the formation of isopentenyl diphosphate and its conversion to geranylgeranyl diphosphate. The synthesis takes place by terpenoid pathway. Enzyme terpene cyclases are involved in the synthesis of geranylgeranyl diphosphate and its conversion to *ent*-kaurene. The latter is synthesized by the two-step cyclization of geranylgeranyl, by way of the intermediate copalyl diphosphate. In these two steps, enzymes copalyl diphosphate synthase (CPS), and ent-kaurene synthase (KS) are required. ent-kaurene is oxidized to bioactive C_{19}-GAs by way of GA_{12}-aldehyde. The cytochrome P450 monooxygenase, and α-ketoglutarate-dependent dixoygenases catalyzes oxidation of GA_{12}-aldehyde. Conversion of GA_{12}-aldehyde to C_{19}-GAs is a multistep pathway. It proceeds via either the `early 13-hydroxylation pathway', or the `non-13-hydroxylation pathway'. Both sets of reactions include 7-oxidation, 13-hydroxylation, 20-oxidation, and 3β-hydroxylation steps. The differences in the two pathways are that early 13-hydroxylation pathway leads to GA_{20} and GA_1, while the non 13-hydroxylation pathway produces GA_9 and GA_4. C_{20}-GAs undergoes successive oxidation at C-20 that result in the loss of CO_2 and the formation of C_{19}-GAs. The oxidation is carried out by GA 20-oxidases. In some cases gibberellin synthesis is controlled by light via phytochrome. Spinach plants can be transferred from short day plant to long day plant. A spinach leaf GA 20-oxidase oxidizes both GA_{53} and GA_{19}. The activity of this enzyme is increased during long day and decreased in short day. The long day spinach plants also show enhanced ent-kaurene synthesis. The activity of GA_{53} oxidase and GA_{19} oxidase is regulated by photoperiod.

Gibberellin promotes wide-ranging growth of many intact plants. They play role in stimulation in shoot elongation, especially in the internode of the stems. Most dicots and some monocots respond by growing faster when treated with gibberellin. Only GA_1 controls stem elongation in maize, rice and peas. A distinct effect of gibberellin (GA_3) is in rosette plant is that it induces to shoot up for the formation of flowers, and they also regulate flowering. They are capable to bring about flowering in plants that ordinarily require a photoperiodic signal or cold treatment for flowering (vernalization). GAs can induce formation of cones in conifers. Michniewiez and Lang (1962) tested gibberellins 1-9, for blotting and flowering effects. They found GA_7 to be the most effective floral inducer. Preformation of fruits and the stimulation of their growth is an additional function of gibberellin. Gibberellins help to overcome bud dormancy especially in evergreen and deciduous trees and shrubs growing in temperate zone. It terminates seed dormancy, probably by softening of the seed coat, and facilitates seed germination by the expression of genes for necessary enzymes. Soon after seed germination, gibberellins convert the storage polymers (of seed) into sucrose and mobile amino acids and amides. Thus, GA stimulates the mobilization of foods and mineral elements in seed storage cells, which can easily move in the seedlings, especially at the age when it is unable to absorb mineral nutrients from soil. GA-induced de novo synthesis of α-amylase and other enzymes in the aleurone layer of barley. In grass seeds, including barley, gibberellins are synthesized in the scutellum an also in other parts of embryo. Scutellum is more important than the aleurone layer. It also secretes enzymes that cause digest endosperm reserves in several species.

CYTOKININS

Cytokinins stimulate cytokinesis; R. Horgan (1984) defined cytokinins as substances which, in the presence of optimal auxin, induce cell division in the tobacco pith or similar assay system grown on an optimally defined medium. Gottlieb Haberlandt discovered that a compound is present in vascular tissues of various plants. It stimulated cell division that caused cork cambium formation and wound healing in cut potato tubers. However, he could not identify this compound. Johannes van Overbeek observed that the milky endosperm from immature coconut contains compounds that promote cytokinesis. Folke Skoog and his coworkers while trying to investigate auxin-stimulation of plant growth in tissue cultures, observed that cells in pith sections from tobacco stems divided much more rapidly if a piece of vascular tissue was placed on top of the pith. They tried to identify this compound. Later it was found that an adenine-like compound in yeast extracts that was highly active. Carlos Miller (1954) discovered an active compound formed by partial breakdown of aged or autoclaved herring sperm DNA. It was named as kinetin. This compound was not found in plants. In D.S. Letham (1974) identified zeatin and zeatin riboside. Zeatin (Z) and its riboside [9R]Z are biologically active cytokinin. Cis isomers of Z {*cis*-Z, *cis*-[9R-5'P]Z and cis-[9R-5'-P]Z} are also naturally occurring cytokinins. Structurally they are adenine derivatives with an isopentenyl-based side chain attached to the N6 amino group. The amino group of adenine is linked with the hydroxylated isoprene residue in the *trans*-position. The side chain of cytokinins is an essential structural requirement for biological activity.

Cytokinins are adenine derivative, they have a basic structure of purine ring with a side-chain rich in carbon and hydrogen attached to the nitrogen protruding from the top of the ring. Each cytokinin can exist in the free-base form or as a nucleoside, in which a ribose group is attached to the 9' nitrogen atom. When the 5'-carbon of ribose is esterified with a phosphate, nucleoside is converted into nucleotide.

Zeatin and zeatin riboside are transported by xylem, but sieve tubes also contains cytokinins. Experiments with detached leaves showed that movement is probably through the phloem. Results from other experiments revealed that cytokinins are not readily distributed in the phloem. Young leaves, fruits, and seeds that are transport sinks do not willingly transport their cytokinins elsewhere via either xylem or phloem. It is tentatively concluded that except for delivery from roots via the xylem, transport of cytokinins within the shoot is rather limited.

Cytokinins have different mechanisms of action in different tissues. There is a common primary effect. The secondary effect may of different types, which depends upon the physiological state of the target cells. Cytokinins are present in very low concentration i.e., 0.01 to 1 ?M. Thus, it needs amplification at the time of initial effect. It is presumed that cytokinin has some promotive effect on RNA and enzyme formation. Stimulation of cell division is by increasing the transition of cells from G_2 to mitosis, and it is due to the increase in the rate of protein synthesis. Some of these proteins could be enzymes needed for mitosis. Cytokinins affect the translation stage of protein synthesis by increased polysome level. Salisbury and Ross studied the effect of cytokinins in the cell expansion of detached cotyledons. It was observed that when radish cotyledons are grown in weak light, cytokinins increase the internal production of reducing sugars. Glucose and fructose are formed in the form of reducing sugars, which act osmotically to cause the water uptake that drives growth. Cytokinins also loosens the cell wall, which results in faster expansion irreversibly under the existing turgor pressure. Wall loosening is much more important for growth promotion than enhanced turgor pressure. The loosening of wall do not takes place due to acidification. The enzymatic system responsible for loosening is yet to be discovered.

Cytokinin synthesis primarily takes place in the meristematic tissues. The principal mechanism by which these cytokinins are produced in plants is de novo biosynthesis. Zeatin is synthesized from AMP and dimethyallylpyrophosphate (DMAPP). In 1978 it was revealed that 5'-AMP is a direct precursor of [9R-5'P]iP. This conversion is catalyzed by dimethylallyl diphosphate:5'-AMP transferase (or isopentenyl transferase). DMAPP is the only acceptable donor for the side chain moiety. The next step is trans hydroxylation of the side chain catalyzed by *trans*-hydroxylase. In this reaction, [9R-5'P]iP is converted into [9R-5'P]Z. the penultimate reaction is dephoshporylation of [9R-5'P]Z to [9R]Z, catalyzed by 5'-nucleotidase. By the involvement of adenosine nucleosidase, [9R]Z is cleaved to form Z. in an alternative pathway; [9R-5'P]iP is dephsophorylated to yield [9R]iP. Ribose is removed from the latter to yield iP. Metabolism of [9R]iP and iP can then yield [9R]Z and Z, respectively. Some tRNAs contain nucleotides in which bases have been modified to form cytokinin molecules. Base modification of these nucleotides can generate cytokinin moieties. These cytokinins are mainly [9R]iP, *cis*-[9R]Z, and their 2-methylthio derivatives. Nucleotide cytokinins are dephosphorylated to nucleoside cytokinins. This yields free base cytokinin after the removal of ribose. Cytokinin oxidase degrades cytokinins. This enzyme system removes the five-carbon side-chain and releases free adenine.

Cytokinin in conjugation with auxin enhances cell division. A high cytokinin/auxin ratio promotes shoot production. Cytokinins increase cell expansion in dicot cotyledons and in leaves. The main cause of growth is due to water uptake that drives cell expansion, because the dry weight of the tissues does not increases. The role of cytokinins is to promote chloroplast development and chlorophyll synthesis. It is also known to induces opening of stomata, and suppress auxin-induced apical dominance. In dicots, cytokinins promote lateral bud development. Another important function of cytokinin is inhibition of senescence of plant organs, especially leaves. The delay of senescence by cytokinins is natural, and partially root-controlled. Cytokinins cause transport of many solutes from older parts of the leaf.

ABSCISIC ACID

In early 1950s Thomas Bennett-Clark and Ned Kefford, while studying endogenous auxins in plant extracts, discovered an acidic compound, which inhibited coleoptile elongation. This compound was referred as β-inhibitor complex. In 1960s plant physiologists isolated abscission-accelerating compound and dormancy inducing compounds. They were named as `abscisin II' and `dormin', respectively. In 1965 Frederick Addicott and his collegues reported the structure of abscisin II and called it as `abscisic acid' (ABA), this had the same property like dormin. ABA is an inducing factor of abscission of leaves and fruits. However, later it was recognized that the major function of this phytohormone is maintenance of water balance and induces seed and bud dormancy.

ABA can be transported by the transpiration stream via the xylem vessels from the roots to the leaves. The movement of ABA in parenchyma cells is similar to that of gibberellins.

The 1' carbon of ABA is chiral. Naturally occurring ABA is a (S)-enantiomer, while commercial available ABA is a racemic mixture. The 2-cis double bond of both enentiomers

[(S)-ABA and (R)-ABA] is isomerized by light to the 2-*trnas* isomers [(S)-2-*trans*-ABA and (R)-*trans*-ABA].

The effect of ABA depends on the tissue involved. There are three major effects (i) effect on the plasma membrane of roots, (ii) inhibition in transcription and (iii) inhibition in translation. Root membranes become more positively charged. This effect is probably involved in the rapid loss of K^+ ions from guard cells, and perhaps in the ability of ABA to rapidly inhibit auxin-induced growth.

ABA synthesis takes place where a water shortage is directly monitored, i.e., leaves and also in roots. It is a product of isoprenoid metabolism. A 15-carbon sesquiterpenoid is synthesized in chloroplasts and other plastids by the mevalonic acid pathway. 1'-deoxyABA was the immediate precursor of ABA in plants. However, results from gas chromatography have revealed that in water stressed leaves 1'-deoxyABA are not the immediate precursor of ABA. These results indicated that ABA was synthesized from a preformed precursor with oxygen atoms at the 1' and 4' positions. The synthesis of ABA proceeds via oxidation of violaxanthin. Robert Creelman and Jan Zeevaart (1984) reported that C_{40} xanthophylls are intermediates in the ABA biosynthesis pathway. Some concepts of ABA synthesis came with the studies of the ABA-deficient viviparous (vp) mutants of maize. It showed blockage at various points in the terpenoid and carotenoid biosynthesis pathways. Geranyl-geranyl diphosphate synthase catalyzes the three successive condensations with isopentenyl diphophate. It converts dimethylallyl diphosphate to the C_{40} compound geranyl-geranyl diphosphate. After metabolism of β-carotene to zeaxanthin, the latter is converted to all-*trans*-violaxanthin via antheraxanthin. This is completed by two epoxidation steps. All-*trans*-violaxanthin is converted to 9'-*cis*-neoxanthin via all-trans-neoxanthin, 9-*cis*-violaxanthin, or both. The oxidative cleavage of 9'-*cis*-neoxanthin yields the first C_{15} intermediate, xanthoxin. The one before the last reaction of the ABA biosynthesis pathway is conversion of xanthoxin to ABA-aldehyde. In this reaction, 4'-hydroxyl group is oxidized to a ketone group. Other changes in this reaction involve 2'-3' desaturation and opening of the 1'-2' epoxide ring. Oxidation of the 1-CHO group of ABA-aldehyde, catalyzed by ABA- aldehyde oxidase, leads to the formation of ABA.

As mentioned earlier, the role of ABA is maintenance of water balance. This is due to induction of nitric oxide (NO) at the closure of the stomata during water shortage. ABA also prevents vivipary.

ETHYLENE

The affects of ethylene was first established by Dimitry N. Neljubow. Ethylene is involved in the induction of senescence. Under this physiological condition, the degradation initiates. Protein is degraded to amino acids. These amino acids along with some ions (e.g., Mg^{2+}) move out via phloem and are reutilized by plants. The annuals reutilize these substances for seed formation, while the perennials store it in the stem, or in the roots. Ethylene production is observed in numerous mechanical and stress effects. Gentle rubbing of stem or leaf, increased pressure, pathogenic microorganisms, insects, water logging and drought are some condition were high ethylene production is observed.

Auxin greatly stimulates ethylene formation. High amount of ethylene is produced in seedling and shoot apex. Methionine is converted to S-adenosylmethionine (SAM), is the precursor for the synthesis of ethylene. This reaction needs ATP. SAM is involved in the biosynthesis of polyamines. It also participates in a wide range of methylation reaction. The enzyme aminocyclopropane carboxylate synthase (ACC synthase) form a cyclopropane. ACC synthase is present in several isoforms, and has seven highly conserved regions, out of which, the fifth includes a lysine residue that reacts with SAM and pyridoxal 5'-phosphate at the active site. In this reaction the positive charge of the sulphur atom in S-adenosylmethionine is cleavaged. Subsequently, ACC oxidase catalyzes the oxidation of the 1-aminocyclopropane-1-carboxylic acid (ACC) to ethylene, CO_2, HCN, and water. Biosynthesis of both ACC and polyamines involves incorporation of the aminopropyl group from SAM. HCN is immediately detoxified by conversion to β-cyanoalanine. The latter is further metabolized to asparagine or γ-glutamyl-β-cyanoalanine. During fruit ripening ACC oxidase activity was found markedly increased. At the time of senescence and ripening induced increase in ACC oxidase is due to increased transcription.

The effect of ethylene depends on the type of physiological phenomenon. An increased enzyme synthesis occurs in the target tissues. When ethylene is involved in abscission, it increases synthesis of cellulase and other cell wall degrading enzyme in abscission layer. If a tissue is wounded, due to the effect of ethylene, phenylalanine ammonia lyase appears. The latter is an important enzyme in formation of phenolic compounds, which are involved in wound healing. A major function of ethylene is in fruit ripening, and necessary enzymes are produced in fruit cells during ripening. Plants metabolize small amount of ethylene and a small amount is bound. As a result of metabolism, CO_2, ethylene oxide, and ethylene glycol are produced. It is concluded that there is a correlation between metabolite production and binding activity. Thus binding of ethylene oxidation of metabolites, and physiological action occur in a dependent sequence.

BRASSINOSTEROIDS

Brassinosteroids (BRs) are steroid phytohormones, which were isolated from pollen and regulate plant development. More than 40 BRs have been identified. The most effective member of the brassinosteroids is brassinolide (C_{28}-BR). It is reported in algae, ferns, gymnosperms and angiosperms but not in microorganisms.

Brassinosteroids were first isolated from bee-collected pollen grains of *Brassica napus*. The chemical structure of BRs is C_{27}-, C_{28}-, and C_{29}- steroid with different substituents on the A and B rings and side chain.

Plant sterols are synthesized from cycloartenol. A characteristic feature of plant sterols is alkylation at C-24 with C_1 or C_2 substituents. The brassinolide synthesis pathway was studied in the cultured cells of *Catharanthus roseus*. Campesterol (C_{28}) is the precursor of BRs. Campesterol is converted to 5α-campestanol via three intermediates. By the C-6 hydroxylation to 5α-campestanol, 6α-hydroxycampestanol is synthesized. The latter is oxidized to yield 6-oxocampestanol. It is still not established, that how 6-oxocampestanol undergoes C-22 hydroxylation and forms cathasterone. Cathasterone is hydroxylated at C-23 to form teasterone. Teasterone is metabolized to typhasterol by way of 3-dehdroteasterone. As a result of successive oxidation and reduction 3α-hydroxyl is epimerized to a 3β-hydroxyl group. Typhasterol may be converted back to teasterone by way of 3-dehydroteasterone. 6-Deoxybrassinosteroids (6-deoxy-BRs) is present in plant tissues in high concentration. It lacks oxygen at B ring. 6-deoxocastasterone is converted to castasterone and brassinolide in

Catharnathus cells. It also synthesizes 6-deoxocastasterone from 6-deoxoteasterone by way of 3-dehydro-6-deoxocastasterone. The biochemical pathway of synthesis of castasterone from 5α-campestanol is known as late C-6 oxidation pathway. Studies on mutant of *Arabidopsis* and *Pisum sativum* have established the existence of this alternative pathway to brassinolide biosynthesis. At the end of this pathway, the 6-oxo group of castasterone is oxidized to a lactone (6-oxo-4-oxa), yielding brassinolide.

Studies carried on mutants of *Arabidopsis, Pisum sativum* and tomato enabled to understand the function of brassinosteroids. The mutant studied had developmental defects like dwarf growth, reduced apical dominance and lowered fertility. BR biosynthesis and sensitivity indicates that BRs biosynthesis and sensitivity indicates that BRs promote the growth or elongation of plant cells. The possible role of BRs in photomorphogenesis was studied in Arabidopsis mutants, grown in light and dark conditions. However, it is still not clear about the involvement of BRs in photomorphogenesis.

Brassinosteroids stimulate shoot growth, folding of leaves, and differentiation of the xylem. They retard root growth, and the formation of anthocyan. Application of BRs induces proton pump activation, reorientation of cellulose microfibrils, xylogenesis, and enhanced ethylene production.

JASMONIC ACID

Jasmonic acid (JA) and other related compounds (methyl jasmonate) are considered as growth inhibitors and involved in defense responses. 12-O-β-glucoside of tuberonic acid (12-hydroxyjasmonic acid) was characterized as a potato tuber inducing factor. It was isolated from a fungus *Botryodiploidia theobromae*. JA is derived from fatty acid. The (+)-7-isojasmonate, and methyl (+)-jasmonate are stereoisomers of (-)-JA and (-)-methyl jasmonate, respectively. They are active agents in plants.

A C_{18} poly-unsaturated fatty acid α-linolenic acid is precursor of JA. The former is synthesized from membranes. The pattern of reactions and enzymes involved in the synthesis of α-linolenic acid show similarities with lipase like enzymes involved in fatty acid metabolism in mammals. The key enzymes involved are lipoxygenase, allene oxide synthase (AOS), and allene oxide cyclase. Enzyme lipoxygenase catalyzes oxidation of α-linolenic acid to 13(S)-hydroperoxylinolenic acid. AOS converts 13(S)-hydroperoxylinolenic acid to 12,13(S)-epoxylinolenic acid. In the next step enzyme allene oxide cyclase cyclizes 12,13(S)-epoxylinolenic acid and 12-oxo-*cis*-10,15-phytodienoic acid (12-oxo-PDA) is formed. It is a cyclopentenone derivative. As a reduction, 12-oxo-PDA is converted to 3-oxo-2 (*cis*-2'-pentenyl)cylcopentane-1-octanoic acid. This reaction is catalyzed by enzyme 12-oxo-PDA-reductase. 3-oxo-2(*cis*-2'-pentenyl)cylcopentane-1-octanoic acid is converted to (+)-7-isojasmonic acid by three rounds of β-oxidation. (+)-7-isojasmonic acid is converted to (-)-JA. However, the details of enzymes involved are not clear. (-)-JA is converted to methyl (-)-jasmonate.

Application of (-)-Jasmonic acid to leaves decreases expression of the nuclear and chloroplast genes involved in photosynthesis. (-)-JA induces fruit ripening and pigment formation. (-)-JA is present in high concentrations developing reproductive tissues of plants. Thus, it can be concluded that it has significant role in the formation of flowers, fruit, and seeds. The most important role of JA is in plant's resistance to insects and disease. Insuction of protease inhibitors appear to be targeted primarily toward certain insects. It is toxic to insects because they impair their digestion of proteins. Accumulation of (-)-JA is observed in wounded plants and also in plants treated with pathogen-derived elicitors (oilgosaccharides) or the peptide systemin. (-)-JA activates expression of antifungal proteins (osmotin and thionin). It induces the phytoalexin-related enzymes chalcone synthase, phenylalanine ammonia lyase, and hydroxymethylglutaryl-CoA reductase.

POLYAMINES

Polyamines are essential for growth and development of cells. They are much more abundant in plants as compared to GAs and cytokinins. Putrescine, spermidine, and spermine are polyamines, frequently found in plants. Polyamines stimulate many reactions involved in the synthesis of DNA, RNA, and proteins. They are polycatonic in nature. Thus, they have high affinity for anionic constituents like DNA, RNA, phospholipids, and acidic proteins. They also show affinity for anionic groups in membranes and cell walls. There is limitation in translocation of polyamines due to its polycationic character.

Polyamines contain two or more amino groups, including the amino acids lysine and arginine. The chemical formulas of some polyamines are as following:

Putrescine	$(NH_2(CH_2)_4NH_2)$
Spermidine	$(NH_2(CH_2)_3NH(CH_2)_4NH_2)$
spermine	$(NH_2(CH_2)_3NH(CH_2)_4NH(CH_2)_3NH_2)$
Cadaverine	$(NH_2(CH_2)_5NH_2)$

The mechanism of action polyamines is not well understood. It is believed that their positively charged amino groups cause them to combine with negatively charged phosphate groups in DNA and RNA in the nucleus and ribosome. It is due to this combination transcription and translation is increased. Polyamines stabilize membranes, which may be due to direct combination with proteins or phospholipids in the membranes. It still needs confirmation.

L-arginine is a precursor of putrescine. The synthesis may occur by two routes. Enzyme arginine decarboxylase catalyzes conversion of l-arginine to agmatine. Enzyme ornithine decarboxylase converts L-ornithin to putrescine. Ornithine decarboxylase is less active than arginine decarboxylase. There is usually a positive correlation between ornithine decarboxylase activity and cell division. Conversion of putrescine to spermidine is thought to be important in controlling the rate of cell division. Spermidine synthase catalyzes the conversion of putrescine to spermidine, and spermine synthetase regulates the conversion of spermidine to spermine. These synthase enzymes transfer an aminopropyl group from decarboxylated S-adenosyl-L-methionine (SAM) to their respective substrates. L-lysine is precursor of cadaverine synthesis. Enzyme lysine decarboxylase is involved. S-adenosyl-L-methionine decarboxylase catalyzes reaction in which SAM is converted to dSAM. Polyamindes are derived from SAM.

In true sense, polyamines are not hormones, but they appear to participate in several key metabolic pathways Polyamine are elicitors of diverse physiological responses. These processes are cell division, tuber formation, root initiation, embryogenesis, flower development, and fruit ripening.

SALICYLIC ACID

Salicylic acid (SA) has a key role in disease resistance. It can retard senescence of petals and induce flowering, it regulates thermogenesis. Thermogenesis is a unique phenomenon in plant mitochondria. At this stage much of the electron flow is diverted from the cytochrome respiration pathway to the cyanide-insensitive nonphosphorylating electron transport pathway. In this alternative pathway, the energy released in the form of heat. According to Adriaan van Herk (1937) in the voodoo lily thermogenesis was initiated by a water-soluble substance "calorigen". This compound is produced in the staminate flower primordia positioned between the upper and lower spadix. This substance was later identified as salicylic acid..

Biosynthesis of salicylic acid was studied in tobacco. Increased level in free and conjugated SA is observed in tobacco inoculated with TMV. SA is synthesized from trans-cinnamic acid, involving a side branch of the phyenylpropanoid pathway. During biosynthesis, *trans*-cinnamic acid to benzoic acid, the latter activates a putative NADPH-dependent cytochrome P450 monooxygenase. 2-hydroxylation of benzoic acid forms SA. Benzoic acid 2-hydoxylase catalyzes this reaction. SA is further converted to salicylic acid-2-O-β-glucoside. It is a SA conjugate that accumulate after TMV infection.

A major role of SA is its involvement in plant disease resistance. The mechanism may be understood by the fact that plants restrict the spread of pathogen by producing necrotic lesions around the initial point of penetration. Five or more families of serologically distinct pathogenesis-related (PR) proteins are synthesized during SA acquired resistance. Two of these families encode the hydrolytic β-1,3-glucanases and chitinases. The expression of other families is not yet been identified, but they are all associated with disease resistance for a large number of pathogen.

Phytohormones and Signal Transduction Chains

Auxin binding protein may function as an auxin receptor. These unknown auxin signal transduction chains are probably so-called Aux/IAA-proteins. The Aux/IAA-proteins are repressor proteins, which bind to further regulator proteins. These proteins are so-called auxin responsive factors (ARF). In *Arabidopsis* 25 different genes have been identified which encode for Aux/IAA-proteins, and 20 genes encode for ARF through signal transduction chain. Auxin causes conjugation of Aux/IAA-repressor proteins with ubiquitin, whereby the repressors are labeled for degradation through the proteasome machinery. In plants ubiquitin are found as highly conserved protein with identical sequence of 76 amino acids. The carboxyl end of the molecule contains a glycine residue. This end is exposed to the outside. The lysine residue of the target protein forms a so-called iso-peptide link between the carboxyl group of the glycine residue of ubiquitin. The degradation of target proteins through ubiquitin takes place by three different enzymes. The ubiquitin-activating enzyme (E1), ubiquitin-conjugating enzymes (E2), and ubiquitin-protein ligase (E3). Proteasomes are multienzyme complex and have a highly conserved machinery for controlled protein degradation.

Gibberellin receptors are present in the plasma membrane. However, the nature of such receptors is unknown. Gibberellins influence gene expression. The signal transduction from the receptor to gene function probably involves a heterotrimeric G-protein, cGMP, Ca^{2+}, calmodulin, and a cascade of protein kinases. The latter is possibly of the MAPK and CDPK types. They modulate the transcription factors for various genes. One effect of GA is to reduce the action of repressor proteins, known as DELLA proteins, which suppress growth. It is due to GA-dependent protein kinase cascades; DELLA proteins are phosphorylated and thus labeled for proteolytic degradation by proteasome machinery as described in auxin section.

Cytokinin receptors are dimeric histidine kinases. As the name suggests, the latter has two histedine residues and two aspartyl residues. The location of this receptor in the plasma membrane shows that the receptor site is directed to the extracellular compartment and the kinase is directed to the cytoplasm. When cytokinin binds, the two histidine kinases phosphorylate their histidine residues reciprocally. This mechanism is known as autophosphorylation. Subsequently, the phosphate groups are transferred to histidine residues or aspartyl residues of transmitter proteins. The latter passes into the nucleus and functions as transcription factors and thus regulate the expression of many genes.

The signal transduction chain of the ABA-action engages G-proteins, protein kinases, protein phosphatases. The synthesis of the messenger substance cyclic ADP-ribose (cADPR) is also involved. Yet in another signal transduction chain of ABA action, Ca^{2+} ions are released through phosphoinositol pathway. This is mediated by a phospholipase C. The rise in the cytosolic Ca^{2+} level, results in the activation of ion channel involved in the opening of stomata. Increase in Ca^{2+} level may be due to cADPR or phosphoinositol triphosphate. The protein kinases cascades modulate transcription factors.

The ethylene receptor is expected to have a very high affinity. It consists of a dimmer of histidine receptor kinase, after autophosphorylation, phosphate group is transferred to histidine residues or aspartyl residue of target proteins. Copper-cofactor takes part in binding of ethylene to the receptor dimer. At this stage, kinase is inactivated and autophosphorylation is prevented. Signal transmission is governed on the phosphorylation state of the ethylene receptor. The transmission takes place via protein kinase cascades, in which MAPKK and MAPK participate to transcription factors, which control the expression of certain genes.

In plants, brassinolide receptor interacts with a receptor-like kinase (RLK). These receptors may be identical to RLK. It is possibly connected to a MAP kinase cascade, regulating transcription factors.

SA acts as the primary long-distance signal that is translocated in the phloem. SA moves from the site of infection to uninfected parts of the plant. In the healthy tissues, it initiates the development of the systemic acquired resistance response. Endogenous SA plays a key role in the signal transduction pathway that leads to activation of genes encoding PR proteins. Thus, hypersensitive response and systemic acquired resistance is established.

Application of Phytohormones in Agriculture

Investigation of plant genome sequence and the associated genomic technologies will enable to understand the genes and enzymes involved in the regulation of plant hormone homeostasis. For proper application of phytohormones it is essential to understand the perceptive of different hormones, their interactions, etc. A proper knowledge about how the quantities of phytohormones are affected by environmental factors such as photoperiod, temperature, water and nutrient supply. All this knowledge will enable to establish a model system, which will be necessary to transfer the understanding to crop plants to facilitate improvement of traits of agricultural and economic importance.

Some compounds synthesized by chemist cause many physiological responses. These are known as growth regulators. Naphthaleneacetic acid (NAA), indolebutyric acid (IBA), 2,4-dichlorophenoxyacetic acid (2,4-D), 2,4,5-trichlorophenoxyacetic acid (2,4,5-T), and 2-methyl-4-chlorophenoxyacetic acid (MCPA) are the best known growth regulators. Synthetic auxins such as 2,4-dichlorophenoxyacetic acid (2,4-D) and naphthalene-1-acetic acid are used extensively in horticulture to induce rooting. It also promotes the set and development of fruit.

2,4-D, NAA, and certain other related compounds are effective herbicides. Four auxin herbicides, 2,4-D, 2,4,5-T, MCPA and derivatives of picolinic acid are more effective on dicots than on monocots. This selectivity towards broad leaves has proved beneficial to kill broad-leaf dicot weeds in cereal grains. The herbicides are formulated as salts of weak bases such as ammonia (amines), as emulsifiable acids, or as esters, and are mixed with an oil or detergent to facilitate spreading and penetration of the waxy cuticle after spraying on the foliage.

Application of gibberellins in cereal crops and a dramatic yield gave birth to green revolution. Retardants (growth inhibitors) are inhibitors of gibberellin biosynthesis. If such compounds are sprayed on cereal fields it decreases the growth of the stalks. This enhances the strength of the cereals stalk at the same time increases the proportion of total biomass in seeds. GA is also applied for the production of long, seedless grapes, which are developed by parthenocarpy. GA_1 causes extension of the cells.

SUMMARY

A plant hormone is an organic compound (signal molecules) synthesized in one part of a plant and translocated to another part where, in very low concentrations, it causes a physiological response. Auxin (Indole Acetic Acid) generally found in apical meristems, derived from indole, it is associated with apical dominance and tropism. It generally influences cell division and differentiation, and stimulates shoot elongation, induction of cambial cell division, and root initiation. Gibberellins derivatives of gibberellane, induce elongation growth of internodes. Cytokinin is a prenylated adenine and stimulates cell division. Abscisic Acid which is formed from carotenoids, regulates the water balance. The most universal response of cells to ABA is growth inhibition. Ethylene in an endogenous plant growth regulator and has major function in fruit ripening. Jasmonic acid a derivative of fatty acids and is involved in disease resistance. It is involved in reproductive development and enhances senescence. It is involved growth inhibition. Brassinosteroids is steroid phytohormone with a key function in the regulation of cell development. Peptide hormones regulate plant development.

Chapter - 15

PHOTOMORPHOGENESIS

Light is an important factor, which controls plant growth, development from germination to the formation of flowers. The control of morphogenesis by light is called photomorphogenesis. Photomorphogenesis of higher terrestrial plants usually dependents on light of long waves length. A surplus of light like the strong UV radiation in high mountains causes drastically reduced elongation of the internodes, a decrease of the assimilating surface, often to a strongly enhanced production of anthocyanes and usually also to intensely coloured flowers. A plant development that is impaired by the absence of light is called etiolation, the plants are known as etiolated.

As a start in understanding, it is important to realize a difference between the action of light in causing photomorphogensis and in causing photosynthesis. In photosynthesis, light in low doses sometimes acts like a trigger to initiate a developmental process that depends upon photosynthetic products for its completion. In other examples, light acts as a modulator. Some light induced reactions show inhibition and increase in some physiological phenomenon. Light shows increase in (i) surface area of cotyledons (ii) negative geotropic reaction of the hypocotyls (iii) synthesis of chlorophyll (iv) synthesis of carotenoids (v) RNA and protein synthesis in cotyledons, while light inhibits (i) elongation of the hypocotys, and translocation from the cotyledons. Light induces development of hypocotyyls hook and hairs at the hypocotyls, development of primary leaf and mature leaf primordia, and xylem elements. It intensifies storage fat and protein breakdown.

Phytochrome

A phytochrome is a photoreceptor, a pigment that plants use to detect light. It absorb light (red and far-red) and effective in causing photomorphogenesis. Many flowering plants use it to regulate the time of flowering. It is present in most organs of photosynthetic organisms. Phytochromes are involved when light initiates the germination of seed up to the greening of the seedling and in the adaptation of the photosynthetic apparatus of leaves to full sunlight or shade. Phytochrome is present throughout the cytoplasm and is also associated with or is within the plasma membrane and the membranes of chloroplasts.

Phytochromes have two isoforms Pr and Pfr, which absorb light at 660nm and 730 nm, respectively.

The phytochrome pigment was discovered by Sterling Hendricks and Harry Borthwick. The pioneer studies on phytochrome were carried out by Harry A. Borthwick, Briggs, Sterling B. Hendricks. W.W. Garner and H.A. Allard were first to observe about the affect of duration of light and dark period in flowering. Cocklebur is a short day plant. If the dark period is interrupted by light, no flowering takes place. It was further observed that red light proved much more effective than the other wavelengths. This effect was not only for interrupting the long nights that proved cockleburs and Biloxi soybean to flower but also exhibited expansion of pea leaves. In certain long day plants, when dark period was interrupted by red light, stimulated flowering. Borthwick and Hendricks observed relation of red light in germination of some dormant seeds. They prepared a spectrograph and obtained an action spectrum with a peak in the red for promotion of seed germination of Grand Rapids lettuce seeds. In few other experiments it was realized that a blue pigment absorbed red light. In 1959 Warren Butler (a biophysicist) and Harold W.Siegeliman (a biochemist) identified phytochrome pigment using a spectrophotometer. During the sixties and seventies phytochrome was purified from several plants and photo reversibility of this pigment was demonstrated by many workers not only *in vivo* but also *in vitro*. Peter Quail and Clark Lagarias (1983) reported the chemical purification of the phytochrome molecule. Hershey and Quail (1987) reported the chemical purification of the phytochrome molecule.

Phytochrome is produced in darkness in the form of Pr, and is accumulated until a certain level is reached. P_r is converted to P_{fr} on exposure of light of the wave length of 660 nm. At wavelength, 730 nm P_{fr}; is reconverted to P_r. the transformation of P_{fr} after exposure to red light is a fast process. P_{fr} is biologically active, while Pr is inactive form. The ratio of P to the total amount of phytochrome in both forms is denoted by $\varphi$and is expressed as following formula

$$\varphi = P_{fr}/P_r + P_{fr} = P_{fr}/P_{total}$$

In far-red region (above 720 nm) there is no P_{fr}, thus φ is zero, while at 667 nm of red light, 80% of phytochrome is converted into P_{fr}, hence φ is 0.8. In bright sunlight, where the red component is stronger than the far-red component, the value of φ is 0.6.

Phytochromes are soluble dimeric proteins with a prosthetic group bilin chromophore, usually called phytochromobilin. It is closely related to phycobilin and to the bile pigment bilirubin. The monomer consists of an apoprotein (Fig. 15.1). The molecular weight of phytochrome varies from 120,000 to 127,000g mol^{-1}. The monomer has a sulfydryl group of a cytseine residue linked to a prosthetic group, functioning as a chromophore. It is an open chain tetrapyrrole, attached to each monomer in the N-terminal part. It absorbs light and causes phytochrome responses. The autocatalytic binding of the tetrapyrrole to the apoprotein results in the formation of a phytochrome P_r (r=red). Phytochromes have absorption maxima in red wavelengths at 660 nm for the red-absorbing bright blue form (P_r) and at 730 nm for the far-red-absorbing green form (Pfr). The absorption of 660 nm wavelength of light, results in a changes from the *trans*- to the *cis*-configuration. This results in conformational change of the protein. The P_{fr} (far red) is the changed and active form of P_r. When P_r is converted to P_{fr} by red light; conjugation of double bonds in the tetrapyrrole is altered. The conjugated double-bond system of P_{fr} has one less double bond than that of P_r.

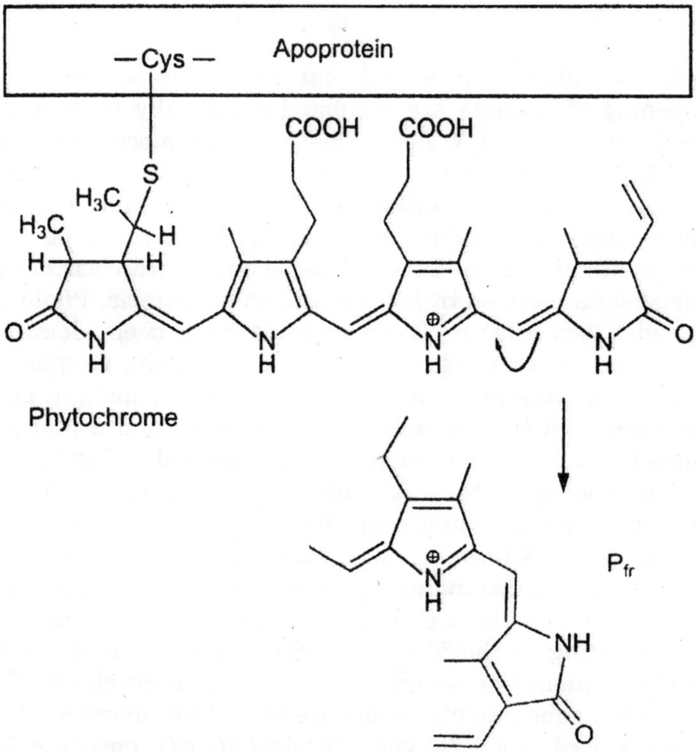

Fig. 15.1. The chromophore of phytochrome

Application of recombinant DNA technology and analysis of genomic libraries of several plants has proved one especially interesting outcome that there are multiple phytochrome genes in higher plants as against one or two lower plants (Mathews and Sharrock, 1997). *Arabidopsis thaliana* is a model plant. Five types of phytochrome (*PHYA* through *PHYE*) have been studied in this plant. The N-terminal region is by and large similar among them as judged from the reduced amino acid sequences, but the C-terminal ends are somewhat variable. Phytochrome A monitors light in the far-red region, whereas phytochrome B monitors red light. The phytochrome C, D, and E fulfill special functions. The details about these functions are still not clear. The amino acid sequences of PHYA, PHYB, and PHYC are equally divergent and demonstrate about 50% homology to each other.

Several forms of phytochrome response are classified as the high irradiance reaction (HIR) and the induction reaction. This is based on the influence and timing of irradiation. Studies on absorption spectrum of phytochrome pigments and action spectrum of various physiological responses have concluded that phytochromes causes some physiological responses as is evident by the similarities of the absorption spectra of phytochrome and action spectra for plant responses. Another significant evidence is that the responses caused by red are almost always nullified by an immediate subsequent exposure to far-red. Very low irradiance levels of either red or far-red are capable of interconverting phytochrome from one form to another also causes these responses. Light absorbed by chlorophyll alters both the action spectra for phytochrome responses caused by P_{fr} and the amount of light required for the response. This required light is absorbed by P_r.

Cryptochrome

A pigment exists that absorb violet and blue light, it was named as cryptochrome. Phototropism and opening of stomata are regulated specifically by blue light. Plants also have photoreceptors for blue and UV light. The name cryptochrome was coined in the seventies because of the cryptic (hidden) nature of the pigment responsible for eliciting blue light mediated responses. Wald and DuBuy (1934) proposed that the chromophore was probably a carotenoid. According to Galston (1949) a flavin could be the photoreceptor. Cryptochrome has sequence similar to DNA photo-lyases, a rare class of flavoproteins that catalyze blue light-dependent electron transfer reactions in bacteria. Photo-lyase is known to bind both a flavin and a deazaflavin. Three proteins have been identified as blue light receptors, viz. cryptochrome 1, cryptochrome 2, and phototropin. Cryptochrome contains a flavin and a pterin, while phototropin contains one flavin as a blue light absorbing pigment. Malhotra *et al.*,(1995) observed the presence of both a flavin and a pterin chromophore in *Sinapis alba*. Koornneef *et al.*, (1980) isolated a long-hypocotyl mutant (hy4) of model plant *Arabidopsis*, which was insensitive to B light for hypocotyl growth inhibition. Ahmad and Cashmore (1993) worked on this mutant and identified *CRY1* gene encoding for the blue light photoreceptors with conserved N-termini, but displaying variations at the C-terminal end. This gene encodes a protein with 681 amino acid sequence and molecular weight of 75.8 kDa. Thus, cryptochrome is encoded by the *CRY1*gene. It is a member of a family of receptors. The most crucial evidence relating to the identity of chromophores associated with the *CRY1* apoprotein has come by sequence homology search in protein database. The N-terminus of the protein encoded by this gene has high sequence identity to microbial DNA photo-lyases. Lin *et al.*, (1996) discovered a *CRY2* gene in *Arabidopsis* encoding a protein whose N-terminus region has high homology with that of *CRY1*, but not at C-terminus. CRY2 protein is very light labile. Huala *et al.*,(1997) observed the existence of two special so-called LOV domains, at the N-terminal end of NPH1. It has been known to bind flavins as also sense redox and voltage changes in proteins of several organisms.

Cryptochromes may also modulate kinase activity. Evidences from some physiological studies like blue light induced opening of anion channels, and proton extrusion from guard cells has led to understand on how cryptochrome work. Khurana and Poff (1989), Konjevic *et al.*,(1992) obtained two types of mutants of *Arabidopsis* (i) the `null' mutant that seemed to lack phototropic activity totally towards unilateral blue radiation (the JK218 mutant) (ii) mutants in which phototropic activity was altered partially (JK224). Reymond *et al.*,(1992) made some basic observation employing the wild-type *Arabidopsis* and the JK224 mutant. In wild type a 120 kDa protein in the plasma membrane fraction from the hypocotyl of wild-type plants underwent phosphorylation by blue irradiation. However, the mutant JK224 was considerably impaired. Further, the addition of irradiated membrane fraction to non-irradiated fraction brought about a higher degree of phosphorylation than with irradiated membranes alone. This suggests that not only the 120 kDa protein underwent autophosphorylation, but phosphorylation was caused by similar proteins-non-irradiated-indicating that the 120 kDa protein was a kinase capable of phosphorylating other substrates. Liscum and Briggs (1995) prepared a phototropic mutant (*nph*) of *Arabidopsis* by bombardment with fast neutrons. The gene coding for putative kinase has been cloned employing this mutant. The mutant *nph*1 is allelic to JK224 and lacks the functional 120 kDa protein. Mapping studies have shown that *NPH*1 gene is located on chromosome III. This gene apparently codes for a protein of 996 amino acid residues, as deduced from the cDNA sequence. Comparison of the cDNA sequence with GeneBank sequences and other databases shows that the end corresponding to the C-terminus of the coded protein is similar to that of a serine-threonine protein kinase. The latter has 11 sequences and belongs to PVPK family. Studies were carried out to understand that how does *NPH*1 code for a light-regulated kinase. Ahmad *et al.*,(1998) observed that in the *cry*1/*cry*2 double mutant, there is neither any phototropic response nor phosphorylation of the 120 kDa protein. However, in either a single mutant, there was not much effect on phototropism. This implied that the energy of a photon captured by the pterin or the flavin moiety in *cry*1or *cry*2 it transferred to the kinase bringing about conformational change in it. The LOV domains as observed by Huala, exhibits the existence of a flavin binding domain in the 120 kDa *NPH*1 protein could facilitate intermolecular transfer of signal between the two molecules, one of which could be considered primarily a photoreceptor and the other primarily a kinase.

The Chimeric Photoreceptors

A chimeric receptor means functions of the two photoreceptors are combined in one molecule. In *Adiantum* chimeric sensory photoreceptors were discovered. Nozue et al., (1998) suggested that this novel phytochrome can mediate actions brought about by both blue and red/far red receptors. The N-terminal end is like true phytochromes, but the C-terminal is almost identical to the full-length NPH1 homologue. A mutant of this plant is nonphototrophic hypocotyls 1, called *nph*1. It fails to show a phototropic response in blue light. Protein kinase activity of NPH1 was established by sequence analysis and biochemical studies. LOV1 and LOV2 are specific amino acid domains of NPH1. They share similarity with proteins from bacteria and eukaryotes. Light, oxygen, or voltage regulates the activity of LOV (light, oxygen and voltage sensitive) proteins, which participate in redox sensing.

Purple bacteria (*Rhodospirillum centenum*) and slime molds, have photoreceptors (a phytochrome) with the central region where the bilin chromophore is normally attached and the histidine-like C-terminal region as in the blue-green algae. A p-hydroxycinnamic acid moiety is covalently attached at N-terminus sequence. It absorbs blue light instead of red light. It is a chimeric receptor, where functions of the two photoreceptors are combined in one molecule.

Mechanism of Photoreceptor Action

Different plant physiologists have conducted experiments to work out the possible mode of action of phytochrome. Various opinions have come forward to understand the possible mechanism of phytochrome action, but still a theory which may provide satisfactory evidence is still to come up. In many plants, the orientation of the chloroplast is controlled by phytochrome to enable it to intercept light optimally. Haupt (1970) studied the effect of microbeam irradiations on chloroplast orientation of the *Mougeotia*. It was observed a critical response in irradiation in this green alga. The irradiation was found in the peripheral regions of the cell, rather than of the chloroplast itself. However, if polarized red or far red light was used, it was observed that the polarizer had to be turned by an angle of 90° in order to achieve red/far red reversibility effects. Thus, it was concluded that the phytochrome molecules are intimately associated with a membrane, which may be plasma membrane. These pigments reorient when converted from one form to the other while still associated with the membrane. In contrast to this localization of phytochrome in plasma membrane, some scientists are of the opinion that they are distributed largely in the cytoplasm, or in some organelles like mitochondria, chloroplasts and nuclei. Phytochrome may control membrane permeability, e.g., transport of specific ions, through modulation of activity of ion channels. The Ca^{2+} ion has received maximum attention. Studies were carried out on protoplasts isolated from dark-grown seedlings and influx of Ca^{2+} induced by red light and efflux of Ca2+ in response of far red wavelength. This phenomenon of rapid influx and efflux is known as phytochrome transformation. There is also evidence for cGMP as a mediator of certain phytochrome controlled response. However, it is still not clear that how does phytochrome alters levels of cGMP or Ca^{2+}. It is possible that phytochrome may interact with a G-protein, which could then activate guanyl cyclase or phospholipase C. The interaction of phytochrome with a G-protein could in turn open a Ca^{2+} channel (Fig. 15.2).

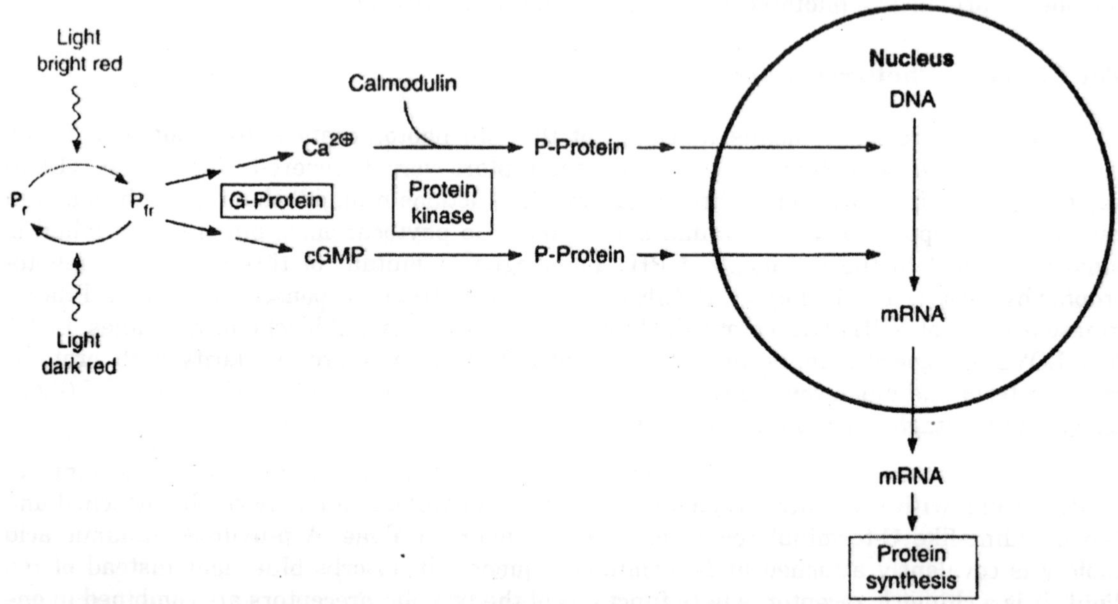

Fig. 15.2. Signal Transduction chains for the effect of phytochrome on gene expression.

Datta *et al.* (1985), observed that red irradiation caused phosphorylation of three nuclear proteins. The effect of red light was reversed by far red light. Thus, it was concluded that some protein kinase activity is closely associated with phytochrome transformation. Wong and Lagarias (1989) are of the opinion of autophosphorylation of phytochrome and a resultant protein kinase activity. They found that phytochrome itself is a phospho-protein containing ~ 0.5 mol P per dimmer and there is a cluster of 8 serine residues at the N-terminus. It is likely that in higher plants, instead of histidine, a serine is phosphorylated. According to Lapko *et al.* (1997), it appears that Ser-7 is phosphorylated by red light. Besides, Ser-7, it seems certain that red light causes the phosphorylation of Ser-598 more specifically. Phosphorylation at Ser-7 in the N-terminal extension region (NTE) is similar in both P_r and P_{fr}, whereas Ser598 in hinge region is phosphorylated in a P_{fr}. According to Park *et al.* (2000), Kim *et al.* (2002) as phosphorylation of Ser598 occurs in Pfr only, thus it is suggested that the phosphorylation and dephosphorylation of Ser598 serves as a switch in phytochrome signaling. The other serines being phosphorylated equally in both dark- and light-grown seedlings.

Details of mode of action of photochrome were studied by genetic engineering technique. mRNA was translated into cDNA and cloned; the nucleotide sequences of photochrome-encoding genes, and that of monoclonal antibodies against certain domains of the protein was studied. *Arabidopsis thaliana* has smallest genome. Five genes encode phytochrome the results of the protein analyses show that two different types of photoreceptor exist: type I, the phytochrome of etiolated tissues and type II, that of green tissues. Type I is encoded by phyA gene; it is termed as phytochrome A. Type II phytochrome is rather heterogenous. It consists of at least two polypeptide chains that are immunologically not related. The polypeptide chains of phytochromes consist of 1,110-1,172 amino acid residues, and both the chains are linked to a chromophore and folds into two domains. The chromophorous group causes the phytochromes to react to the light signal, while the different protein structures allow them to forward the signal to different receptors inducing a number of different physiological reactions.

The Signal Transduction Chain

Photoreceptors probably affect the activity of transcription factors via signal transduction chains, involving the change of cytoplasmic Ca^{2+} concentration and protein kinase. A protein kinase phosphorylates a protein changing its conformation and works in close conjugation with a protein phosphatase which removes the phosphate group. The signal transdcution chains for cryptochromes and phototropin are still not clear. Phototropin is a receptor kinase. Upon exposure to blue light, it is autophosphorylated by its protein kinase activity. Cryptochrome undergoes blue light dependent phosphorylation. The signal function of phytochrome A can be terminated by conjugation of P_{fr} with ubiquitin, which marks it for porteolytic degradation by the proteasome pathway. Phytochrome A affects the transcription of 10% of all genes in *Arabidopsis*. Red light-dependent phosphorylation of NDPK shows its possible role in phytochrome-mediated signaling. In this plant, Nucleoside-diphosphate kinase (NDPK)2 has been identified as a phytochrome-interacting protein by using the C-terminal domain of phytochrome A (PhyA). Domain studies revealed that PER-ARNT-SIM domain A in the C-terminal domain of phytochrome is the binding site for NDPK2. Phytochrome recognizes both the NDPK2 C-terminal fragment and the NDPK2 hexameric structure to fulfill its binding. His-197-surrounding residue mutants were studied, and it was found that the H-bonding with His-197 inside the nucleotide-binding pocket is critical for NDPK2 functioning. His-197 accelerates NDPK2 autophosphorylation and the phospho-transfer between the phosphorylated NDPK2 and its kinase substrate. NDPK is a ubiquitous enzyme. The major function is to catalyze the transfer of the phosphate from NTP to NDP. Thus, enzyme used to maintain nucleoside triphosphate levels in organisms. Now it is recognized as a multifunctional enzyme. It participates in the regulation of growth, development, and signal transduction processes. NDPK is capable to supply GTP; hence it is engage in G-protein mediated signaling. It serves as a guanine nucleotide exchange factor as well as a GTPase-activating protein. The active phytochrome Pfr influences gene expression via transcription factors. Phytochrome controls the expression of many genes. Thus, different types of signal transduction chains are involved. Phytochrome A and B are present in cytosol, and enter the nucleus in activated form (P_{fr}) (Fig. 15.3). In nucleus each of them can associate with a phytochrome interacting factor 3 (PIF3). PIF3 is a basic helix-loop helix transcription factor that exhibits phytochrome-mediated binding to promoter regions of various light-activated genes. The gene regulation occurs due to binding of complex PiF3 and Pfr to the promoter region of genes. In certain plants, protein kinase activity was observed. Thus, it may be concluded that phytochromes are linked to protein kinase cascade regulating transcription factors. In a mutant of tomato, cells are unable to green due to lack of phytochrome. The micro-injection of Ca^{2+} ions and calmodulin into single cells of these mutants stimulated the action of phytochrome. This induces the synthesis of light harvesting complexes. These results indicated that Ca^{2+} is involved as a messenger in signal transduction from phytochromes. Phytochrome also induces the enzyme chalcone synthase. Chalcone synthase may be induced by injection of cGMP in the mutants as described earlier. Hence, it may be concluded that cGMP participate as another messenger to Ca^{2+} in Pfr-induced signal transduction chains. G-protein is involved in both transduction chains.

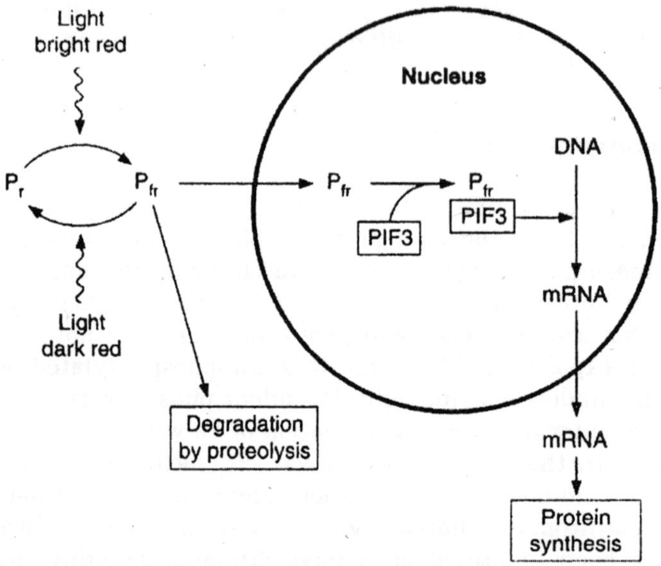

Fig. 15.3. Effect of phytochrome A and B on gene expression.

Phytochrome meshes with the control of transcription; this fact has led the scientist to find out the specific DNA recognition sequences causing a light-induced transcription. Morelli *et al.* (1985) showed that a sequence of 33 base pair is essential for the light-induced control of gene expression. It includes TATA-Box. This gene precedes the gene for small subunit of RuBP carboxylase. Now light-induced transcription of several proteins/ enzymes has been studied.

Seed Germination

Seeds that require light for germination are said to be photodormant. There is an interaction between light and temperature in a germinating seed. In 1907 Kinzel studied 964 species and reported that seeds of 672 species showed enhanced germination in the presence of light. He also found that seeds of 258 species showed inhibited germination by sunlight relative to darkness. The far-red and blue wavelengths are inhibitory. The far-red light decreases the amount of Pfr in the seed to a level below that needed for germination. Blue light effects the blue complex and thus it is inhibitory. The amount of Pfr required for seed germination is important as the value of φin promoting seed germination may be as low as

0.01 and high as 0.06. The chemical reactions controlled by P_{fr} are quite temperature sensitive. At high temperature, the level of P_{fr} decreases as it is reverted to P_r.

A general feature in photodormant and non-dormant seeds is imbibition of water, which results in hydration of seeds, and it gets swelled up. When the seeds are imbibed, light can break the dormancy. The imbibition depends on the permeability of seed coat and fruit coat. In a sufficiently hydrated seed, P_r is transformed in P_{fr}. Some seeds remain viable in soil for many years. These seeds wait for a proper combination of light, temperature and moisture so that P_r may be transformed to P_{fr}. In such types of seeds phytochrome remains stable for many years in dry seeds. The chlorophyll content in the embryos, when seed is ripening is an important factor. It decides whether the seed will be photodormant. Embryos covered by maternal tissues at the time of ripening contain a high amount of chlorophyll. The chlorophyll absorbs the red wavelengths and prevents P_{fr} formation. Such seeds require light for germination, so that it can absorb red wavelengths to form Pfr that promotes germination. It is also evident that long day plants have photodormant seeds. P_{fr} is formed in the hypocotyl-radicle tissue, and also in cotyledons. It increases the growth potential of the radicle cells. It is especially active in the cells of elongation region, by decreasing their water potential so that they more easily absorb water from soils and germinate. In photodormant seeds, the radicle is unable to grow because it cannot apply sufficient force to break through the layers that surround it.

Auxin does not promote germination of photodormant seeds. It is also not effective for nondormant seeds. Gibberellin is the most effective substitute of light in germination of photodormant seeds. In certain cases cytokinins are also helpful but not up to that extent. Thus, it may be concluded that P_{fr} might break photodormancy by causing synthesis of a gibberellin or a cytokinin.

Seedling and Vegetative Growth

The grasses and cereal grain seed germinates and the coleoptile elongates. This elongation goes on until the tip breaks through the soil. The first internode present between the scutellum and the base of the coleptile is known as mesocotyl. The elongation of mesocotyl is very much sensitive to light. The mesocotyl elongates so that the leaves come out of soil surface. Much more adventitious roots come out just above the mesocotyl. Elongation of the mesocotyl, coleoptile, and leaves enclosed by the coleoptile is necessary to carry leaves into the light. Leaves enclosed in coleptile elongates after germination of seed. At this stage elongation of coleptile should be equal or exceed the elongation of leaves, otherwise leaves would grow out of the coleoptile and probably be broken off in the soil. At soil surface the rate of elongation of leaves is promoted by light. Thus, the first emerging leaf breaks coleoptile tip and stops coleoptile elongation. It may be because this stops auxin transport from the tip to elongating cell below. These phenomenons of growth (elongation of leaf and inhibition of coleoptile) are phytochrome responses of sunlight. The seeds of maize are sown near the soil surface. Mesocotyl elongates very little. The first two leaves are rolled up inside the coleoptile. When exposed to light, they began to unroll. Unrolling is caused by more rapid growth of cells on the concave side. It is established that Pfr promotes gibberellins synthesis in rolled leaves. It results in unrolling of leaves. The role of cytokinins is yet not clear.

Seed germination in dicots is generally epigeal. However, pea has hypogeal germination of seeds. In dicot plants, when seed germinates, a hook is formed the stem apex. Soon after germination, this hook is formed due to the effect of ethylene and unequal growth of the two sides of the hypocotyl or epicotyl. Phytochrome is involved in straightening of the epicotyl or hypocotyls hook of dicot seedlings. As the hook emerges from the soil, red light acting through P_{fr} promotes opening of the hook. Light increases leaf blade expansion, petiole elongation, chlorophyll formation, and chloroplast development.

High irradiance reaction (HIR) causes light promoted leaf growth. Pfr initiates synthesis of delta-aminolevulinic acid (ALA) from glutamic acid. ALA is converted to chlorophyll only at higher irradiance of red or blue light. Chloroplast development depends strongly on chlorophyll formation. When cotyledons and leaves start photosynthesis, the elongation of stem is inhibited by light.

Effect of light on Pigments

Light effects on chloroplast arrangements. At high irradiance of light, the chloroplast aligns themselves along the radial walls of the cell and overlaps each other. Such arrangement protects them from light damage. When light intensity is weak, chloroplasts are separated in two groups distributed along the walls nearest to and farthest from the light source. Such arrangement helps to absorb maximum light. Such movement as directed by light shows phototactic nature of chloroplast. In such movement, phytochromes are not involved; rather cryptochrome takes part in it. It is not essential that chloroplast to come in direct light. Instead of this, light is absorbed elsewhere in the cell causes chloroplast movements through effects on cytoplasmic streaming. The streaming is a result from interaction between microfilament and microtubules.

Light promotes synthesis of anthocyanin and other flavonoid pigments. These pigments are synthesized in specialized cells in one or more organs of a plant. Anthocyanin synthesis is increased by light acting photosynthetically in leaves or green apple fruit skins. The action spectra shows that red, far red and blue light gives maximum response for anthocyanin synthesis. Flavonoid synthesis requires phosphoenol pyruvate and erythose-4-phosphate. These sugars provide carbon atoms needed for the B ring and as a source of acetate units required for the A ring. The accumulation of flavonoids in many autumn leaves during senescence exhibits a relationship between protein hydrolysis, phenylalanine appearance, and the use of phenylalanine in ring B formation. This reaction is catalyzed by phenylalanine ammonia lyase and it is promoted by light in various organs of many plants. Phenylalanine ammonia lyase also takes part in lignin synthesis.

Light promotes formation of xylem, and its differentiation from the vascular cambium, lignin formation in xylem cell walls.

SUMMARY

Light is an important factor, which controls plant growth, development from germination to the formation of flowers. The control of morphogenesis by light is called photomorphogenesis. A phytochrome is a photoreceptor, a pigment that plants use to detect light. It absorb light (red and far-red) and effective in causing photomorphogenesis. Phytochrome is produced in darkness in the form of P_r, and is accumulated until a certain level is reached. P_r is converted to P_{fr} on exposure of light of the wave length of 660 nm. Phytochromes are soluble dimeric proteins with a prosthetic group bilin chromophore, usually called phytochromobilin. Several forms of phytochrome response are classified as the high irradiance reaction (HIR) and the induction reaction. A pigment exists that absorb violet and blue light, it was named as cryptochrome. The name cryptochrome was coined in the seventies because of the cryptic (hidden) nature of the pigment responsible for eliciting blue light mediated responses. Cryptochromes may also modulate kinase activity. Chimeric receptor means functions of the two photoreceptors are combined in one molecule. In many plants, the orientation of the chloroplast is controlled by phytochrome to enable it to intercept light optimally.

Chapter - 16

PHYSIOLOGY OF FLOWERING

Flower is a modified stem and all the floral parts are modified leaf. Flowers are important in agriculture. In many plants flower formation is governed by internal factors; in other plants it is controlled by precise environmental conditions. During flowering, the meristematic tissue determinate shoots from which additional shoots do not arise. Flowers can be formed from the apical meristem of the main shoot or solely from the meristem of lateral buds or even from buds that grow out from the bark of the tree trunk. When flowers are born apically, it stops the growth of the shoot, while in lateral flowers; the main shoot continues to grow. Flowering is controlled by photoperiod, light quality, and temperature. Besides this, a transmissible signal moves from photoinduced leaves to the meristems, where flowers are produced. At the site of flowering signal (hormone) stops making leaves and start making flower. Once the flower is formed, further development is also often strongly influenced by photoperiod. Sex expression and seed filling is also effected by photoperiod.

Photoperiodism

Darwin reported the effect of light on phototropism. In 1920 Garner and Allard discovered and elucidated photoperiodism. Photoperiod is the length of exposure to light, and photoperiodism is the response to the timing of light and darkness. Photoperiod regulates flowering in a wide variety of plants. It was found that day length (light period) is not that important for plants. As a matter of fact it is dark period, which is decisive for stimulation of flowering. Actually, dark period exert both a stimulating and an inhibiting influence on flower formation. Based on photoperiod, plants are classified in three categories. (1) *Short day plants (SDP)* flowering is promoted when they are grown under long-night/short day conditions (2) *Long day plants (LDP)* flowering is promoted in short-night/ long-day conditions. *Arabidopsis* is a facultative long-day plant, and (3) *Day neutral plants (DNP)* flowering is not affected by photoperiod. Some plants have a double photoperiodic requirement: these are LSDPs that require LDs followed SDs. Such plants are described as dual-day length plants,e.g., *Bryophyllum*, *Cestrum*. These plants stay vegetative if grown on continuous long day or continuous short day, but flower if exposed either first to long then short days (long-short-day plants), or vice versa (short-long-day plants). The interaction between the temperature and photoperiod is obvious. Some plants initiate flowering after having undergone exposure to a period of cold. There are two physiological stages the photophilic and skotophile. In photophilic stage the physiological activities take place with the onset of light period. After a prolonged light period (i.e., 9-12 hours) inhibition starts and the plant enters its skotophile, i.e., its darkness-loving phase. The next photophilic phase begins just a few hours later, long before dawn and thus quite independent of the actual conditions of alternation of day and night.

Role of Phytochrome

Floral morphology influences both an individual's breeding system and reproductive assurance. The evolutionary strategy of an organism so that it can adopt and match their phenotype to local conditions is designated as phenotypic plasticity. Developmental information at the genetic level can inform evolutionary studies of plasticity. Genetic and environmental factors affecting floral morphogenesis are relevant to individual fitness.

Karl C. Hammer and James Bonner (1938) discovered night interruption phenomenon. In a series of experiments they observed that dark interruption of the day had little or no effect, but night interruption by light inhibited flowering in SDPs and promoted flowering in LDPs.

H. Mohr and P. Schopfer (1978) differentiated SDPs and LDPs in view of the effect of phytochrome as following:

Long-Day Plants: short day + enough PFR (during the middle of the skotophile phase) > production of flowering hormones.

Short-Day Plants: short day + enough PFR (during the middle of skotophile phase) > no production of flowering hormones.

It became apparent that red light was more effective than other wavelengths. The Action spectra of this phenomenon are typical of those for other phytochrome responses. The preliminary conclusion may be that phytochrome apparently detects the light break, and its effectiveness depends upon the time of irradiation. Null response technique was applied to understand an indirect measure of the ratio of the two forms of phytochrome at any given time. In this technique, plants are exposed to mixture of red and far-red light. The mixture that provides no response must be a measure of P_{fr}/P_{total} at the time of irradiation. If the irradiation has no effect, that ratio must have already been present at the time of irradiation. The results obtained from experiments on different plants, it can be a general conclusion that the HIR could be operating, phytochrome intermediate could play a role in night interruption inhibition in flowering of short-day plants and promote flowering of long-day plants. Red light is more effective with short-day plants and a mixture of red and far-red with long-day plants.

Phytochromes and red:far red wavelengths affect floral phenotypes. This was established by a comparative account of (1) wild-type *Arabidopsis thaliana* and *Brassica rapa* plants experiencing high red:far red characteristic of sunlight vs. low red:far red typical of foliar shade and (2) wild-type and phytochrome-deficient A.*thaliana* plants exhibits some characteristic features in floral morphology. Wild type exposed to low red:far red had diminished petal and pistil lengths but longer filaments for a given petal size than plants exposed to high red:far red. *Brassica rapa* plants had qualitatively similar responses. Mutants of A. *thaliana* lacking phytochrome A had smaller flowers i.e., smaller petals, pistils and filaments, whereas phytochrome B-deficient mutants had long filaments. All PHY genes of A.*thaliana* are also active in flowers. Phytochrome A and *PHY*B are expressed in all the floral organs other than petals. Both *PHY*D and *PHY*E are expressed in sepals. PHYE is further expressed in filament tips.

To understand the effects of red: far red and phytochromes on floral morphogenesis, growth-chamber and green house experiments were also carried out. The buds of wild type of A.*thaliana* and *Brassica rapa* were exposed to both high and low red : far red conditions. These conditions which are similar to sunlight and foliar-shade environments, respectively. Wild type of *A. thaliana* was compared with the monogenic mutants deficient in phytochrome A, B, or D. In comparison of flowers of wild type of *A. thaliana* and *B. rapa* opening under high red:far red, and opening under low red:far red exhibited significant differences in either petals, pistil, or filament lengths. Floral development also differed between wild-type individuals and phytochrome-deficient plants.

Florigen

Chailakhyan (1930) coined the term florigen (La *flora* flower; Gr *genno* to begtet). He proposed a hypothesis, that action of a hormone produced in the SD-treated leaf and then transported to the apical meristem where the hormone caused the meristem to stop making leaves and to start making flowers. Lona (1946) and Thurlow (1948) observed that in SDP (*Xanthium strumarium*) if a photoinduced single leaf was cut off and grafted in another plant permanently kept in the non inductive LD conditions, the grafted leaf would induce flowering in the host plant. This was concluded that the grafted leaf would have produced florigen even when it was kept in LD conditions. Many different varieties or species representing different response types have been grafted to each other and caused flowering. In a graft experiment induced *Xanthium strumarium* (a SDP) was grafted in Silene armeria (a LDP) caused flowering. This experiment exhibits that florigen produced by one response type (SDP), when grafted to other response type (i.e., LDP) is capable to induce the latter. In another graft experiment, plants from several species that require only a single inductive cycle were defoliated at various times after that cycle, and the level of flowering some time later was plotted as a function of the time of defoliation. In SDPs, the leaves were cut off immediately following the inductive long dark period, plants remain vegetative and no flowers appeared. The same result was observed in LDPs when defoliated after long light period.

Flowering is controlled by multifactorial processes. Many molecules promote flowering, but none is sufficient to trigger flowering universally. One of the important molecules which promote flowering is gibberellic acid (GA). Studies carried out on a long day plant *Sinapis alba* showed that flowering is regulated by several molecules, including plant hormones and other common metabolites. GA, cytokinins, sucrose and polyamines are flower promoting in *Arabidopsis* and S.*alba*. A mutant of *Arabidopsis* called *gal*-3 is a GA-deficient. When this mutant was grown in long days, flowering occurs later in the mutant than in the wild type. Furthermore, when plants are grown in noninductive short days, flowering does not occur at all in the mutant, unless exogenous GA is applied. The flower promoting signals are transported by phloem.

A. Lang, MSU-DOE (1984) made following conclusions

1. In the leaves of plants is a hormone-like substance (florigen) is produced, which may be also a complex of substances that is conducted to the shoot meristems in order to stimulate them to pass from vegetative growth to flower formation.
2. The florigen is not species-specific. It can be transferred to members of the same species, to members of one genus and to members of one genus and to members of different genera. The transferability seems to be limited only by the acceptability of the graft. A good connection between the phloem of the two partners seems to be important. Grafting of monocots is very difficult to do and a florigen transfer by grafting has not been successful, but it could be shown with other techniques that are required in monocots, too.
3. Florigen is physiologically not specific. It can without difficulty be exchanged between short-day, long-day and day-neutral plants. It is very likely that it is identical in all plants the main difference between short-day and long-day plants is that the florigen production occurs only under a certain (inductive) light program that differs in the two types.

There are some inhibitors also which effect flowering. The transferable substance produced under short-day conditions and suppress flower formation is known as *antiflorigen*. Dostl and Hosek (1937) observed that in Circaca if auxin is added, it could inhibit flowering. Bonner and Thurlow (1949), von Denffer (1950), proposed that auxin is the inhibitor for flowering. Both promoters and inhibitors influence flowering. If a plant is induced by exposing only one leaf to suitable photoperiodic conditions, flowering is frequently inhibited by the noninduced leaves. According to Lang (1961) the release from auxin inhibition was not the prime controlling mechanism for flowering. However, auxin could modify the flowering process. Lang *et al*. (1977), grafted plants of both SD and LD cultivars to tobacco on a DN cultivar. This cultivar was grown under various light conditions. When exposed to SDs , the Sd graft partners caused the DNPs to flower earlier, and under LDs. LD partners also promoted earlier flowering of DN partners. Thus, it shows the presence of a positive acting flowering hormone.

The Genetic Aspect

After the results of grafting experiments in several examples, scientist made further studies and found opposite reactions to photoperiod. It was found between varieties of the same species, indicating that photoperiodic reaction might be controlled by a small number of genes. Brener (1931), Little and Kantor (1941), Laibach (1947) found that a single gene controls the photoperiodic reaction. It was later proved that many genes are involved in the production, transmission, or perception of signals that induce flowering.

It is essential to understand the genes influence flowering. Studies on mutants help to identify genes involved in the production of floral signals in the leaves, and to identify the genes which are involved in the sensitivity of floral signals within meristem. A simple observation that a the mutants of *Arabidopsis* flowers earlier or later as compared to wild types, gave a conclusion that flowering is genetically controlled by many factors. Genes that repress flowering can be identified in those mutants who show earlier flowering, while in mutants with late flowering show genes which act as promoter. In *Arabidopsis* two late-flowering phenotypes correspond to mutant alleles of the genes *CONSTANS (CO)* and *LUMINIDEPENDENS (LD)*. *CO* is a recognized transcription factor with similarity to zinc-finger proteins. In long days it is up-regulating flower promoter. The ectopic expression of *CO* rapidly induces the expression of flowering genes and promotes flowering, i.e., it is involved in activation of flowering genes. The dominant allele of gene *FRIGIDA (FRI)* governs late flowering in *Arabidopsis*. In natural conditions, FRI regulate flowering by sensing cold temperature. The analysis of this gene shows that prolonged cold treatments may also regulate flowering at the shoot apex. Their late flowering phenotype is reversed with vernalization treatment. A combined expression of *FRI* and *FLOWERING LOCUS C*, modulates a requirement for vernilization. Obviously, FRI regulates flowering by sensing cold temperatures. Cytosine demethylation is promoted by vernalization. In *Thlaspi arvense*, *KAH* gene encodes the enzyme ent-kaurenoic acid hydoxylase, essential for GA biosynthesis. In this plant, a correlation was established between the transcriptional activation of *KAH* and cytosine demethylation. This indicates that veranlization may activate expression of *KAH* and possibly other genes involved in the promotion of flowering. Gene *SUPER-MAN (SUP)* defines the identity of various floral organs. It has been confirmed that methylation is involved in controlling this gene. Genes *LEAFY (LFY)*, *APETALAI (API)*, and *CAULIFLOWER (CAL)* are involved to define the identity of floral meristem. These genes are ultimate targets of the signals that promote flowering. It is still not clear, whether the products of these genes interact with transmissible flower-promoting signals directly or indirectly. *LFY* gene carries two functions. It plays role in defining floral meristem and in affecting the time to flowering. The *lfy* mutants of *Arabidopsis* have green flowers and consist of whorls of sepal-like and carpel-like organs. The plants have more inflorescence branches as compared to wild type of plant. This gene shows conversion of indeterminate shoots to determinate shoots. On the other hand, the ectopic expression of *LFY* promotes early flowering and the conversion of shoots to flowers. Gene *TERMINAL FLOWER 1(TFL1)* determines the meristem. The recessive mutant (*tfl1*) lacks lateral branches with reduced number of flowers. It shows early flowering and the primary inflorescence branches produce terminal flowers. This gene exhibits conversion of determinate shoots to indeterminate shoots. *TFL1* is exhibited in the shoot meristem. The recessive mutants of genes *tfl1* and lfy express opposite phenomenon. There are three more genes in *Arabidopsis* which influence the determinacy of the floral meristem. These are named as *AP1, APETALA2 (AP2)*, and *CAL*. *LFY, AP1*, and *TFL* play some early role in floral meristem specification. The expression of gene *LFY*, and *AP1* is exhibited in floral meristem before the organ primordia develop. *LFY* is expressed in low amounts during the vegetative phase, and it is increased with floral induction by photoperiod treatment or in response to GA. *AP1* is expressed in flower primordia until one to two days after floral induction begins. *AP2* is expressed throughout the life cycle. *TFL1*, is expressed throughout development in cells that lie just below the apical dome of the primary shoot meristem. The recessive mutants *ap1* and *ap2* enhance the lfy phenotype. *cal ap1* mutants are more highly indeterminate. This strikingly overproduce floral meristem. As a result, there is formation of cauliflower like inflorescence.

The CONSTANS and The Phloem Mobile Signal

Brian G. Ayre and Robert Turgeon (2004) discovered that the plant protein, CONSTANS plays an important role in generating the signal-florigen. According to Suarez-Lopez (2001) CONSTANS mediates between photoperiod perception and the transition to reproductive growth, and transcript levels accumulate under inductive, long day conditions. Studies were carried out on *Arabidopsis*, in which all CONSTANS proteins had been abolished. This led to the conclusion to understand that CONSTANS is a signal involved in flowering. A copy of CONSTANS gene (CO) was introduced under the control of the galactinol synthase promoter. The latter causes the protein to be synthesized only in the leaf. This signal had dramatic effect on flowering. It was concluded that there are two possibilities regarding the role of CONSTANS. One possibility is that it moves throughout the plant to the site of flowering. The translocation takes place through phloem. The other way is that CONSTANS interacts with another factor inside the phloem that is transported to the site of flowering. The further evidence came from grafting *Arabidopsis* plants that contained no CONSTANS protein on plants synthesizing CONSTANS in their leaves. The explanation of the results is that CONSTANS, or another factor to which, it interacts, was able to move through the graft junction to signal flowering in the parts of the plant. These parts were formerly devoid of any of the protein. It is still doubtful whether CONSTANS is in fact the flowering hormone. According to Ayre "It may be interacting with another downstream factor that moves to the site of flowering action. It is clear now that CONSTANS is an important factor in generating this signal". Ayre (2005) further studied on florigen and a genetic approach to Long-Distance signaling through phloem.

Putterill et al., (1995) observed that in *Arabidopsis* the homozygous mutant (CO) flower late under long-day conditions, whereas plants over expressing CO flower early. They expressed that CO from the CmGAST promoter in transgenic *Andriopsis* to establish whether CO acts upstream or downstream of the phloem mobile signal. In transgenic wild type and co-1, CO expression in the minor veins accelerated the transition to flowering under non-inductive, short-day conditions. It was suggested that CO participates in generating a phloem mobile stimulant. FLOWERING LOCUS T (FT) is a downstream target of CO. It was demonstrated that in cells unable to express CO, the correlation between FT and CO expression pattern shows that CO does not directly promote gene expression in cells.

A potential signaling compounds was introduced to the phloem translocation stream. Later, an observation was made to understand the effect in the tissues where this compound is unloaded. This provides an alternate approach to identify the compounds involved in long-distance signaling. In this regard, the companion-cell-specific promoters are used. It expresses the genes involved in generating potential signals. In *Arabidopsis* AtSUC2 is used as phloem-specific promoter. A promoter CmGAS1 gene encodes for Galactinol synthase was isolated from *Cucumis melo*. It is unique and is specific for the companion cells of minor veins. This plant transports significant quantities of the oligosaccharides (raffinose and stachyose), and phloem loads by the polymer trap mechanism. Galactinol is the galactosyl donor for the synthesis of raffinose and stachyose. According to Kim *et al*. (2001) the presence of LeT6 transcription factor mRNA in the phloem correlates with morphological changes in recipient tissues in tomato. In 2002 they opined that transcription factor proteins can pass through plasmodesmata despite predominant nuclear localization. The same opinion was put forward by Wu *et al.* (2003).

SUMMARY

Flowers can be formed from the apical meristem of the main shoot or solely from the meristem of lateral buds or even from buds that grow out from the bark of the tree trunk.

Flowering is controlled by photoperiod, light quality, and temperature. Besides this, a transmissible signal moves from `photoinduced' leaves to the meristems, where flowers are produced. Photoperiod is the length of exposure to light, and photoperiodism is the response to the timing of light and darkness. Phytochrome apparently detects the light break, and its effectiveness depends upon the time of irradiation. Phytochromes and red:far red wavelengths affect floral phenotypes. Chailakhyan (1930) coined the term `florigen' in the leaves of plants which is a hormone-like substance (florigen) produced, may be also a complex of substances, that is conducted to the shoot meristems in order to stimulate them to pass from vegetative growth to flower formation. The florigen is not species-specific. Studies on mutants helps to identify genes involved in the production of floral signals in the leaves, and to identify the genes which are involved in the sensitivity of floral signals within meristem. The plant protein, CONSTANS plays an important role in generating the signal-florigen. A potential signaling compounds is introduced to the phloem translocation stream. In *Arabidopsis* AtSUC2 is used as phloem-specific promoter. A promoter CmGAS1 gene encodes for Galactinol synthase was isolated from *Cucumis melo*.

SECTION - III :

BIO TECHNOLOGY

Chapter - 17

TRANSFORMATION AND TRANSPOSONS

Recombinant DNA technology has been extensively used with microorganisms. It is also applied for gene manipulation in plants. This technology has helped to develop plant varieties which have high productivity with enhanced nutritional value. Even if a single plant cell is genetically engineered, and if the gene transfer is efficient, transient gene expression can be quantitatively measured. This is generally in terms of protein encoded by transient gene rather than measuring mRNA to show the success of gene transfer. If the transgenes become integrated in plant cell genome, it will be inherited. If this transgene carrying cell is cultured under artificial conditions, it will develop into a plant. This is because of totipotent nature of plant cells. The transgenic cell thus obtained will develop into a transgenic plant. If this transgenic plant is fertile, it will produce viable seeds, and thus the desired trait is passed to successive generations. There are a number of effective DNA-delivery systems and expression vectors that work in a range of plant cells.

Transformation is the phenomenon by which the plant cells uptakes the transgenes. In other words, stable integration of a foreign gene into the genome of a plant regenerated from DNA-treated protoplasts or intact cells. These transgenes may be transformed in a plant cells by using plasmid vector. The most developed systems for higher plant transformation is derived from the tumor inducing mechanism of the soil bacterium *Agrobacterium*. *Agrobacterium tumifaciens* causes crown gall disease, and *Agrobacterium rhizogenes* causes hairy root. Transgenes may also be transformed by physical methods.

TRANSFORMATION USING PLASMID OF AGROBACTERIUM

Molecular Biology of the Crown Gall

Agrobacterium is a gram-negative, non-sporing, motile, rod-shaped bacterium. On the basis of biochemical tests and utilization of carbohydrates, the strains of *Agrobacterium* are classified in three biovars. Two species of this bacterium have plasmid viz. Ti in A. *tumifaciens* and Ri in A. *rhizogenes*. *Agrobacterium tumifaciens* causes crown galls tumors in plants. Galls are formed due to genetic transformation, which interfere with the normal growth of an infected plant. In 1970s Jeff Schell found that virulent bacterium contains a plasmid, which has the potential to infect plants. The tumor is formed by the genetic expression of Ti-plasmid of this bacterium. Ti-plasmid (tumor inducing) contains a specific gene known as T-DNA (transferred DNA). It integrates into the plant cell genome and the proteins synthesized by it forms tumor. A. *tumifaciens* causes crown gall disease in dicot plants only. It enters inside the plants through wounds, often at the base of the stem. The motile cells of bacterium are attracted to wound sites by chemotaxis. The infection occurs at the crown, and infected cells proliferate at the site of infection, forming a tumor or gall. It is also observed that the susceptible plants excrete some phenolic compounds (acetosyringone and hydroxyacetosyringone) (Fig. 17.1) and this soil inhabiting bacterium is attracted to it. Ti plasmid containing bacterium able to recognize wound phenolic compounds, hence they are attracted strongly towards the wound. Acetosyringone attracts bacterium even if present at a very low concentration. These compounds act to induce the activity of vir genes. The latter are present on Ti-plasmid (Fig. 17.2) and are responsible for virulence. *vir* region has 25 genes with 8 operons, out of them six (excluding *vir*A and *vir*G) encode for various proteins involved in T-DNA transfer. Genetic analysis of the various vir genes suggested that *vir*A and *vir*G comprised a two-component system necessary for the expression of the vir regions.

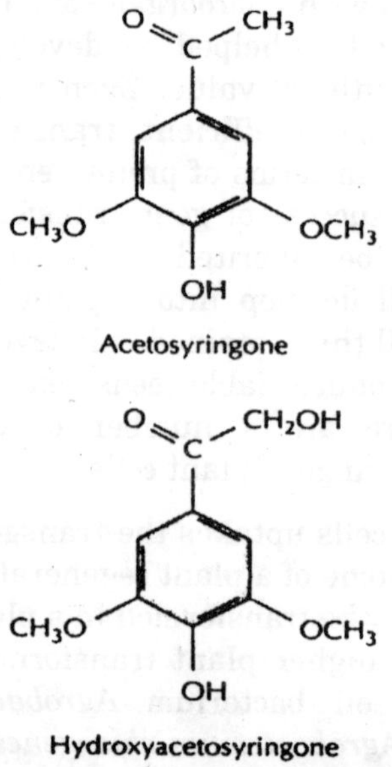

Fig. 17.1. Structure of Acetosyringone, Hydroxyacetosyringone

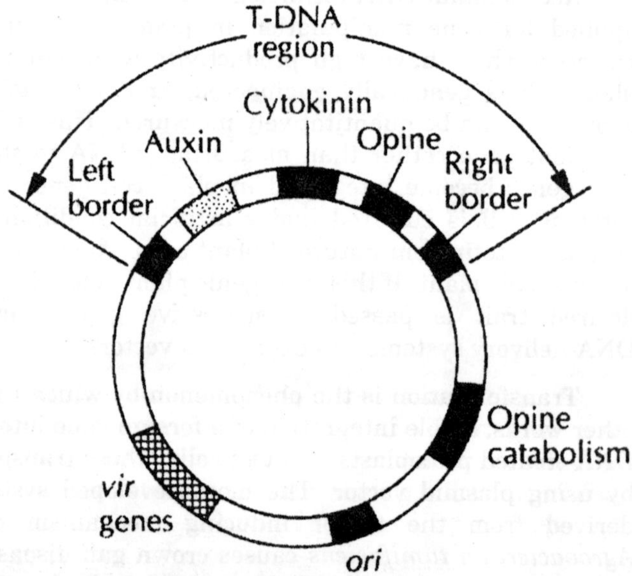

Fig. 17.2. The Ti plasmid

The genes of vir region can not transfer themselves; they only induce the transfer of T-DNA. The vir genes Two vir genes, *vir*A and *vir*G, are constitutively expressed and are involved in mediating the induction of the other genes. *vir*A protein was postulated to be a sensor of the pheolic-inducing molecules. This gene shows positive regulatory mechanism. *vir*A has been shown to be autophosphorylated at one specific histidine residue under appropriate induction conditions. It is a membrane-associated protein which monitors the presence of the induction compounds. The presence of inducers is then signaled to the positive regulatory protein and it results in induction of other vir genes. After induction of virA gene, phosphate is then transferred to virG product increasing the transcriptional activating activity of *vir*G. VirG protein is a transcriptional activator of the vir genes. The *vir*D codes for at least two polypeptides called VirDI and VirDII. VirDII has shown to be required for the nicking activity and single-strand synthesis and encodes an endonuclease activity. VirDI has topioisomerase activity while VizDz is an exonuclease. VirD, binds to the right boarder sequence, and relaxes supercoiling which facilitates the action of protein Vir Dz. Vir Dz nicks at the right boarder and covalently binds to the 5' end generated as a result of nicking. It remains bind to it during the T-DNA transfer. The 3' end produced at the site of nick serves as a primer for DNA synthesis in the 5' - 3' direction. It results in displacement of T-strand from the DNA duplex. VirE is not absolutely required for T-DNA transfer, but it has shown to behave as a single-stranded DNA binding protein. The VirE and VirDII proteins may provide protection to the T-strand intermediate during the transfer process. VirE appears to be a nonspecific single-stranded binding protein that coats the T-strand. It is presumed that it protects from endonucleolytic catalysis. The VirDII protein attaches covalently at 3' end of the T-strand. It protects T-DNA from exonucleolytic degradation. Another function of VirDII is that its N-terminal portion has been shown to direct a fusion protein to the plant nucleus protein. Another product of vir gene is VirCI protein that has been shown to have DNA binding activity. VirCI binds specifically to the overdrive sequence of the octopine-type A6 T-DNA. VirC like VirE is not absolutely required for T-DNA transfer and tumorgenesis on some plants. It is suggested that VirCI and overdrive facilitate the interaction of VirDII with the border nick site.

The *vir*B locus encodes 11 potential polypeptides, some of which appear to be membrane spanning proteins. The function of *vir*B gene product in tumorigenesis process is not clear. After the induction of activity of *vir* gene T-DNA (Transfer DNA) is transferred to the host cell. This transfer has by somewhat similar mechanism as found in conjugation in bacterium. T-DNA it enters into the host cell in the form of a linear, single-stranded molecule. The formation of the single-stranded form of T-DNA is initiated by strand-specific cutting at both borders of the intact T-DNA region. After entering inside the host cell T-DNA, integrates in the plant chromosomal DNA. The 5' end of the single-stranded T-DNA carries the right border sequence and the left border sequence is at the 3' end. The right border of the T-DNA has repeating units of 25bp, and it is believed that this ends makes the integration successful. The left border also contains 25 base repeats; however, it has no role in integration of T-DNA in plant cell genome. The genes which are located in the T-DNA region contain the necessary structural features for expression by the plant. The mRNA is polyadenylated and contains a 7-methyl-guanosine cap at the 5' end. Willmitzer *et al.* (1983) reported that total of eight mRNAs in plant tissue containing the TL-DNA of the octopine-type Ti plasmid. According to Salomon *et al.* (1984) five mRNA species have been mapped to the TR-DNA. These transcriptional units do not have introns. Two TL- DNA genes control the synthesis of indole-3-acetic acid from tryptophan. These genes viz iaaM and iaaH are also known as *tms*1 and t*ms*2, respectively. Besides these two genes, T-DNA also contains tme gene, which also known as *ipt*-gene. These genes code for synthesis of plant growth hormone. *iaa*M codes for the enzyme tryptophan 2-monooxygenase, which converts tryptophan to indole-3-acetamide, *iaa*H codes for indole-3-acetamide hydrolase, which converts indole-3-acetamide to indole-acetic acid.

While, gene *ipt* codes for enzyme isopentenyl transferase. The latter adds 5'-AMP to an isoprenoid side chain to form the cytokinins isopentenyladenine and isopentenyladenosine. These two compounds are hydroxylated by plant enzyme and to generate cytokinins. The cytokinins are Transzeatin and Transcribosylzeatin. Transzeatin is formed form isopentenyladenine, while isopentenyladenosine forms Transcribosylzeatin. Formation of auxin and cytokinin in excess causes gall formation. T-DNA also carries a gene, which synthesizes opine. The taken is formed by the unusual condensation of an amino acid and a keto acid, or an amino acid and a sugar, i.e., octopine is formed by the condensation of amino acid arginine and keto acid pyruvic acid, and nopaline is formed by the condensation of argenine and ketoglutarate. Octopine and nopline are not present in normal healthy cells. Opines (Fig. 17.3) are unique molecules synthesized in the crown galls and then secreted. Opine production is controlled by T-DNA genes which encode the enzymes for opine synthesis. The gene for nopaline synthase (nopaline dehydrogenase) has been the most widely used opine gene and was derived from the nopline plamsid pTiT37. Ti-plasmid carrying A. *tumifaciens* uses this compound as carbon and sometimes as nitrogen source. Ti-plamsid has a specific gene that can catabolize a particular opine. This gene is present in Ti-plasmid but not in the T-DNA region. The different opines specified by Ti plasmid are octopine, nopaline, succinamopine and leucinopine.

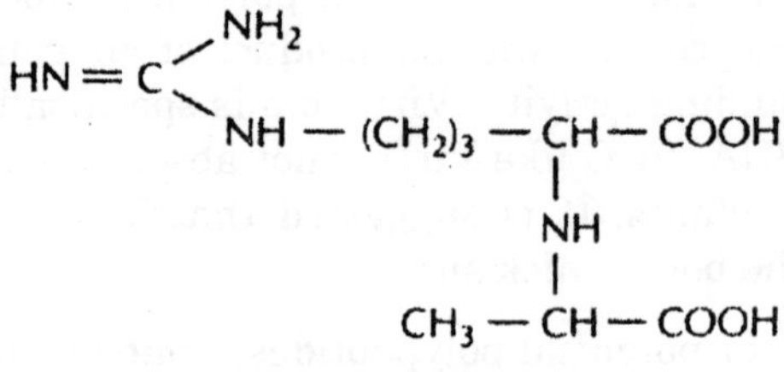

Octopine

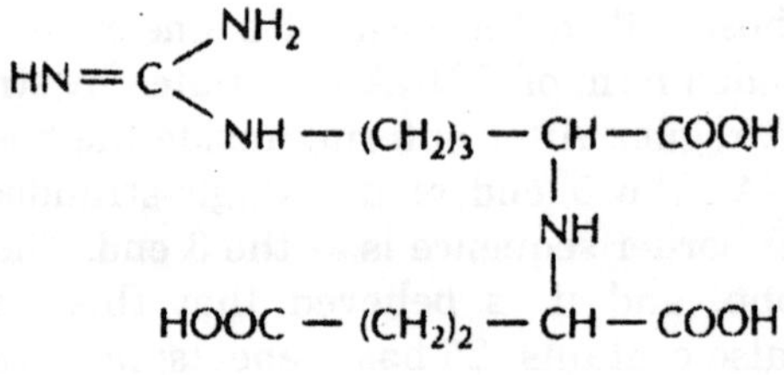

Nopaline

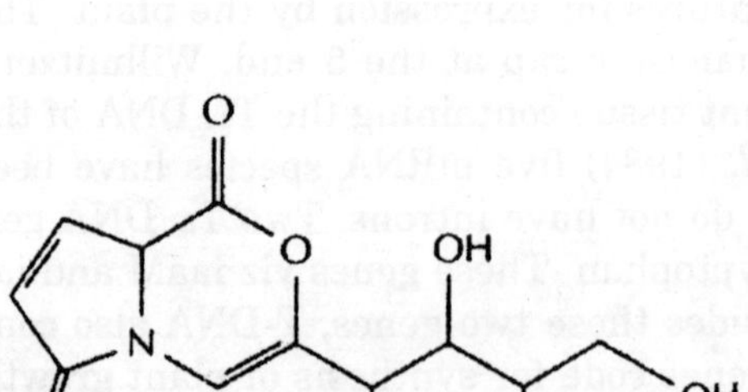

Agropine

Fig. 17.3. Chemical structure of octopine, nopaline, and agropine.

Ti-plasmid-derived Vector Systems

The length and number of T-DNA regions vary in Ti and Ri plasmids. Ti plasmid is a large, circular molecule with about 200 bp. In plant genetic engineering techniques, Ti-plasmid is used as a vector. In Ti-derived vector system, the desired DNA sequence is inserted into the T-DNA region and then this plasmid is used to deliver and insert this gene into the new susceptible host plant cell. Methods for inoculation of plant tissues vary depending on the plant species and the *Agrobacterium* delivery system. There are some serious limitations in gene cloning using this plasmid. Some steps are necessary to overcome these limitations. They are removal of genes viz. auxin and cytokinin gene, opine-synthesis gene. Besides this large segments of DNA that are not essential for a cloning vector must be removed. Ti plasmid is not capable to replicate in *E.coli,* so in developing Ti-plasmid -based vectors, an origin of replication should be added so that this plasmid can be used in *E.coli.*

T-DNA of wild Ti plasmid vector contains *iaa*M, *iaa*H and *ipt* genes, which are considered as oncogenes. They code for phytohormones which causes a disorganized growth in transformed cells. Transformed cells (with wild Ti plasmid) growing in culture loose regeneration potential, which prevents them to form mature plant. Hence, the genes governing auxin and cytokinin production are deleted from T-DNA region of Ti plasmid. This technique is known as disarming and plasmid is known as disarmed plasmid. These genes are replaced by pBR 322 sequences. The modified T-DNA region was non-tumorous, and has the potential to produce plantlets.

Two strategies are applied to clone vectors in recipient plants. After cloning, the vector should by themselves affect the transfer and integration of the T-DNA region into recipient plant cells. The first approach is binary vector system and the second is cointegrate vector system. The **binary cloning vector** (*E.coli-A.tumifaciens* shuttle) lacks *vir* gene (Fig. 17.4). It has a pair of plasmid of *E.coli* and *A.tumifaciens* origins. All the cloning steps are carried out in *E.coli* before the vector is introduced into *A.tumifaciens*. The modified Ti plasmid contains a complete set of vir genes but lacks portions, or the entire T-DNA region. This binary cloning vector consists of a pair of plasmid of which one plasmid contains disarmed T-DNA sequences (with at least one boarder must be present), and the second plasmid of the pair contains vir region. This is also known as defective plasmid and it codes for *vir* gene products. The plasmid containing disarmed T-DNA is called mini-Ti or micro-Ti, and the plasmid with *vir* gene (and lacking T-DNA) is the helper plasmid. The products of *vir* region mobilize the T-DNA region of the binary cloning vector plasmid. Thus infected Ti plasmid acts as a helper plasmid, enabling the T-DNA to be inserted into the plant chromosomal DNA. In **cointegrate vector system** (Fig. 17.5) - the genes to be transferred into the plant are first cloned in plasmid of *E.coli*. This modified plasmid is intermediate vector. This cloning vector recombines with a disarmed Ti plasmid. For such recombination, *E.coli* cells containing modified plasmid are made to conjugate with *Agrobacterium*. The cointegrate cloning vehicle is a product of recombination of modified *E.coli* plasmid and disarmed Ti plasmid. The co-integration of the two plasmids is achieved within Agrobacterium by homologous integration. Thus Ti plasmid and *E.coli* plasmid should have some base pair common so that recombination can take place. Normally these sequences lie inside of the T-DNA region of Ti plasmid. Ti plasmid provides vir genes to cointegrate vector. The latter is necessary for the transfer of the T-DNA to the host plant cells. This cloning vector is maintained in *A.tumefaciens* as a part of a cointegrated structure. Through this vector genetically engineered T-DNA region can be transferred to plant cells.

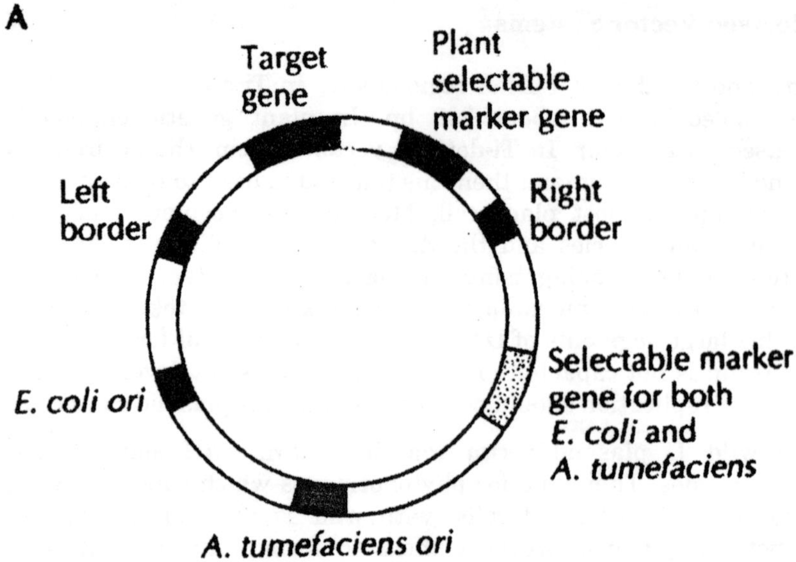

Fig. 17.4. Two Ti plasmid derived cloning vector systems.

Markers

Markers ascertain how effectively transforming DNA is introduced. *Selective markers* are generally used for stable transformation. These markers give a dominant phenotype on transformed cells. Traits which can be used to select transformed cells from those that are untransformed are grouped in two classes: traits giving viability or lethality and traits with some distinguishing characteristic. The latter class has a negligible effect on the survival of cells. Selective markers of first category include Herbicide and Antibiotic Resistance Markers, Auxotrophic Markers, Lethal or Conditionally Lethal Markers. The second category of selection markers i.e., visible markers include Histological markers, Morphological Markers, Pigmentation markers. There are certain criteria in choosing an appropriate selective agent. These are 1. selective agent should be readily available and soluble in plant tissue culture media, 2. it should also be inexpensive and nontoxic to the investigator. *Assayable markers* are most commonly used to quantify gene expression in plants. These markers direct the expression of a readily quantifiable product, which is not necessarily distinguishable in living cell. These products may be enzyme or an antigen. Some markers can also be assayed both qualitatively and semiqualitatively. The choice of appropriate assayable marker or reporter gene depends on the type of assay which is most suited to the experimental aims of the investigator. Sensitivity and reproducibility of the assay method are most crucial factors for quantitative experiments. When a gene is transferred, it is necessary to assess marker gene expression. This gives the picture of stable transformation. Therefore, assayable marker are frequently used to optimize DNA delivery in the early stages of developing a transformation procedure. Integration of marker gene into the target cell genome, and cell division following integration are the basics of stable transformation. There are some important factors which affect transient gene expression. These parameters are the number of transformed cells, the number of active templates inside those cells which are transformed, and the viability of the transformed cells following DNA delivery.

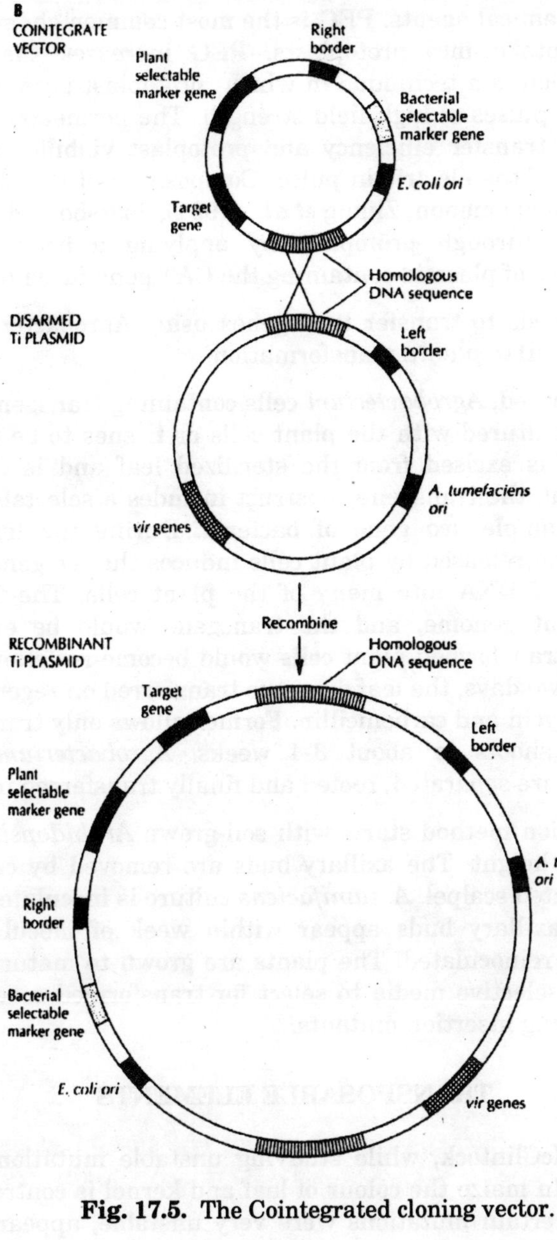

Fig. 17.5. The Cointegrated cloning vector.

The gene transfer

Gene transfer or DNA uptake is a process which moves a specific piece of DNA into a protoplast or a cell. The piece of DNA is usually ligated with bacterial plasmid. Basic techniques for gene transfer are methods for transformation of plant protoplasts, cells, and tissues. The technique of transformation of protoplast means transfer of foreign genes into plants using protoplasts as the starting material. Protoplast capable of gene integration and plant regeneration may be referred to as competent. Another method of gene transfer into protoplast is by using chemical agents. PEG is the most common chemical treatment used to stimulate late DNA uptake into protoplasts. PEG increases the permeability of cell membrane. Electroporation is a technique in which protoplast permeability is increased by applying short electrical pulses of high field strength. The permeability generated facilitate DNA uptake. The gene transfer efficiency and protoplast viability are influenced by the amplitude and duration of the electric in pulse. Composition of the electroporation medium also influences on such phenomenon. Zhang *et al.*, (1991), Joersbo and Brunstedt (1990) have reported transformation through protoplast by applying a brief exposure of 20 KHz ultrasound in the presence of plasmid containing the CAT gene fused to the 35S promoter.

There are two methods to transfer transgenes using Agrobacterium viz. (i) co-culture with tissue explants and (ii) *in planta* transformation.

In the co-culture method, *Agrobacterium* cells containing transgenes in the T-region of a disarmed plasmid is co-cultured with the plant cells or tissues to be transformed for about two days. A small disc is excised from the sterilized leaf and is cultured with infected bacterial cells. In general, the transgene construct includes a selectable reporter gene. This reporter gene is for example *neo* gene of bacteria. During the leaf disc-*Agrobacterium* co-culture, acetosyningone released by plant cells induces the *vir* genes. This results in the transfer of recombinant T-DNA into many of the plant cells. The T-DNA would become integrated into the plant genome, and the transgene would be expressed. Due to the expression of *neo* gene, transformed plant cells would become resistant to kanamycin. After an incubation period of two days, the leaf discs are transferred on regeneration medium. This medium contains kanamycin and carbenicillin. Former allows only transformed plant cells to divide and regenerate shoots in about 3-4 weeks. *Agrobacterium* cells are killed by carbenicillin. The shoots are separated, rooted and finally transferred into soil.

Planta Transformation method starts with soil-grown *Arabidopsis* plants with primary bolts of around 3 cm in height. The axillary buds are removed by cutting deeply into the rosette with a sharp-pointed scalpel. *A. tumifaciens* culture is inoculated in this wound. Some new inflorescence and axillary buds appear within week of inoculation, which is again removed and plants are reinoculated. The plants are grown to maturity, and the seeds are collected and plated on selective media to select for transformed seedlings. This procedure may be useful in generating insertion mutants.

TRANSPOSABLE ELEMENTS

In 1951 Barbara McClintock, while studying unstable mutation in maize discovered Transposable elements. In maize the colour of leaf and kernel is controlled by genetic traits. McClintock found that certain mutations were very unstable, appearing and disappearing from one generation to the next, or even during the lifetime of an individual plant. On the basis of these observations, she concluded that there are certain genes which can move from one place in a chromosome to an entirely different site, and effects on phenotypic expression. These movable genes and the phenomenon were named as transposable elements and transposition, respectively. In late 1960s molecular biologists working on bacteria discovered some DNA sequences, which can move from one place to another in the genome, and these transposable elements were named as Transposons. Thus, transposons are discrete sequences in the genome that are mobile i.e., they are able to transport themselves to other locations within the genome. They may provide the major source of mutation in the genome. The simplest transposons are called insertion sequences (IS). The IS elements are normal constitutes of bacterial chromosomes and plasmids. Transposon promotes DNA rearrangement.

The isolated plant transposons have the length of 1.4 to 1.7 kb. Mu transposon is 1367 nucleotide long with terminal inverted repeats of 213 and 215 nucleotides. It creates a duplication of nine nucleotides upon insertion. Studies on base sequences of transposons revealed that these elements have repeated sequences at the opposite ends, but in opposite orientation. In simple language transposons have inverted repeats of base sequences. These inverted repeats are significant as they are recognized by transposases, which brings out integration of these elements into target sites. The maize transposons have inverted repeats at the ends and short direct repeats in the adjacent target DNA.

Several transposable elements are reported in maize and *Antirrhinum majus*. These transposable elements have been isolated and analyzed at the DNA level. They are categorized in two groups. Group one has similar inverted termini. During insertion, it created duplication of 8 nucleotides. It includes Ac/Ds system of maize and the Tam 3 element of *Antirrhinum majus*. The second group has inverted termini of 13-15 nucleotides and during insertion duplication three nucleotides are created. The examples of this group are En (Som) system of maize, the Tam 1 and Tam 2 elements of *A. majus*, Tgm 1 sequence of soyabean is also included in this group.

The molecular studies with Ac/Ds group and the En (Spm) of maize have revealed that these elements can undergo different structural changes in their sequences. They can cause a variety of chromosomal alterations at or near the insertion site. Genetic analysis shows that Ac and En(Spm) are autonomous elements i.e., they can move in the genome. The `receptor' elements are also present. They are dependent on the presence of either Ac or En (Spm) for transposition. Reverent sector arises by somatic excision of such receptor elements. These sectors are alike in size. Excision occurs at a certain frequency in a given time which is determined by the receptor element itself.

Transposition

Transposition is joining the transposon to its target site. It was also discovered that transposition is encoded by a protein. The latter was named as Transposase. Transposition occurs by replicative and nonreplicative mechanisms. In a general mechanism of transposition, the transposon is nicked at both ends, and target site is nicked on both strands. The nicked ends are joined crosswise. This produces a covalent connection between the transposon and the target. The transposon is joined to the protruding ends, and the gaps are filled. Due to the asymmetric nature of the nick, there are a few bases, which may be present as single strand. Such SS gaps which may be present on both the strands are filled up by the polymerase. Transposition occurs by replicative and nonreplicative mechanisms. In *replicative transposition*, the element is duplicated. The transposing unit is a copy of the original element. One copy remains at the original site, while the other inserts at the new site. Transposase and resolvase enzymes are required for replicatve transposition. Transposase acts on the ends of the original transposon. The recombination reaction is called resolution and the enzyme resolvase is involved in such action. The 3' ends of the strand transfer complex are used as primers for replication. The fusion of the two original molecule results in the formation of a cointegrate. The latter has two copies of the transposon, one at each junction between the original replicons oriented as direct repeats. A homologous recombination between the two copies of the transposon releases two individual replicon. Out of them one is original donor replicon, while the other is target replicon. Recombination between the trasnposons copies regenerates the original replicons. The recipient has gains a copy of the transposon. In *non-replicative transposition* the transposing element does not replicates rather it moves as a physical unit directly from one site to another, and is conserved. Thus, the element is inserted at the target site and lost at donor site. There are two mechanism of movement, the composite transposon Tn10 and Tn5 requires transposase. This mechanism involves the release of the transposon from the flanking donor DNA during transfer. The second mechanism utilizes the connection of donor and target DNA sequences and share some steps with replicative transposition. Conservative transposition is a sort of non-replicative transposition. The element is removed from the donor site and inserted into a target site by a series of events. In this mechanism every nucleotide bond is conserved.

Transposition pathway of phage Mu was studied in details. When phage infects a host cell, Mu integrates into the genome by non-replicative transposition. While during the subsequent lytic cycle it behaves in a replicative manner and the number of copies are produced. MuA transposase takes charge in the initial manipulation of the phage DNA. Three MuA-binding sites are located at each end of Mu DNA. These sites are of 22 bp consensus and are recognized as L1, L2, and L3 at the left end and R1, R2, and R3 at the right end. A complex is formed when MuA binds to the both ends at the internal site of phage genome. Joining the Mu transposon to a target site passes through the three stages synapsis, cleavage, and strand transfer. MuA subunit bound to only the two sites closest to each end of the trasposon and form a tetramer. This results in synapsis of the two ends of the trasnposon. Following this functioning of tetramer brings coordinated reaction on both ends of Mu DNA. The latter has two sites for manipulating DNA. The active site cleaves the Mu DNA strands at positions adjacent to the L1 and R1 sites. It is unable to cleave DNA sequence that is adjacent to the consensus sequence in the consensus-binding site, but it cleaves appropriate sequence on a different stretch of DNA. MuA subunit cleaves the ends of the transposons by trans mode, i.e., the monomers actually bound to L1 and R1 do not cleave the adjacent sites. One of the monomer bound to the left end nicks the site at the right end and vice versa. Another protein MuB prefers transposition at target immunity site, i.e., >10-15 kb away from the original insertion. In *vitro* experiments in a plasmid with Mu containing donor and Mu deficient target, MuA and MuB proteins, *E. coli* HU proteins, Mg^{2+} and ATP shows that the presence of MuB and ATP restricts transposition exclusively to the target plasmid. When MuB binds to the MuA-Mu DNA complex, MuA causes MuB to hydrolyze ATP, after which MuB is released.

Transposition in Maize

In maize, break in one chromosome causes loss of dominant allele, and as a result of transposition the effects of the chromosome breaks generates `controlling elements'. These elements often insert near genes that show phenotypic effect, but this effect is non-lethal. Maize exhibits clonal development and, therefore, the occurrence and timing of a transposition event can be pictured. A sector of tissues is formed due to the presence of mitotic progeny of a given cell in the same location. As a consequence changed phenotypic characters appear within the tissue of the original phenotype. This is known as variegation. The number of division present in a lineage will give rise to a site, which represents the timing of the change in genotype. Insertion of a controlling element may affect the activity of adjacent genes. Chromosome breakage is a general outcome of the presence of some elements. The elements transpose and promote genetic rearrangements at characteristic times and frequencies during plant development. The characteristic behavior of controlling elements in maize is typified by the Ds elements. This element is the site of chromosome breakage. Breakage at Ds generates acentric fragment. This acentric fragment carrying the dominant markers. Being acentric it is lost during mitosis. Thus the progeny will have only the recessive markers carried by the intact chromosome. Fusion between the broken ends of the chromosome generates dicentric chromosomes. During the segregation of dicentric chromosome in mitosis spindle, each of two centromeres pulls toward an opposite pole. The tension breaks the chromosome at a random site between the centromere. The fusion-breakage-bridge cycle continue through further cell generations. It is due to such cycle genetic changes continue in progeny.

Transposon Families

A transposon family consists of a single type of autonomous elements accompanied by many varieties of nonautonomous elements and there is interaction between such elements. In maize genome, there are transopon families of several controlling elements. They are grouped into two categories with autonomous and nonautonomous controlling elements. The autonomous controlling elements code for proteins that permit them to transpose, while the nonautonomous controlling elements have the capacity to catalyze transposition, but they need necessary proteins supplied to them by autonomous controlling elements. The autonomous controlling elements are well capable of transposition; hence its insertion at any locus creates an unstable or `mutable' allele. Such autonomous elements turn into stable allele if they undergo mutation. Autonomous and nonautonomous elements can undergo genetic as well as epigenetic changes. One of the major changes is that autonomous elements may be converted into nonautonomous elements. Nonautonomous elements may undergo further changes.

The autonomous controlling families in maize are *Ac* (activator), *Mp* (modulator), *Spm* (suppressor-mutator), *En* (enhancer), *MuDR* (mutator). The nonautonomous controlling families are *Ds* (dissociation), *dSpm* (defective Spm), *I* (inhibitor), and Mu.

TRANSPOSON TAGGING

Plant transposable elements have already been used to isolate genes from indigenous species. Transposon tagging is a very attractive alternative method of insertional mutagenesis. The molecular isolation of transposable elements has enabled to clone of genes in which the element is present. It is advantageous because those genes whose functions are not known can be cloned. A transposable element must first be cloned and characterized at the molecular level before it can be used in gene tagging experiment. To isolate gene mutations two genetic strategies are in practice, targeted and non-targeted tagging. *Targeted tagging* is a technique of isolation of one specific gene. To isolate such specific gene a cross is made between a plant containing transposons and a homozygous for a recessive mutant phenotype. A vast majority of plants of the progeny which will inherit an active copy of the gene from the parent containing the transposon will not exhibit mutant phenotype. Rare gametes from this parent will contain an insertion of a transposon within the gene of interest. These progeny will be apparent. They exhibit the mutant phenotype in the first generation after the cross. In *non-targeted tagging* attempt is made to identify and recover new mutations for a variety of different plant characteristics. In this method, plants are isolated which have inherited a germinal transposon excision event. Many of these individuals will contain a transposon at a new location within the plant genome. If it is located within a gene, the transposon will probably inactivate it. However, as the insertion will almost certainly be heterozygous in this generation, no mutant phenotype will be detected.

These plants are self fertilized and mutants are screened in subsequent generation. Transposon tagging is a stepwise procedure. The first step is to isolate an interesting mutation which is the result of the insertion of transposon within a gene, i.e., a mutant for a specific trait is identified. Isolation of genes by transposon tagging is itself a stepwise procedure. First a trasposon is inserted within a gene. Such insertion results in induction of mutations which may be detected phenotypically. A transposon sequence probe is used to isolate DNA fragment containing the transposon. Sequences flanking the transposon can be used to screen a library made from non-mutant plants. A genomic library is created after the homozygous mutant is recovered. It is screened with a radioactive labeled probe of the transposable element sequences. The screened selected clone showing homology to the transposable element probe is isolated. It will contain the element in the clone. Sequences for the mutated gene will lie adjacent to the element, as the mutation is generated by the integration of the element within the gene sequence. A subclone is developed containing sequences from the gene is then developed from the non-transposable element DNA of the original clone. This clone is then used to screen a genomic library containing DNA from a normal plant. In this manner, any clone that is selected should contain a full, normal copy of the gene. This system is very powerful and hence it is applied in introducing elements from corn and *Antirrhinum* into other species using transformation techniques. It has been demonstrated that these elements can be induced to move from one location to another in the new species. If this movement is coupled with the appearance of mutant phenotype, then the gene responsible for the phenotype can be cloned in that particular species. These techniques are applied in transposon tagging in those plant species where active transposable elements have not been identified.

SUMMARY

Recombinant DNA technology has been extensively used with microorganisms. It is also applied for gene manipulation in plants. *Agrobacterium* is a gram-negative, non-sporing, motile, rod-shaped bacterium. Two species of this bacterium have plasmid viz. Ti in *A. tumifaciens* and Ri in *A. rhizogenes*. *Agrobacterium tumifaciens* causes crown galls tumors in plants. The tumor is formed by the genetic expression of Ti-plasmid of this bacterium. vir region has 25 genes with 8 operons, out of them six (excluding *virA* and *virG*) encode for various proteins involved in T-DNA transfer. Ti-plasmid is used as a vector. Two strategies are applied to clone vectors in recipient plants. There are two methods to transfer transgenes using *Agrobacterium* viz. (i) co-culture with tissue explants and (ii) in *planta* transformation.

There are certain genes which can move from one place in a chromosome to an entirely different site, and effects on phenotypic expression. These movable genes and the phenomenon were named as transposable elements and transposition, respectively. Transposition is joining the transposon to its target site. Transposition occurs by replicative and nonreplicative mechanisms. Transposition pathway of phage Mu was studied in details.

Chapter - 18

TRANSGENIC PLANTS

Transgenic plants carry transgene, i.e., DNA introduced into them by other means and not through gametes. The complete exercise to introduce a desirable gene in a plant needs four components 1 vector, 2 selectable markers, 3 efficient techniques for transformation and 4 regeneration. Interesting genes are identified, isolated, modified, and transformed in plants by T-DNA or transposable elements. This technical effort results in genetically modified plants, called transgenic plants and the entire technique is known as transgenesis. The genes introduced in new host plant should integrate in it and thereby inactivate the native gene of this host plant. Transgenesis has proved in developing plants (especially crop plants) to resist to viral disease, insects pests, herbicides and extending the shelf- life of fruits.

Transgenic Plants Resistant for Pest and the Herbicide- Tolerant Plants

Herbicide-Tolerant Crops Weeds are hazardous in crop productivity. Herbicides are used to control weeds. Several classes of herbicides are quite effective for broad spectrum weed control. Such herbicides are either nonselective and kill crop plants or they significantly injure some crops. Most of the modern herbicide interferes with amino acid biosynthesis in plants. Chemical safeners can be used to reduce herbicide toxicity to a crop. To develop herbicide tolerant crops is a way to protect crop from herbicide damage. Two general methods have been used to generate herbicide-tolerant crops: selection and direct herbicide-tolerant gene insertion using genetic engineering techniques. There is some phenomenon that provides crop plant selectivity. This includes metabolic detoxification of the herbicide, internal compartmentalization and sequestering of the herbicide within crop plant cells, and physical and physiological barriers in the crop plant which interfere with herbicide uptake, translocation, or target-site interaction. Herbicide selective pressure during plant cell culture has been used successfully to produce imidazolinone and/or sulfonylurea- tolerant cell lines and plants.

Genetic resistance is developed by the use of mutagens and/or herbicide selection pressure during *in vitro* cell culture or seed germination. Mutagenesis, prior to seedling growth or plant regeneration, has also yielded herbicide-tolerant mutants. *Arabidopsis* mutant for gene *csrI* and *imr* produced resistance to chlorsulfuron and imidazolinone, respectively. Resistance to isoxaben was obtained from two unlinked mutations generated after *Arabidopsis* seed mutagenesis. However, mutagenesis and genetic selection techniques may not allow identification of herbicide-tolerant crop genotypes for certain herbicides. Genetic engineering of herbicide tolerance is based on inserting genes to express proteins which give herbicide tolerance. Three mechanisms are applied (1) target-site modification-herbicide tolerant proteins are introduced. Such mechanism gives tolerance to sulfonylureas, imidazolinones, glyphosate, and atrazine. (2) target-site overproduction approach is applied for glyphosate and phosphinothricin tolerance (3) metabolic inactivation of the herbicide i.e., genetic insertion of an herbicide inactivation enzyme. This mechanism is applicable to produce phosphinothricin, bromoxynil, or 2,4-dichlorophenoxyacetic acid. *DL-Phosphinothricin* (PPT) is a nonselective, broad-spectrum, postemergent herbicide. It is a potent inhibitor of the enzyme glutamine synthetase (GS). It can detoxify ammonia released by nitrate reduction, amino acid catabolism, and respiration. Two approaches have been applied for PPT tolerance. First mechanism is the overproduction of the target enzyme. Transgenic tobacco plants have shown PPT tolerance.

This plant showed over expression of an alfalfa cytosolic GS gene regulated by either the cauliflower mosaic virus 35S promoter or the potato ST/LS1 promoter. Metabolic inactivation of herbicide is another mechanism of PPT tolerance. L- Phosphinothricin is acted by enzyme phosphinothricin acetyl transferase (PAT). bar gene isolated from *Streptomyces hygroscopicus* encodes for this enzyme. Transgenic tobacco, tomato, and potato plants, expressing the bar gene under the regulation of the CaMV 35S promoter, were tolerant to 4-10 times the field rate of application of the herbicide. The *bar* gene was used successfully as a selectable marker for the transformation. *Bromoxynil tolerance* this herbicide is used for selective, postemergence control of seedling broad leaf weeds in wheat, barley, and seedling turf grasses. The mode of action is based on inhibition of photosynthetic electron transport . It binds a component of the quinone-binding complex of Photosystem II. A soil inhabiting bacteria *Klebsiella pneumoniae* subsp. *ozaenae*, has nitrilase activity, which can detoxify bromoxynil. The latter is hydrolyzed to 3,5-di-bromo-4-hydroxybenzoic acid. The bromoxynil nitrilase gene (*bxn*) is present in plasmid. This gene was transformed to tobacco plant. As a result of expression of this gene, nitrilase was found in the cytosol of the transgenic plant. The biochemical reaction was driven by the light-regulated ribulose-bisphosphate carboxylase small subunit promoter. Such plants have the quality to resist bromoxynil. In other transgenic plants different level of nitrilase activity was observed. *Glyphosate* an active principle of Roundup herbicide, which is used as a non-selective post emergence herbicides. It is highly effective against the majority of annual and perennial grasses and broad- leaved weeds. This compound inhibits aromatic amino acid biosynthesis. These amino acids are required for meristem function. It works by competitive inhibition of the enzyme5-enol-pyruvylshikimic acid 3-phsophate (EPSP) synthase. The latter catalyzes the synthesis of 5-enol-pyruvylshikimic acid-3-phosphate from phosphenol-pyruvate and shikimate-3-phosphate. This is an important biochemical reaction of pathway of synthesis of some amino acid like phenylalanine, tyrosine and tryptophan. Overproduction of EPSP synthase provides resistant to this herbicide. This is encoded by gene aroA, which has been isolated and studied in *E.coli*, *Salmonella typhimurium* and *Arabidopsis thaliana*. Base pair sequence in resistant and sensitive isolates have shown that in resistant mutant there is a change in single base pair (C to T). This simple change resulted in a change from proline to serine at position 101 of protein.

Higher plants (*Petunia hybrida* and *Corydalis sempervirens*) showing over production of EPSP synthase exhibit glyphosate resistance. There are two approaches for the production of genetically engineered glyphosate resistant plants (a) transfer of glyphosate sensitive EPSP synthase (EPSPS) under the control of a powerful promoter resulting in over expression of this enzyme, in other words it is overproduction of wild-type plant EPSPS and/or (b) transfer of a mutated gene encoding glycophosate resistant EPSPS. The glyphosate-resistant petunia cell line was used to isolate EPSP cDNA gene and a chimeric gene under the control of the 35S cauliflower mosaic virus promoter was constructed. This was transferred to Petunia cells, which demonstrated 40-folds increase in EPSP synthase activity. The field results were appreciable. There is a second method for isolation of a mutated glyphosate- resistant EPSP synthase encoding gene from *S. typhimurium* and its insertion and expression in higher plants. The mutated gene was fused to an octopine synthase promoter so that it can express itself in plants. Regenerated transgenic plants showed glyphosate tolerance. However, impaired growth behaviour was also observed. In addition to the petunia EPSPS cDNA, a full length cDNA clone of tomato EPSPS and the genes for petunia and *Arabidopsis* have been isolated. These genes have seven introns and there is a high degree of conservation in the amino acid sequences of the mature EPSPS of these plants. Transgenic tobacco plants were produced which contained the mutant *aroA* gene driven by either of two constitutive promoters. The gene was transferred using *Agrobacterium* vectors. The promoters were octopine synthase (octopine dehydrogenase) and mannopine synthase (mannopine dehydrogenase). The plants showed partial tolerance for roundup. Transgenic plants with improved glyphosate tolerance have been obtained by using other mutant EPSPS genes with chloroplast transit peptide sequences. The transit peptide i.e., the amino-terminal sequences direct the precursor proteins into the chloroplast. These precursors proteins located in the plastids. They are coded by nuclear genes and synthesized on cytoplasmatic ribosomes. EPSPS itself is also synthesized as a precursor protein which is enzymatically active and binds glyphosate. Recently, it has proved possible to transfer higher plant gene for EPSPS to *E.coli* and obtain glyphosate-resistant enzymes by site-directed mutagenesis a selection of resistant strains. *E.coli* B was grown on medium containing glyphosate.

The highly glyphosate-insensitive mutant EPSPS was isolated. The *aroA* gene which encodes this mutant enzyme (SM-1) was cloned and contained a single amino acid substitution. The mutated DNA sequences were incorporated into transgenic plants. Ti plasmid was used as vector. Gene constructs were used that ensured the synthesis of EPSPS with an N-terminal transit peptide for transport into the chloroplasts. Transgenic tobacco plants have been produced which express glyphosate tolerant bacterial and plant mutant EPSPS genes. The gene expression was regulated by the CaMV 35S promoter. The latter caused constitutively high-level expression. *Sulfonylurea and Imidazolinone* sulfonylurea is an active principle of sulfomenturon methyl and chlorsulfuron, and imazapyr. While, American Cyanamide heribicide "Arsenal" has Imidazolinones as active principle. Both of these groups interfere in biosynthesis of essential amino acid in plants. Both classes of herbicides achieve their effect by inhibition of the same enzyme i.e., acetolactate synthetase (ALS) or acetohydroxy acid synthase (AHAS). ALS is involved in the synthesis of the branched chain amino acids valine, leucine and isoleucine. Bacterial, plant, and yeast ALS enzymes have been identified which are tolerant to sulfonylureas. The heribicide-insensitive ALS gene was obtained from tobacco plants. Smith *et al.*,(1988) isolated two mutant tobacco ALS genes, the Hra and C3 lines. The Hra and C3 lines had two amino acid and one amino acid substitutions, respectively.

The substitutions in Hra line are proline to alanine substitution at amino acid 196 and a tryptophan to leucine substitution at amino acid 573. In C3 line there is only one substitution i.e., proline to glutamine at amino acid 196. The sulfonylurea-resistant *Arabidopsis csrI* gene, under the regulation of its own promoter, has been expressed in transgenic tobacco plants. The transformation was carried out by *Agrobacterium*-mediated method. *2,4-Dichlorophenoxyacetic acid* (2,4-D) is highly active against many species of broad-leaved weeds and is widely used as a weed control agent on cereal crops. Bacterium *Alcaligenes eutrophs* has the ability to degrade 2,4-D. the degrative pathway comprised of 5 enzymes that convert 2,4-D to chloromalylacetic acid, which is subsequently metabolized to succinic acid. The gene (*rfdA*) encodes a 2,4-D monoxygenase which converts 2,4-D into 2,4-dichlorophenol by cleavage of the aliphatic side chain. Lyon *et al.* (1989) cloned *rfdA* gene of *A. eutrophus* and placed in a CaMV 35S promoter expression vector. Transgenic tobacco plants were produced via *Agrobacterium*-mediated transformation. Transgenic progeny plants survived up to 8 times the usual field application rate.

Pest Resistant Transgenic Plants

World is facing a major loss in agricultural and other crops due to damages caused by insects. Plants have evolved an impressive array of defense for insect predation. The basic mechanisms are temporal avoidance, physical defenses and chemical defenses. A simple approach to develop insect resistant crop, is transforming genes in plants which makes them insect resistant or insect tolerant. The functioning of genes transferred is encoding insecticidal enzymes or toxins.

Bacillus thuringiensis is a biological insecticide. This fermented bacterium showed high specificity toward lepidopteran pests. It is rather environmentally safe. Spray is still very limited because of low stability of the protein in the field. Schnepf *et al.* (1985) Schnepf and Whiteley (1985) first reported toxin gene DNA sequence of this bacterium. On the basis of similarity between cry (crystal protein) gene sequence homologies, at least four unique classes of insecticial genes are identified.

The currently accepted classification is lepidopteran-specific, *cryI*; lepidopteran- and dipteran-secific *cryII*; coleopteran-specific *cryIII*; and those active against dipteran species, *cryIV*. Many strains of this bacterium contain additional genes encoding alternative toxin peptides. They may have gene encoding enzymatic pathways for biosynthesis of other metabolites. These metabolites are like nucleoside analogues that have broad cytotoxic activities against organisms other than insects. Out of the four genes, *cryI* gene is thoroughly studied. It encodes proteinaceous protoxins of 130-140 kDa, which are localized within bacterial spores as bipyramidal crystal inclusions. The guts of lepidopteran insects have alkaline pH (9-12). The crystals of δ-endotoxin and associated peptides disperse at this medium. The proteolytic activity is then capable of the high molecular weight protoxins to smaller, insecticidally active peptides of 65-70 kDa. A second class of δ-endotoxin with coleopteran toxicity, the cryIII genes has been studied. It is rhomboidal crystal inclusions in spores, and has been found in three morphologically different isolates of *Bacillus thuringiensis*. The DNA sequences of the protoxin genes of these strains are identical. However, the precise character of the crystallized protoxin or the final insecticidal fragments may differ. This difference is due to the potential for alternative proteolytic processing occurring within the several hosts. The genes each encode a 72-kDa peptide. It is processed to an approximately 66-kDa toxin by the proteolytic cleavage of 57 amino-terminal residues.

The tendotoxin has the single gene nature of the protein and the advantages to take the task for transferring this protein to plant. Schnepf and Whiteley reported that a truncated form of the gene, encoding only the amino- terminal portion of the protoxin, retained essentially all the insecticidal activity of the intact gene.

The range of expression of these genes was typical of chimeric gene expression. The variation in level of expression was due to insertion into differing chromosomal locations. Transformed calli was well able to regenerate the plants, which appeared to be as healthy as control. The endotoxin was produced in tobacco plants transferred with a shortened by 2 gene from *Bacillus thuringiensis* strain Berliner 1715 encoding the amino terminal half of the *cryI*A(b) protein. Expression of the transgene was under control of the *Agrobacterium tumefaciens* mannopine synthase gene promoter which confers constitutive transcription. High protection was observed with transgenic plants when kanamycin was used as a selecting agent during the regeneration of transformants. The transgenic plants expressing the condensed toxin as a fusion protein with neomycin phospho transferase, as a selectable marker. The transgenic tobacco and tomato plants expressing the amino terminal half of the *cryI* A(a) *cryI*A(c) endotoxin, respectively obtained from *Bacillus thuringiensis* var ***kurstaki*** strains. The experiments showed transcription of the toxin gene was driven by CaMV 35 S promoter. However, it is doubtful that all pest species will be eradicated by *Bacillus thuringiensis*. Much more, the feeding habit of other pests may intake the toxin of this bacterium.

Protease Inhibitor Approach protease inhibitors are widely distributed within the plant kingdom. They are particularly abundant in seeds and storage organs. Synthesis of protease is one of the mechanisms, which makes crop insect tolerant. The effects on susceptible insects are generally seen as an increase in mortality, decrease in growth rate, and prolongation of developmental period of the larvae. Experiments conducted with insects raised on artificial diets into which inhibitors had been incorporated have demonstrated the protective role for protease inhibitors. Studies of the wound-induced synthesis of inhibitors in potato and tomato plants have established a direct role off protease inhibitors in plant protection. Protease has anti- metabolic activity against a wide range of insects. These enzymes have no effect on endogenous plant metabolism, but show specificity towards animals.

The primary effect of protease inhibitors is assumed to be to cause a critical shortage of essential amino acids and their effectiveness. In plants the main types of proteases are serine proteases, thiol protease, and metalloprotease. Plant kingdom has at least eight non-homologous families of protease inhibitor. There is tremendous variability in the primary structure of homologous proteins. The insect tolerant transgenic plants had transformed chimeric gene constructs for plant expression, using the coding sequence of PI-II gene coding sequences from tomato and potato. These genes were fused to the CaMV35S gene constitutive promoter and introduced into transgenic tobacco. Experiments on predator insect horn-worm (*Manduca sexta*)of tobacco showed that the hornworm caterpillars were shown to be significantly reduced on transgenic plant expressing PI-II proteins. The degree of protection afforded by CpTI alone is probably insufficient to be commercially acceptable in developed, high-input agricultural systems. An approach of package of gene concept is introduced for the sake of increased protective efficacy, spectrum of activity, and durability of resistance. In this regard, a package of genes CpTI and pea lectin was introduced in tobacco plant. The product of package i.e., two genes (CpTI and pea lectin) transformed tobacco plants together showed an additive protective effect against tobacco budworm caterpillars.

Transgenic Plants with Enhanced Resistance to Microbial Pathogens

Plants have their own defense mechanism to face plant pathogens, specially bacteria, virus and fungi. The host plant undergoes a number of physical and biochemical changes as a part of defense response.

Fungal Resistant Crop Chitin is a major component of fungal cell wall. This is hydrolyzed by a lytic enzyme, known as chitinase. The level of this enzyme goes high in plants facing fungal invasion. However, in many cases, the chitinase levels are increased on treatment with the phytohormone ethylene or oligosaccharide elicitors. Chitinase have been purified from several plant species. Broglie *et al.* (1986) reported that in bean, chitinase is encoded by a small multigene family consisting of at least three members. Using cloned cDNAs as hybridization probes, the level of chitinase mRNA in bean seedlings was found to increase 50 to 75-fold following ethylene or elicitor treatment. Nucleotide sequence analysis of isolated cDNA clones indicate that at least two of the chitinase genes are transcribed in response to ethylene treatment. In transgenic tobacco plant, the expression of these genes was named as CH5A and CH5B. A 43 base pair sequence is precisely conserved between the CH5A and CH5B genes may play a role in transcriptional regulation of chitinase gene expression by ethylene. The induction of the CH5B gene in response to attack by several pathogenic fungi was studied. β-glucuronidase was used to assess the temporal and spatial pattern of CH5B expression during fungal infection.

A chimeric gene was constructed by fusing a 1.7-kb fragment bearing CH5B upstream sequences to the coding region of β-glucuronidase followed by the termination signals of the noplaine synthase gene. A novel chitinase gene was constructed by replacing the native, inducible promoter with the high-level, constitutive 35S promoter of cauliflower mosaic virus. Western blot analysis technique was used for assay of chimeric gene expression in transgenic plant. It showed that high levels of the bean protein are produced in healthy, unstressed plants. Further, the polypeptide from transgenic plant and bean chitinase had the same size. This gives the conclusion that the precursor polypeptide was correctly processed in the heterologous system and localized in the correct subcellular compartment. The tobacco plants exhibiting high-level constitutive chitinase expression would be more resistant to attack by fungus *Rhizoctonia solani*. With a number of experiments, it appears that growth of transgenic plants with high-constitutive levels of chitinase may be beneficial in reducing the indigenous levels of *Rhizoctonia* in the soil by interfering with their ability to produce successful infection.

According to Flor's "gene to gene hypothesis" there is a gene in host which provides defense, while the pathogen has a gene for virulence. The genetic inter-dependence of host and pathogen determines the outcome of the interaction i.e., disease. Different approaches are being pursued for isolation and molecular characterization of resistance genes including map- based cloning and transposon tagging. The resistant gene is isolated and transferred to other cultivars/species. This facilitates resistance to specific races of fungal pathogen. The disadvantage is that the resulting transgenic plant will remain susceptible to other races of the pathogens.

Stirpe *et al.* (1992), have observed plant ribosome- inactivating proteins. Such proteins inhibit protein synthesis in target cells by N-glycosidic cleavage of 28S rRNA. They do not interfere with the functioning of self ribosomes but exert inhibitory activity towards fungal ribosome.

Viral Resistant Crop The principal to develop viral resistant crop is by introducing of virus-derived or a virus- targeted sequence into the genome of a potential plant. These sequences interfere with specific stages in the viral infection cycle. Virus genome has either RNA or DNA. RNA may be messenger sense (plus strand) or ambisense sense (minus strand). DNA may be single stranded or double stranded. All viruses encode proteins involved in the replication of their nucleic acid. Single- stranded DNA viruses contain gene for enzymes necessary for replication via complementary DNA. In 1939, Mc Kinney observed that if a plant is infected with mild strain of a virus, it is able to resist infection by other related strains. This is called as cross infection. Experiments on cross protein-mediated to viruses has proved successful in plant genetic engineering. The degree of resistance is correlated with amount of intact coat protein expressed in transgenic plants. Beachy (1986) introduced the TMV coat protein gene into the genome of tobacco plants. This transgenic plant expressed high levels of coat protein, and the plant infected by this virus showed delayed disease development. Genes encoding for coat protein have been isolated from different single-stranded (plus-strand) RNA plant viruses. These genes have been transferred into various crop plants. The genes are cloned and transformed to plant using *A. tumifaciens* as vector. The coat protein approach seems to completely prevent virus infection at low inoculum concentrations and often delays the appearance and extent of virus symptoms at higher concentrations.

Satellite is small extragenomic RNA species that depend on a helper virus for replication. However, they themselves are not required for the replication of the helper virus. Much more, they have no appreciable sequence similarity with the genome of their helper virus. Till now one or more viral satellites or satellite- like replicating RNAs have been found in association with some 30 plant viruses. Kassanis (1962) was the first to observe that viral satellites have ability to modify the disease symptoms of their helper viruses. Some satellite RNAs are able to attenuate the symptoms of their helper virus. Transgenic plants that express viral satellite RNA sequences can resist crop diseases induced by their specific helper viruses. Certain satellites show autocatalytic cleavage and ligation reactions. Cucumber mosaic virus (CMV) infects over 800 plant species. A vast study is carried out on the satellite RNAs of cucumber mosaic virus (CMV) in disease protection. Other viral satellites and secondary RNA agents such as DI-RNAs have similar properties. The apparent advantages of satellite-mediated tolerance are that it seems to be independent of both the strength of the challenge inoculum and the amount of RNA expressed from the inserted gene. Disadvantages of the satellite-mediated approach are that some plant tissue is infected and tolerance is delayed. In a few cases certain satellite RNAs in combination with specific helper viruses and host plants can aggravate disease symptoms.

A combination of protein coat approach and satellite helper technique in transgenic plants could provide two lines of defense against viral diseases. Coat protein can prevent or delay the onset of viral disease, provides the first line of defense. While, satellite RNA, provides a late developing tolerance independent of the strength of the inoculum will provide a second line of defense.

Deletion mutants of genomic viral sequences, which depend on their parent virus for replication, are known as defective interfering particles. Like satellite RNAs, these particles can attenuate the disease symptoms of their helper virus by interfering with its replications. The effort to construct artificial defective interfering particles of some virus is being practiced. Antisense constructs have generated protection against RNA and DNA viruses.

Antisense coat protein genes mediate high level or resistance in PLRV. Huntley and Hall (1993) reported an antisense transcript of BMV RNA 3 intercistronic region interferes with viral RNA replication. Ribozyme are RNA molecules catalyzing endonucleolytic cleavage reactions. Some viriods and satellite RNAs can undergo self-cleavage reactions. Mostly, the cleavage site consists of a consensus structure. These consensuses are "hammerhead" motif. The nucleotide region directing the catalysis of cleavage reaction could be separated from the region where the cleavage occurs and the recognition of the target could be modified by changing the nucleotide sequence of the regions flanking the cleavage site.

Bacterial Resistant Pathogen one of the approach to produce transgenic plant for bacterial disease is based on the transformation of gene encoding a toxin-detoxifying enzymes from pathogen itself. A dipeptide tabtoxin is responsible for chlorosis in wild fire disease on tobacco. The causal organism is *Pseudomonas syringae pv. tabaci*. In plants, tabtoxin is converted to tabtoxinine-B-lactam. The latter inhibits glutamine synthetase leading to an accumulation of cytotoxic ammonia. The pathogen protects itself against the toxin by expression of tabtoxin resistance gene, ttr. This gene encodes an enzyme that acetylates tabtoxin. Anzai *et al.* (1989), observed that the transgenic tobacco plants constitutively expressing the ttr gene under control of CaMV 35S promoter did not produce the chlorotic halo as it is observed in diseased plants.

Enzyme ornithine carbamoyl transferase takes part in citrulline bio-synthesis. The gene encoding this enzyme was isolated from *Pseudomonas syringae pv. phaseolicola* and was introduced in the genome of the tobacco plants. The transgenic plants expressed an ornithine carbamoyl transferase which was insensitive to phaseolotoxin produced by this bacterium. The progeny of these plants displayed a reduction in disease symptoms.

Murein is a structural component of bacterial cell wall. It undergoes hydrolytic cleavage by lysozymes. Genes encoding lysozyme has been isolated from other sources than plants and is used for plant genetic engineering.

SUMMARY

Transgenic plants carry transgene. Interesting genes are identified, isolated, modified, and transformed in plants by T-DNA or transposable elements. Weeds are hazardous in crop productivity. Herbicides are used to control weeds. To develop herbicide tolerant crops is a way to protect crop from herbicide damage. Two general methods have been used to generate herbicide- tolerant crops: selection and direct herbicide-tolerant gene insertion using genetic engineering techniques. Plants have evolved an impressive array of defense for insect predation. A simple approach to develop insect resistant crop, is transforming genes in plants which makes them insect resistant or insect tolerant. *Bacillus thuringiensis* is a biological insecticide. Protease inhibitors are widely distributed within the plant kingdom. They are particularly abundant in seeds and storage organs. Synthesis of protease is one of the mechanisms, which make crop insect tolerant. Transgenic plants have been successfully prepared which show resistance towards plant pathogens like fungi, bacteria and viruses.

Chapter - 19

TISSUE CULTURE

Plant tissue culture is commonly used to describe the *in vitro* and aseptic growth of any plant part on a nutrient medium.

Historical background

The history of plant tissue culture begins with the `cell doctrine' proposed by Schleiden (1838) and Schwan (1839), according to this theory cell is a functional unit, which means that cells are autonomous. doHenri-Lousi duhamel duMonceau's (1756) performed pioneering experiments on wound healing in plants and demonstrated spontaneous callus formation on the decroticated region of elm plants.

Gottlieb Haberlandt isolated mesophyll cells and cultured it using Knop's nutrient solution, supplemented with sucrose, asparagines and peptone. Haberlandt cultured mesophyll cells of *Lamium purpueum* and *Eichhornia crassipes*, epidermal cells of *Ornithogalum* and hair cells of *Pulmonaria*. The cells survived 3-4 weeks but without cell division, there was increase in cell size. Pierre Roger Gautheret could develop callus form piece of cambium cut from trees (*Acer pseudoplantanus, Sambuscus nigra* and *Salix capraea*) cultured on solid media. He also worked on application of yeast extracts (B vitamin) and the auixn (IAA) in cambium culture. In 1939, he reported the propagation of carrot as the first plant tissue culture of unlimited growth. In 1942, Gautheret observed development of buds in tissue culture of *Ulmus* and plantlets tissue culture in Chicory. Georges Morel developed the method of meristem culture for the elimination of viruses. He enriched natural extracts like yeast extract and coconut milk in culture media, and obtained indefinite growth of monocotyledonous tissues. He successfully multiplied tropical orchids through division of protocorms. In 1955 Morel and Martin recovered virus free potato. In 1971, Morel isolated octopine and nopaline chemicals from the dicot plants infected by *Agrobacterium tumifaciens*. This bacteria posses Ti plasmid, and when this plasmid integrates in the host genome produces these chemicals. Philip R. White (1934) reported root tip culture of tomato and obtained indefinite growth of roots. He used Knop's solution with sucrose and three vitamins (pyrodoxine, thiamine and nicotinic acid) instead of yeast extract.

Some Basic Terms and Definitions

Androgenesis : *in vitro* development of haploid plants originating from totiopotent pollen grains through a series of cell division and differentiation.

Callus : An unorganized growth of plant cells, *in vitro* on culture medium.

Clonal Propagation : A sexual reproduction or vegetative propagation of plants considered to be physiologically man/or genetically uniform and which originated from a single individual or explant.

Clone : Population of plants derived from a single individual by vegetative propagation or by regeneration of plants from in vitro cultures.

Population of cells mitotically derived from a single cell.

Caulogenesis : The development of adventitious shoot bud from the callus tissue during organogenesis. It was observed that if in the culture medium concentration of kinetin (a hormone) is more than auxin, caulogenesis takes place.

Cytodifferentiation : In plant tissue culture during growth and maturation of callus tissue, few dedifferentiated cells are redifferentiation of vascular tissue, particularly tracheary elements. The ultrastructural study of cytodifferentiation shows that a chain of intra-cellular degradative changes (autodestruction of cell organelles, like chloroplast, endoplasmic reticulum, dictyosome, ribosome, mitochondria) are associated with cyto differentiation.

Dedifferentiation : when a mature cell reverts to a meristematic state and forming undifferentiated callus tissue.

Embryogenesis : Initiation and development of embryo from zygotes, or of bipolar structures from somatic cells that parallel the developmental path of zygotic embryo. In the later case, such embryo may be either adventitious of *de novo* in origin.

Explant : the tissue taken from a plant or seed and transferred to a culture medium to establish a tissue culture system or regenerate a plant.

Meristemoid : a localized group of meristematic cells that arise in the callus tissue. Adventitious roots and shoots develop from these cells.

Organogenesis : the development of adventitious organs or primordial. The process of initiation and development of a structure shows natural organ, the initiation, of which is temporally separated from the initiation of other organs. Such organs may be adventious or *de novo* in origin.

In some cultured tissues, due to error, some anomalous structures are formed. These structures develop directly from the periphery of the callus tissue and not from organized meristemoid.

Production of Haploids Haploid set of chromosome of a plant is obtained by different techniques. It facilitates to conduct various gene manipulations and to study mutation and other genetic studies. Gynogenesis is production of a haploid individual by the development of an unfertilized egg-cell. Haploids may be produced by inactivating egg nucleus and introducing a male nucleus in it. This method is known as ovule androgenesis. Semigamy also results in a haploid chimera. It is a condition wherein the nucleus of the egg-cell and generative nucleus of the germinated pollen grain divide indecently. In intergeneric or interspecific crosses, a condition of selective elimination may arise, where after fertilization;

genome of only one parent develops. This results in embryo formation with only haploid set of genome. Physical techniques used to obtain haploids are chemical treatment, temperature shocks, exposure of X-ray or UV-light.

Redifferentiation : the ability of a dedifferentiated cell to form a whole plant of plant organs.

Rhizogenesis : development of adventitious root in callus tissue during organogenesis. It was observed that if in the culture medium, concentration of kinetin is lower than that of auxin, rhizogenesis occurs.

Somatic Embryo : embryos formed (*in vivo* or *in vitro*) by the sporophytic cells other than zygote.

Somatic Embyrogenesis : is the process of a single cell or a group of cells initiating the developmental pathway that leads to reproducible regeneration of non-zygotic embryos capable of germinating to form complete plants.

Totiopotency : the property of normal cells that they have the genetic potential to give rise to a multicellular or multiorganed individual. Thus, it is characteristic of a cell having the potential for forming all the cell types in the adult organism is retained. Exceptions include terminally specialized cells.

Xylogenesis : differentiation of parenchyma into cells that have localized secondary wall thickenings as seen in xylem of vascular plants.

The Different Culture System

Organ Culture

In this technique an excised organ primordial or whole or part of an organ is cultured *in vitro* and maintained in a way that may allow differentiation and preservation of the architecture and/or function.

Callus Culture

An unorganized and proliferated mass of tissue is called Callus. In plants callus may be formed in natural course either due to deep wound or due to some diseases. In callus culture explant is obtained form such tissue and cultured aspetically *in vitro*, for this purpose culture medium is supplemented with phytochrome.

Cell Suspension Culture

In this type of culture, a single cell or few cells (aggregates) are cultured on a liquid medium. During culture, the medium is kept continuously agitating.

Embryo Culture

The embryo is excised carefully so that the ovule, seed or capsule is not damaged and this excised explant is cultured in a test tube. Studies on embryo culture helps to understand the factors that influence the embryonic growth under controlled conditions.

Anther Culture

The anthers are excised from unopened flower bud. The anthers are collected at a precise and critical stage. During culture, the microspores within the anther develop into callus tissue or embryoid.

Pollen Culture

Pollen of uninucleated stage is cultured, which instead of developing into microspore forms an embryoid or callus tissue. It is collected aseptically from intact anther.

The plantlets developed from Anther or Pollen Culture is either through organogenesis or embryogenesis. They are haploid, and are very important for the study of fundamental and applied aspects of genetics in higher plants.

Protoplast Culture

Protoplast is isolated from any part of the plant by removing cell wall of the tissues. The plasma membrane is present. Such protoplast has potential to regenerate cell wall.

Tissue culture-Basic techniques

The Explant In tissue culture most important is to obtain explant and maintain it under asepsis. This explant is grown on a culture medium under controlled conditions. Any plant part can be used as an explant, and it has to be sterilized before transferring it to the culture medium. Few chemicals are used for sterilization of explants viz. sodium hypochlorite (1-4%), calcium hypochlorite (7%), mercuric chloride (0.1-0.2%), silver nitrate solution (1%), hydrogen peroxide solution (10%), and bromine water(1%). Calcium hypochlorite solution is in general use. Mercury is avoided because of the problem of proper disposal. 70% ethyl alcohol is also used for the purpose of sterilization. The type of a particular sterilant and its concentration to be used and time of exposure depends upon the explant. Teepol, Tween-80, is incorporated into sterilizing solution to enhance the penetration and effectiveness of the sterilizing agent. After sterilization, explant should be thoroughly washed several times with sterilized distilled water so that the sterilant is completely removed and the explant is ready for transferring to the nutrient medium. When explant is from underground plant parts (roots, tubers, rhizomes) are used for tissue culture, it has to be pretreated with fungicide and antibiotics.

The Culture Medium In the early researches, culture media was prepared based on Knop's solution. Later the discovery of plant growth regulators and role of micronutrients and vitamins in plant growth and nutrition invented changes in culture medium. Murashige and Skoog (1962) proposed a combination of a culture medium. This medium is known as Murashige and Skoog medium and is used for all types of tissue culture work. Later, Linsmaier and Skoog (1965B), Gamborg *et. al.* (1968), Schenk and Hildebrandt (1972), Bitsch and Bitsch (1969) and Llyod and McCown (1980) also proposed different types of culture medium used for tissue culture.

A nutrient media used for culturing of plant parts, has basic ingredients like carbon source, inorganic nutrients, vitamins, growth regulators and some organic supplements. The successful plant tissue culture depends upon the choice of nutrient medium. Carbon is essential for life, as it acts as the energy source. In culture medium glucose, fructose, maltose or sucrose is generally used as carbon source. However, 2-4% sucrose has proved to be the best carbon source. The inorganic nutrients are salts, which are used as macro- and micro-nutrients. Plants need potassium, calcium, magnesium and sulphur as macro nutrients, nitrogen is utilized in the form of nitrate salts of metals or in the form of ammonium salts. The micro-nutrients include I, Bo, Mn, Zn, Mo, Cu, Co and Fe. Vitamins viz, Thiamine, pyridoxine and nicotinic acid are used in culture medium. Plant growth regulators are used in culture medium for two purposes, i.e., for cell proliferation and for cell regeneration. 2,4-dichlorophenoxy acetic acid (2,4-D) or 1-napthalene acetic acid (NAA) and a cytokinin (kinetin, benzyl adenosine, 2-isopentyladenosine, zeatin, thidiazuron) are used for cell proliferation, and low auxin (NAA, IAA, IBA and a cytokinin in high amount, but not 2,4-D is used for regeneration. Amino acid (glycine) and casine hydrolysate (nitrogen supplement) are used as organic supplements. Coconut milk, yeast extracts, peptone, and malt extracts are some organic supplement, which may be used in culture medium.

Sterilization the measure to prevent the contamination is called sterilization. Glass vials, metal instruments, and aluminium foil are sterilization by dry techniques. The articles are properly sealed and exposed to hot dry air (160°-180°C) for 2-4 hour in a hot air oven. Autoclaving is the main wet sterilization techniques. Autoclaving is to sterilize things with water vapour under high pressure. The later is defined as 121°C and 15 psi for 15-20 min. The apparatus is a metal vessel. Culture media is sterilized by autoclaving. Certain chemicals like vitamins, hormones or plant extracts may be decomposed by autoclaving. In such cases, filter technique is applied. The solutions are passed through a membrane-filter under positive pressure. The pore size of such filters does not permit bacteria to pass through it. The filters generally used are Millipore or Seitz filter with a pore size of not more than 0.2 μm. Surface sterilization is applied for the plant part used as explant. The plant part is kept in sterilizing solutions. The latter is solution with different concentrations. Solutions generally used for surface sterilization with their dilutions are mercuric chloride (0.1-1.0%), ethyl alcohol (75-95%), hydrogen peroxide (3-12%), bromine water (1-2%), and silver nitrate (1%). The time duration of treatment varies generally, 5-30 minutes. The bacterial and fungal contamination may be removed by this technique. The culture room is washed with detergent, Lysol, zephiram and roccal are effective disinfectant. The surfaces of transfer area are wiped with 2% sodium hypochlorite solution or 95% ethyl alcohol. The culture room and other apparatus used for transfer are exposed to UV radiations. Plant part transfers or other transfers are not conducted under UV radiations.

Application of Tissue Culture

Tissue culture techniques are specially applied in plant breeding, horticulture and forestry. Microporpagation, clonal propagation and other techniques of tissue culture have proved to be fruitful in vast and varied fields.

Haploid plants are used in breeding and improvement of crop species. It has helped in releasing new varieties through F_1 double-haploid system, diploid production, selection of mutants resistant to disease, transfer of desired gene, and production of transgenic plants. Embryo culture technique has practical applications like rescuing embryos from incompatibile crosses, production of monoploids, clonal micropropagation. It has helped in overcoming dormancy and seed sterility. This technique is helpful in shortening breeding cycle. Genetic transformation is transfer and expression of foreign genes into plant cells. It has vast applications especially in introducing genes for herbicide tolerance, and virus resistance. Production of viral disease free plants is the major achievement in plant breeding and crop improvement. Meristem-tip (apical meristem) culture and micrografting techniques are applied for this purpose. However, meristem-tip culture is the only way to obtain a clone of virus free plant. Virus is eliminated from the tissue by thermotherapy. Preservation of germplasm is important to protect and study genetic diversity. It has helped in preserving primitive cultivars and wild relatives of crop plants. Tissue culture technique has provided a germplasm storage procedure. Some natural products, generally called secondary metabolites like vitamins, pigments, essential oils, alkaoids, sweeteners, food flavours, feedstocks, antimicrobial and pharmaceuticals have industrial applications. Plant cell cultures have advantages in metabolite production. *In vitro* conditions facilitate plant cell cultures to produce such metabolites much more than intact plants. Heterogeneity in plant cell culture enables successful selection of cells producing high amounts of secondary compounds. Special techniques have been developed in cell free culture for exploring the heterogeneity. Immobilized plant cell culture technique has optimized the production of such compounds. Immobilzation is carried on gel or membrane entrapment. Another procedure to produce secondary metabolites is two-phase system culture. In this method, cells are kept apart from the product by a separation phase (liquid) placed in a reactor. The product is continuously collected from reactors.

SUMMARY

Plant tissue culture is commonly used to describe the *in vitro* and aseptic growth of any plant part on a nutrient medium. Different types of culture system involve organ culture, callus culture, cell suspension culture, embryo culture, anther culture, pollen culture and protoplast culture. In tissue culture most important is to obtain explants and maintain it under asepsis. Culture media was prepared based on Knop's solution. Later the discovery of plant growth regulators and role of micronutrients and vitamins in plant growth and nutrition invented changes in culture medium. The measure to prevent the contamination is called sterilization. The application of tissue culture is discussed in brief.

Chapter - 20

BIOINFORMATICS

Bioinformatics is a subject in which the computational technology is used to manage/ solve biological information/problems. The basic effort of Bioinformatics is to combine data from various sources. Laboratory automation and computing, has made new advancement in DNA sequencing of very large genomes. Studies on the genome of *Arabidopsis thaliana*, a model plant was funded by the National Science Foundation. One of the most basic operations in bioinformatics involves searching for similarities or homologies between a newly sequenced piece of DNA and previously sequenced DNA segments from various organisms. Bioinformatics covers the use of software, databases and networks for gene and open-reading frame (ORF) identification, Database homology and pattern searching with both DNA and protein sequences; Comparative sequence analysis and multiple sequence alignment; Protein structure prediction with reference to mapping of functional sites; and Protein homology modeling.

Bioinformatics companies and scientists work for (i) collecting and storing data, (ii) searching databases, (iii) classifying data, (iv) analyzing and interpreting the data and (v) distributing biological information derived from sequencing and functional analysis projects.

This chapter will include only the definitions and principles related with bioinformatics.

Genomics is the study of all the genes of an organism. These studies include DNA sequences, structures, gene regulation, interaction and products. The investigations are carried to study the structure and function of very large numbers of genes in a simultaneous fashion. A worldwide initiative aimed at determining a large number of protein structures in a high throughput mode, is defined as *Structural Genomcis*. X-ray crystallography and NMR spectroscopy are important High-throughput tools. Structural genomics enables to understand the molecular events in biological processes. Structural genomics will yield a large number of experimental protein structures and an even larger number of calculated comparative protein structure models. This structural data will be freely available, and evolution in natural ecosystems, agricultural plant genetics, breeding of farm and domestic animals, and human health and disease. Application of Structural genomics in drug discoveries is highly beneficial. *Functional genomics* is the process of determining the function of genes and an understanding of how they affect normal physiology. The basic aim is to supplement genomic sequence data with information about the biological functionality and regulatory control of the product. The best method for understanding gene function is mutagenesis, or the production of changes in the DNA sequence that affect the expression or structure of gene products. Technologies that have emerged from functional genomics are sequencing, positional cloning, knock out, gene expression profiling, model organism development, and developed technology for large-scale protein analysis. *Comparative Genomics* is morphological and physiological studies. It basically involves the comparison of proteomes *i.e.*, the complete protein set of two or more organisms. The nature and significance of differences between genomes also provides a powerful tool for determining the relationship between genotype and phenotype. Comparative genomics is also essential for studying and understanding genetic diversity present in healthy outbreed populations.

One of the major goals of comparative genomics is to try and attempt to predict gene function. Single nucleotide polymorphisms (SNPs) are the points where individuals vary with respect to sequence. The importance of SNPs lies in the fact that these changes in sequence may determine the ways in which subpopulations of a given species are functionally different. The genomic industry has evolved four primary business models: (1) information companies-Structural genomics (2) information companies-Functional companies. Information companies in these two models are involved in sequencing genomic map and deducing gene function. (3) Target drug discovery companies are involved with genetic variability, lead generation and lead optimization, forward integrated drug discovery and pharmacogenomics. (4) Enabling genomic technology companies, to work in the area of sequencers, DNA arrays, High throughput screening, bioinformatics, miniaturization, as reagent companies. *Pharmcogenomics* is an intersection of the fields of pharmacology and genetics. It studies that how genetic variation affect the ways in which people respond to drugs. This field helps in identification and elucidation of genetic variations that will alter the efficacy of a drug or suggest new target drugs. It is predicted that diseases will be reclassified on the basis of biomechanisms and molecular markers. The individuals will be treated for a disease and the medicine tailored to their molecular profile. Such preselections may prove inadequate to support regulatory approval for the patient subpopulation that responds to the drug. The future potentials of pharmacogenomics may be efficacy of administered drugs may be improved, predicting patients best response to a particular drug, powerful, better and safer drug, advanced screening for disease. This will all help in providing better vaccines and decrease in the cost of health care.

Proteome and Proteomics Marc Wilkins coined the term proteome. It is the expressed protein complement of a genome. Leigh Anderson introduced the term Proteomics. It is functional genomics at the protein level, and involves the separation, identification and characterization of proteins in order to understand their functions. Proteomics seeks to provide functional information for all proteins. Thus, it is complementary to genomics. The basic motto of proteomics is to provide functional explanation for the entire proteome. The role of proteome within and among cells cooperates to produce the biochemistry and physiology of individual cells and organisms is defined as *Physiomics*. Study including anatomy and function of cell and whole organism is defined as *Phenomics*. Proteomics yields significant discoveries because this set of proteins is likely to include new enzymes, signaling molecules, and pathways that may be excellent and unanticipated therapeutic targets. Genomics provides the comprehensive sequence and expression data required to advance protein research. Proteomics provides data on the outcome of gene expression.

Proteomics Technology is rather difficult. The DNA molecule is a simple linear molecule and can be amplified. The story is different with proteins. They exhibit complicated structures and cannot be amplified. DNA is basically static: proteins change in innumerable ways even in an individual cell over a short period of time. Thus, it is simple to carry out studies related to DNA and RNA, as compared to study of proteins. The scarce proteins are often critical control elements like protein kinases, or other important enzymes like telomerase. The membrane proteins are hard to separate because they are insoluble without detergents. These proteins are often involved in cell-cell signaling, and make ideal drug targets. Hence, it is critical to detect them.

2-D gel electrophoresis and mass spectroscopy techniques are used in proteomics technology. Proteins are loaded onto an MS plate. Later generating spectra, and searching proteins databases like SWISS-PROT for matches is a time consuming process. To speed up things entire process is automated, i.e., from extracting proteins from the gel through protein identification and quantification. Advances in 2D gel electrophoresis are revolutionizing proteomics by providing a means to separate different proteins for further analysis. 2D gels are transformed into high-resolution digital maps using staining, imaging, and bioinformatics software. Individual proteins are excised from 2D gels: split into polypeptides fragments, and sequenced using mass spectrometry. The amino acid sequence is a "fingerprint" that can be compared with existing sequences in protein and DNA sequence databases. Ruedi Aebersoid has devised an ingenious method for measuring the relative quantities of proteins using mass spectrometry.

Isotope Coded Affinity Tags (ICAT) is one of the most elegant techniques in proteomics. It is applied to make any separation technology, including HPLC and capillary electrophoresis, quantitative, 2-D gel electrophoresis is more useful, as it allows pre-fractionation with quantitation. In this technique, two samples of proteins (out of which, one is control) are labeled with two closely related isotopes. The two samples are mixed together and then proteins are identified using mass spectrometry. Since both samples are labeled with different isotopes, ratios can be calculated. Protein mass fingerprinting technique helps to identify proteins from MS, faster and more efficiently. Proteins fragments are obtained by selectively cutting it with an enzyme (usually trypsin) and the fragment masses are compared to theoretical peptides. The theoretical peptides are available from protein databases, i.e., protein fragments prepared by computer. If mass fingerprinting doesn't fix the protein, then the peptides can be further fragmented and analyzed in a second, "tandem" MS, or "MS/MS". *Human Proteomics Initiative (HPI)* is an out break of the Swiss Institute of Bioinformatics (SBI) and the European Bioinformatics Institute (EBI). These institutes are in effort to annotate, describe and distribute a large amount of highly curate information concerning human protein sequences. The HPI project has two different time-related aspects: one of which is a nine-month "marathon" to catch up with the current state of research, the other one is a long term commitment to keep such a project alive as long as it is necessary. *Protein Modeling:* To find out the location of a gene locus and the three-dimensional model of the protein that it encodes, encompasses, protein modeling. There are few steps in protein modeling: Location of Transcription initiation/termination, Location of Translation initiation/termination, Detection of intron/Exon splices Sites, and the Prediction of 3-D Structure. Once the sequence of primary amino acid is known, the next step is to develop probable 3-S structure of the concerned protein. This requires analysis of different data.

Genetic and Physical Mapping

The genetic and physical maps provide the likely order of items along a chromosome. The former provides an indirect estimate of the distance between two items and is limited to ordering certain maps, while the latter mark an estimate of the true distance, in measurements called base pairs, between items of interest. **Genetic maps**: in these maps, genetic markers are used to guide researchers on their gene hunt. These markers (or genetic markers) are used as landmark in genetic mapping. Any observable variation that results from an alteration, or mutation, at a single genetic locus is described as marker. Markers may within genes that code for a noticeable or not noticeable physical characteristic. DNA-based reagents can also serve as markers. These types of markers are found within the non-coding regions of genes and are used to detect unique regions on a chromosome. There are some commonly used markers. *Restriction fragment length polymorphisms (RFLPs)* markers are defined by the presence or absence of a specific site, called a restriction site. *Variable number of tandem repeat polymorphisms (VNTRs)* is defined by the presence of a nucleotide sequence that is repeated several times. However, the number of times a sequence is repeated may vary. *Microsatellite polymorphisms* are variable number of repeats consist of the nucleotides, or bases, cytosine and adenosine. The number of repeats for a given microsatellite may differ between individuals. *Single nucleotide polymorphisms (SNPs)* are individual point mutations, or substitutions of a single nucleotide, that do not change the overall length of the DNA sequence in that regions. These polymorphisms are found throughout an individual's genome. Genes located in the same chromosome exhibit linkage and during meiosis, crossing over results in recombination.

Application of co-segregation, and statistical technique (the genetic linkage analysis) it is possible to identify markers that lie close to each other and infer a likely crossover pattern. This enables to analyze an order of the makers involved. Co-segregation of alleles refers to the movement of each marker during meiosis. The functions that tend to move only in one direction is known as Monotonic functions. If recombination occurs as a random event, then two markers that are close together should be separated less frequently than two markers that are more distant from one another. **Physical Maps** are of three types: chromosomal or cytogenetic maps, radiation hybrid maps and sequence maps. These maps show difference in their ability to measure the separation of elements that are close together. This property is known as the degree of resolution. Higher the resolutions better the picture. *Chromosomal or cytogenetic map* is based on the distinctive banding patterns observed by light microscopy of stained chromosomes. It can be used to locate genetic markers defined by traits observable only in whole organism. They are considered to be physical maps and exhibit the lowest resolution. *Radiation Hybrid (RH) Maps* are similar to linkage maps. They show estimates of distance between genetic and physical markers. It is able to provide more precise information regarding the distance between genetic markers. In this mapping DNA is exposed to a measured dose of radiation, and while in this course, it is possible to control the average distance between breaks in a chromosome. A varying degree of radiation exposure to DNA results in breaking of two markers which are very close to each other. Thus, almost any genetic marker as well as other genomic fragments may be localized. *Sequence Maps* show genetic markers, as well as the sequence between the markers, measured in base pairs. It gives the most detail of markers. Radiation hybrid maps and sequence maps are similar to linkage maps.

RH mapping is used as a bridge between linkage maps and sequence maps. Sequence tagged site (STS) is a short DNA sequence. In STS mapping, the exact location and order of the bases of the sequence must be known. The common sources of STSs are expressed sequence tags (ESTs) and Simple sequence length polymorphism (SSLPs). ESTs are short sequences obtained by analysis of complementary DNA (cDNA) clones. STSs are significant for unique gene i.e., not belonging to a gene family. A gene family means a set of gene in which all of the genes have the same or similar sequences. SSLPs are arrays of repeat sequences that display length variations. The polymorphic SSLPs and already mapped by linkage analysis are significant. Such SSLPs provide a connection between genetic and physical maps. In STSs mapping the overlapped DNA fragments are digested into smaller fragments by restriction enzymes. The data obtained are analyzed so as to make out which fragment contain which STSs. Molecular cloning technique (a recombinant DNA technology) is adopted for this purpose. DNA fragment is introduced into a suitable host by using a proper cloning vector. In the host cell this foreign DNA fragment reproduces along with the host cell DNA. The daughter DNA molecules are used for experimental study. In the next step, the clones, or copies, are assembled in the order they would be found in the original chromosome by determining which clones contain overlapping DNA fragments. The assembly of overlapping clones is named as clone contig. The information about the clone is processed in database.

In plants physical maps are commonly prepared by three methods. These methods are (1) DNA contigs assembled using large insert DNA clones, (2) to locate genetically mapped DNA markers to specific chromosomal segments using cytogenetic stocks, and (3) to directly visualize DNA sequences on chromosomes by *in situ* hybridization. Physical maps generated through contig assembly have the ultimate resolution. In this method, yeast artificial chromosomes (YACs) and bacterial artificial chromosomes (BACs) are used as large insert DNA clones. Mozo *et al.* (1999) has developed BASC contigs covering the entire genome of *Arabidopsis thaliana*. Different types of cytogenetic stocks are used in physical mapping. These stocks include B-A translocations and oat-maize translocation in maize, deletion stocks in wheat, chromosomal translocation stocks in barley. In such stocks, the resolution is not only dependent on the number of stocks, but also on the accuracy of their cytological characterization. A cytological defined chromosomal fragment can include several megabases of DNA, which could significantly limit the resolving power of the physical maps.

Relation between Genetic and Physical Distance

There is a considerable evidence that there is not a consistent relationship between genetic distance in centimorgans and physical distance in base pairs and that there is variation in this relationship form one part of the genome to another. The differences in the frequency of recombination along the length of a chromosome results in genetically close markers to be far away in terms of base pairs (or vice versa) due to differences in the frequency of recombination along the length of a chromosome. In plants, significant discrepancies between genetic and physical distances have been observed. Schmidt *et al.* (1995); and Mayer *et al.* (1999) studied a YAC-based contig map and complete sequencing of chromosome 4 of *A. thaliana*. The average ratio of physical to genetic distance between markers was 185 kb/cM. Relative hot spots (30-50 kb/cM) and cold spots (0.550kb/cM) were distributed throughout the chromosome. In rice, variation of the ratio of physical to genetic markers was observed. According to Umehara *et al.* (1994) genetic mapping of the ends of large-size YAC clones suggests that the physical distance corresponding to 1 cM varies from 120 to 1000 kb, depending on the chromosomal region. Large size of chromosome and large genome is present in wheat and barley. Recombination is mainly distributed along the distal half of the chromosomes. Most markers are tightly clustered in small-sized physical segments. The 50% length of each chromosome covers centromeric regions. In this region recombination was found to be suppressed. When considering the average length of DNA per unit of recombination, different segments of chromosome should be considered independently. Tanksley *et al.* (1992) observed significant suppression of genetic recombination in the pericentric regions in tomato. Such features were not evidently observed in *A. thaliana*. The degree of recombination suppression in the centromeric region of rice chromosome 10 is similar to that of chromosome 4 of *A. thaliana*. However, it is different from wheat and barley.

Genetic and Physical Mapping of Plants

The very first genome map in plants was reported in maize, followed by rice, *Arabidopsis* etc. using RFLP markers. Integrated genetic and physical maps are prepared with the help of all available genetic and DNA marker resources. Cantrell and Lei used PCR-based DNA markers, to prepare and discuss the integrated genetic and physical map of cotton. The simple sequence repeats (SSRs) were being anchored to physical mapping resource utilizing a large-insert Maxxa BAC library. BAC-PCR pools provide an efficient way for localizing an SSR locus to a BAC address, thus converting the SSR to a sequence tagged site. Wang *et al.* (2001) studied genetic and physical mapping of the evergreen gene. One single recessive evergreen gene (*evg*) controls evergreen trait. The latter was identified as a unique material for studying the process of winter dormancy and cold-hardiness in peach trees. The map was prepared by amplified fragment length polymorphiysim (AFLP) and bulked segregant analysis (BSA) technique. This map consisted of fourteen loci. Four dominant AFLP markers were found to be tightly linked to the evg locus and were converted into STS markers. These markers are named as EAT/MCAC, ETT/MCCA2, EAT/MCTA and ETT/MACC. STS markers were identified and are being used to initiate chromosomal walking to the evg locus. Microsatellite markers in these peach BAC clones were characterized. Rice is a model monocot plant, having 460-Mb genome.

Rice genome is three times larger than that of *A. thaliana*. Cheng *et al.* prepared a physical map of rice chromosome 10 developed by fluorescence in situ hybridization (FISH) mapping of bacterial artificial chromosome (BAC) clones on meiotic pachytene chromosomes. This map was found fully integrated with a genetic linkage map of rice chromosome 10. Meiotic pachytene chromosomes are often more than 10 times longer than somatic metaphase chromosomes. Xi and Earle (1996), Song *et al.* (2000) used these chromosomes have recently been used as a target for FISH mapping. Causse *et al.*, (1994) prepared the RFLP-based genetic linkage maps of rice. It was among the most saturated maps in plants. Kinga *et al.* (2000) prepared the physical and genetic map of grasses (*Lolium perenne* and *Festuca pratensis*). The AFLPs of *F. pratensis* demonstrated: (1) the first large-scale analysis of the physical distribution of AFLPs; (2) variation in the relationship between genetic and physical distance from one part of the *F. pratensis* chromosome to another; (3) that nucleolar organizer regions (NORs) and centromeres greatly reduce recombination; (4) that coding sequences are present close to the centromere and NORs in areas of low recombination in plant species with large genomes; and (5) apparent complete synteny between the *F. pratensis* chromosome and rice chromosome 1.

Markers

The advancement of DNA technologies has added a new chapter to uncover a large number of genetic polymorphisms at the DNA sequence level, and to use them as markers for evaluation of the genetic basis for the observed phenotypic variability. There are two terms polymorphic and polymorphism. *Polymorphic* means the existence of two or more forms of the same gene, or genetic marker with each form being too common in a population to be merely attributable to a new mutation. *Polymorphism* is a useful genetic marker; it helps in distinguishing which allele was inherited. Any stable and inherited variation that can be measured or detected by a suitable method, and can be used subsequently to detect the presence of a specific genotype or phenotype other than itself, which otherwise is nonmeasurable or very difficult to detect is known as genetic marker. The markers revealing variations at DNA level are referred to as the molecular markers. The variations may be at different level, i.e., at the morphological, chromosomal, biochemical or DNA level. These markers are divided in two categories: Hybridization-based markers (includes the traditional RFLP analysis); and PCR-based markers (probe-hybridization step is not required). In Hybridization-based markers DNA fragment to be studied is digested with restriction enzyme and hybridized with a DNA fragment of known origin or sequence (labeled probe). In the second type, a particular DNA molecule is amplified with the help of specifically or arbitrarily chosen oligonucleotide sequences (primers). A thermostable DNA polymerase is used in this technique. The primer sequences are chosen to allow base-specific binding to the template in reverse orientation.

Molecular markers are used as versatile tool in plant genome analysis. Molecular markers include biochemical constituents (e.g., secondary metabolites in plants) and macromolecules, viz. proteins and DNA. DNA markers are more suitable and ubiquitous to most of the living organisms. Techniques have been developed to visualize DNA sequence polymorphism. In 1985 Alec Jaffrey introduced the term DNA-fingerprinting. In this technique, genomic DNA fragment is separated by electrophoresis, and bar-code-like DNA fragment patterns generated by multilocus probes. Presently, DNA fingerprinting/profiling is used to describe the combined use of several single locus detection systems. It is being used as versatile tools for investigating various aspects of plant genomes. It is applied in characterization of genetic variability, genome fingerprinting, genome mapping, gene localization, analysis of genome evolution, population genetics, taxonomy, plant breeding, and diagnostics. The DNA markers are described under two heads, single or low copy probe and multiple copy probe. The single copy probe includes Restriction fragment length polymorphism (RFLP), Restriction landmark genomic scanning (RLGS), Sequence-tagged sites (STS), and Allele-specific associated primers (ASAPs). The multilocus probes include Repetitive DNA, Microsatellite and Minisatellite. Other repetitive DNA-type markers are transposable elements, Alu-repeats, Arbitrary sequence markers. *Microsatellites and Minisatellites* are multilocus probes creating complex banding patterns by simultaneously detecting multiple DNA loci. They are usually non-specific occurring ubiquitously.

Litt and Lutty coined the term Microsatellite for tandem repeats with a monomer repeat length of about 11-60bp. Minisatellite term was introduced by Jeffery, which consists of 1 to 6 bp long monomer sequences that is repeated several time. It is also known as short tandem repeats/simple sequence repeats (STRs/SSRs). These loci contain tandem repeats. A single locus that contains variable number of tandem repeats (VNTRs) or numerous loci containing tandem repeats within a genome generating high levels of polymorphism between individuals i.e., hypervariable regions (HVRs). These markers are dominant fingerprinting markers. Another feature is that they are codominant sequence tagged microsatellites markers (STMS). Both are converted into PCR-based markers. *STMS-PCR* STMS includes DNA polymorphism using specific primers designed from the sequence data of a specific locus. Variation in the number of tandem repeats in a given repeat motif provides polymorphism. Primers complementary to the flanking regions of the simple sequence repeat loci yield highly polymorphic amplification products. The dinucleotides [like (CA)n; (AG)n; and (AT)n] are generally abundant in genomes and have been used as markers. However, Tri- and tetranucleotide microsatellites are more applicable for STMS analysis. The di-and tetranucleotide repeats are present mostly in the non-coding regions of the genome. Trinucleotide repeats are present in or around the genes. The di-and tetranucleotide repeats are present as a clear banding pattern after PCR and gel electrophoresis. Direct amplification of minisatellite DNA markers (*DAMD-PCR*) technique is applicable for generating DNA probes useful for detecting polymorphism. The clones can yield individual-specific DNA fingerprinting pattern. It is the potential as markers for species differentiation and cultivar identification.

DNA Microarrays

An array is an orderly arrangement of samples. Through array technique, it is possible to match known and unknown DNA samples. The unknown sample is identified on the basis of base-pairing rules and automating the process. Arrays are described as macroarrays and microarrays. The samples applied in macroarrays have spot diameter of about 300 microns or larger. The sample spot diameters in microarray are typically less than 200 microns, and these arrays usually contain thousands of spots. The basic principle of DNA microarray technique is base-pairing or hybridization. The DNA microarray is a small, solid supports onto which the sequences from thousands of different genes are immobilized, or attached, at fixed locations. There is a support on which DNA is printed, spotted, or actually synthesized directly. The supports themselves are usually glass microscope slides, the size of two side-by-side pinky fingers, but can also be silicon chips or nylon membranes. This technique promises to monitor the whole genome on a single chip. This provides knowledge to understand a better picture of the interactions among thousands of genes simultaneously. DNA microarray technique is used to identify gene/ gene mutation and to determine the expression level of gene. Before going into the details, it is a must to understand two terms probes and targets. According to the nomenclature recommended by B.Phismister, *Probe* is the gethered nucleic acid with known sequence, whereas a target is the free nucleic acid sample whose identity/abundance is being detected. *Foramt I* and *Foramt II* are the two variants of the DNA microarray technology. This is in terms of the property of arrayed DNA sequence with known identity. *Foramt I* the probe complementary DNA is 500~5,000 long. It is immobilized to a solid surface such as glass using robot spotting and exposed to a set of targets either separately or in a mixture. *Format II* the probe is oligonucleotide (20~80-mer oligos) or peptide nucleic acid (PAN). It is synthesized either *in situ* (on-chips) or by conventional synthesis followed by on-chip immobilization. The array is exposed to labeled sample DNA, hybridized, and the identity/abundance of complementary sequences is determined.

Designing a microarray experiment is based on hybridization probing, a technique that uses fluorescently labeled nucleic acid molecules as `mobile probes' to identify complementary molecules, sequences that are able to base-pair with one another. The basic steps in the design and implementation of a DNA microarray experiments are (1) DNA types-cDNA/oligo with known identity (2) Chip fabrication-putting probes on the chip by photolithography, pipette, drop-touch, piezoelectric (ink-jet), electric process (3) Sample preparation (target)-fluorecently labeled sample the (mRNA) cDNA. cDNA are made from RNA samples of interest. (4) Assay-hybridization, electrophoresis, flowcytometry, PCR-DIRECT, TaqMan (5) Readout (6) Software (informatics)-Robotics control, Image processing.

A microarray is significant as it contain a very large number of genes in a small size. It is useful specially because it is helpful to survey a large number of genes quickly or when the sample to be studied is small. The microarray technology is having a significant application on many fields like, Gene discovery, disease diagnosis, drug discovery (Pharmacogenesis), and toxicological research (Toxicogenomics).

Artificial Chromosome

The artificial or synthetic chromosomes were first created in yeast. It enables a scientist to understand the elements of a natural chromosome. By placing genes onto artificial chromosomes and then inserting them into cell, may give a picture on how a gene's chromosomal surroundings regulate its activity. Artificial chromosomes could also deliver much larger genes as compared to viruses or other vectors. Both yeast and human chromosome appear to contain three main components: the telomeric (TEL- protects the linear DNA from degradation by nucleases), centromeric (CEN it attaches to the spindle fiber during mitotic cell division). It pulls a copy of chromosomes at either end of the cell, and replication origin sequences (ORI the specific DNA sequences that allow the DNA replication machinery to assemble on the DNA and move at the replication forks). The artificial chromosome has mitotic stability i.e., they are reliably reproduced and passed on every time the cell divides. One of the great appeals of using artificial chromosomes is to introduce all the subtleties of the genetic controls that exist at the chromosomal level. A yeast artificial chromosome (YAC) is a vector. YAC contains some specific sequences like **A** and **B: selective markers** it helps in easy isolation of yeasts cells that have taken up the artificial chromosome, **Recognition site** for the two restriction enzymes EcoRI and BamHI.

YAC can be used to clone DNA fragments larger than 100 kb and up to 3000 kb, generally of 400 kb length. Techniques for cloning genomic DNA into YAC makes it possible to analyze very long DNA sequences like human genome. There are few steps in cloning human genomic DNA into a YAC: first the genomic DNA is partially digested by EcoRI (a restriction enzyme). This yields very large DNA fragments. The YAC is digested by the two restriction enzymes EcoRI and BamIII. The genomic DNA fragment and YAC recombine at the EcoRI sites and are covalently linked by the DNA ligase. The vector thus produced by recombination can be used to infect yeast cells and generated an unlimited number of copies. A bacterial artificial chromosome (BAC) is synthesized on a fertility plasmid, generally from *E.coli*. It is used for transforming and cloning in bacteria. It has the capacity to clone fragment ranging from 100 to 300 kbp. It is often used to sequence the genetic code of organisms in genome projects. P1-derived artificial chromosome (PAC) is based on bacteriophage P1 genome. It is used to clone DNA fragments of 100 to 300kbp with average of 150 kbp.

Databases and Bioinformatics Tools

The biological databases are of two types: the Raw databases and Processed databases. The former database is automation of base sequence in DNA or amino acid sequence of proteins, while the database provides analysis and links information from various databases. The organization of biological databases will include (i) genome databases, the totality of genes on all chromosomes in the nucleus of a cell. (ii) transcriptomics (or mRNA) databases, (iii) proteomics databases, (iv) protein sequence databases, (v) protein structure databases, (vi) protein function databases, (vii) protein/protein, DNA/protien interaction databases, and (viii) metabolic databases.

A computer programme for finding genes will contain at least the following elements: Algorithms, Data Tables, Taxonomic differences, and Analysis rule. The Algorithms for pattern recognition are the probability formulae that are used to determine if two sequences are statistically similar. Data Tables contain information on consensus sequences for various genetic elements. Taxonomic differences inclusion on consensus sequences differences in an analysis speeds processing and minimizes error. Analysis rules are the programming instructions that define how algorithms are applied. They define the degree of similarity accepted and whether entire sequences and/or fragments thereof will be considered in the analysis.

The data management tools required in bioinformatics is Software programmes. These programmes are developed by various Universities, organizations. There are certain advantages as well as different limitations in different bioinformatics tools.

DATABASE

A database is collection of data that have some common element or theme and is composed of entries-secrete coherent parcels of information. It is a logically coherent collection of related data with inherent meaning built for certain application. Databases are needed to collect and preserve data, to make data easy to find and search, to standardize data representation and to organize data into knowledge. Database can be searched or cross-referred either over the Internet or using downloaded versions on local computers or computer networks by multiple users. Information available in these databases can be searched, compared, retrieved and analyzed. The primary goals of databases are (i) minimizing data redundancy and (ii) achieving data independence.

The International Nucleotide Sequence Database are (i) National Centre for Biotechnology Information (NCBI), (ii) the European Molecular Biology Laboratory (EMBL), and it contains sequences from direct author submission and genome sequencing groups, and from the scientific literature and patent applications. The format of EMBL entries is consistent with SWISS-PORT format. The database is produced in collaboration with DDBJ and GenBank. The collaborating groups collect the data. These datas are a portion of the total sequence data reported worldwide. The new and updated entries are then exchanged between the groups. (iii) DNA Data Bank of Japan (DDBJ) database is produced, maintained and distributed at the National Institute of Genetics sequence.

Classification and Types of Database

There are many different database types, depending both on the nature of the information being stored and on the manner of data storage. There are two principal types of databases viz. relational and object-oriented. The relational database orders the data to tables made up of rows giving specific items in the databases columns giving the features as attributes of those items. The object-oriented database includes objects such as genetic maps, genes, or proteins which have an associated set of utilities for analysis which help in identifying the relationships among these objects. Databases are broadly classified into generalized database (DNA, protein, carbohydrate or similar databases), and specialized databases (expressed sequence tagged site STS, genome survey sequences GSS, single nucleotide polymorphism SNP). Generalized databases are again of two types sequence databases (contain the individual sequence records of either nucleotides or amino acids or proteins) and structure databases (individual sequence records of biochemically solved structure of macromolecules. e.g., Protein 3D structure).

However, more specifically databases can be classified into three types based on the complexity of the data stored. These classes are primary databases, secondary databases, and composite databases. (i) Primary database contains data in its original form, taken as such from the source. GenBank and SWISS-PORT for gene and protein sequences, respectively are the examples of primary database, (ii) Secondary database is a value added database which contains some specific annotated and derived information from the primary database, e.g., SCOP, CATH, PROSITE and (iii) Composite database amalgamates a variety of different primary database structures into one.

Databases of nucleic acid and protein sequences maintain facilities for a very wide variety of information reclamation and analysis operations. Such operations are recovery of sequences from the data base sequence comparison, translation of DNA sequences to protein sequences, simple types of structure analysis and prediction, pattern recognition and molecular graphics.

Nucleic Acid Sequence Databases: in such databases, entries are made with each entry has the form of a text file. The latter contains data and annotations for a single contiguous sequence. The text of text file can be read by human beings as well as a computer. In 1964, nucleic acid sequence of yeast t-RNA with 77 bases was recorded. There are three International Nucleotide Sequence Databases Collaboration viz. National Centre for Biotechnology Information (NCBI) located in USA, the European Molecular Biology Laboratory (EMBL) of UK, and DNA Data Bank of Japan (DDBJ). These three centers together form the International Nucleotide Sequence Database Collaboration.

GenBank Primarily the DNA sequence database (GenBank), sequences were distributed on floppy disk. The arrangement of data in each entry in the GenBank is self-explanatory. GenBank entry consists of a number of keywords, relevant associated subkeywords, and an optional Feature Table. The source of GenBank is in USA and is accessible through the NCBI portal over the Internet. GenBank from NCBI incorporates sequences from publicly available sources, primarily from direct author submissions and large-scale sequencing projects. Now about 100 billion numbers of bases in GenBank are available.

Protein Sequence Database The Protein sequence Database was developed at National Biomedical Research Foundation (NBRF) at Georgetown University, USA. Now the Protein Sequence Databases have been maintained collaboratively by PIR (Protein Information Resource) at NBRF, the International Protein Information Database of Japan (JIPID), and the Munich Institute of Protein Sequence (MIPS), Germany. The latter is an International association of macromolecular sequence data collection centers. The sequence of amino acid in a polypeptide chain is stored in primary databases as linear alphabets. This denotes the constituent residues.

The PIR maintains databases about protein. Some of the important of them are headed by PIR-PSD (the main protein sequence database), ASBD (annotation and similarity database: each entry is linked to a list of similar sequence), ALN (database of protein sequence alignment), RESID (a database of covalent protein structure modifications).

Another protein sequence database is SWISS-PORT (http.//www.expasy.ch/ sprot). Database is made available by Swiss Institute of Bioinformatics (SIB) in collaboration with the EMBL. It is interlinked to many other resources. SWISS-PORT provides an annotated database of amino acid sequences. It is a curated protein sequence database which strives to provide high level annotations. Structure of protein domains, its post-transitional modification, functions of the protein is available in this database.

Structure and Composite Database the structure database provides three dimensional structures of protein and nucleic acid, while composite database integrates a variety of different primary sources. Through composite database, sequence searching is much more efficient.

SEQUENCE ALIGNMENT

Sequence alignment is the procedure of comparing two (pairwise alignment) or more (multiple sequence alignment) sequences by searching for a series of individual characters or character patterns that are in the same order in the sequences. Sequence alignment is used to look for a common or related pattern amongst them. The common pattern of alignment may be a short stretch of one sequence that is similar to a short stretch of the other sequences or many short similar bits of the two sequences. The biological similarities between the two sequences may be structural, functional or evolutionary relationship. The extracted sequence, for which functional and structural information is available, will help us to predict the structure and function of the query sequence. The changes that occur during divergence from the common ancestor can be categorized as substitutions, insertions and deletions. The basic purpose of sequence alignment is to understand if two sequences display sufficient similarity such that an inference of homology is justified. As genetic information is passed on from one generation to the next, the information gets altered slightly during the process of copying.

SOME BASIC DEFINITIONS/CONCEPTS IN SEQUENCE ALIGNMENTS

Similarity: when two sequences after the best alignment, show identical residues occur at identical positions. The number of such matches indicates the degree of similarity. Similarity could arise by chance, or it could be convergence towards a common sequence and structure and therefore function, through evolution. Sequence similarity searches of database enable us to extract sequences that are similar to a query sequence.

Homology: sequences with common origin point out homology. The homology could also arise from divergent evolution of the two sequences from a common ancestral sequence. In terms of bioinformatics, if the structure of one of the proteins is known, the structure of the other may be built by homology modeling.

Twilight zone: this is a zone of sequence among protein with approximately 10-20% identity, over approximately 300 amino acids within which alignments are not statistically significant, and therefore homology cannot be directly ascertained.

Heterologs: heterologous sequence may be similar but they do not have a common origin. They do not exhibit common function or activity.

Analogous: Sequences with the same function but they lack sufficient similarity to imply common origin. These sequences followed evolutionary pathways from different origins to converge upon the same function and activity. Analogs have homologous function but heterologous origins.

Orthologs: these are homologs that arise by speciation. The two sequences are taken from two different species, but they show similarity because they are both derived from the same ancestral sequences that were present in the ancestral organism.

Paralogs: the homologs which have come out as a result of gene duplication, without this duplication event being followed by a speciation event. Paralogs are homologous sequences that exist in the same organism, and that have different functions.

Xenologs: homologs appearing as a result of horizontal gene transfer. They are an exception to rule that homologs are always descended from a common ancestor.

Global alignment: the comparison of one complete protein or gene sequence with another. In such alignment an attempt is made to align the entire sequence, using as many characters as possible, up to both ends of each sequence. Global alignment considers the similarity across the full extent of the sequence. It helps in understanding of the biological relationship between the two entities that were compared.

```
L G P S S K Q T G K G S S R I W D N
| | | | | | | | | | | | | | | | | |
L N I T I K S A G K G A M R L G D A
```

Local alignment: compares short portions of sequences or a whole library of sequences with short portions of another. It focuses on regions of similarity in parts of the sequences only. In this alignment, stretches of sequence with the highest density of matches are aligned thus generating one or more islands of matches or subalignments in the aligned sequences. It is used to identify local regions of a protein or gene that may have resemblances to other local regions in some other protein or gene.

```
T G K G
| | | |
A G K G
```

ALGORITHM

Algorithm is logical sequence of steps to perform a task. It is a complete and precise specification of a method for solving a problem. There is a set of rules for calculating or solving a problem which normally is carried out by a computer program. An algorithm must stop after a finite number of steps and all steps are precisely defined. Input and output to the algorithm must be specified. Algorithm must be very effective and operation of the algorithm must be basic.

Dynamic Programming: is something like techniques used in a variety of technological, scientific and commercial fields to find the set of parameters that would give the desired optimum solution. It includes all the parameters and find out the best location.

Dynamic Programming (DP) Algorithms: Computer assisted DP algorithm. In an alignment of two sequences effort is made so that there are maximum possible number of matches, the minimum number of mismatches and the minimum number of gaps. To maximize matches, gaps and mismatches are to be ignored. Similarly, to minimize the gaps matches and mismatches are to be ignored. All these procedures are carried out simultaneously, and it is called dynamic programming algorithms.

The Needleman-Wunsch Alogrithm: In 1970, Needleman and Wunsch proposed the first application of dynamic programming to the alignment of sequences. Needleman and Wunsch algorithm is used in global alignment to find similarity between sequences across the entire length. They explained it with the following sequence

Sequence 3: ABCNJRQCLCRPM

Sequence 4: AJCJNRCKCRBP

Sequence 3 is S3(i), i stands for 1,m (m=13), and sequence 4 is S4(j), where j = 1,n (n=12). The first step in the algorithm is to create a matrix, known as similarity matrix F(i.j). One sequence is written along the top of the matrix, and each residue of this sequence labels one column. The other sequence is written vertically along the left of the matrix, and each residue of this sequence labels one row. Thus a pair of residues one from each sequence, labels each element of the matrix. Some features of matrix are lines drawn connecting the cells are called paths through the matrix. A valid path is defined as if a line connects the cell with indices (i.j) to the cell with indices (m,n), then either (i and j) or (im and jn). Thus, lines joining the cells are all diagonal, never vertical or horizontal. If we consider a continuous line going from one corner of the matrix to the other, hopping from cell to cell, then every such continuous line is a valid.

The Smith-Waterman algorithm: In this method portions of one sequence are aligned to similar portions of the second sequence. This is useful in comparing a relatively short sequence with a very long sequence, or a set of long sequences. It is a local alignment program method. The Smith-Waterman algorithm method is applied in cases where, it is possible that a portion or portions of the relatively short query sequence is similar to not just one, but many regions of the long target sequence. It may also happen that portion 1 of the query sequence align with one part of the target, and portion 2 f the query aligns with another part, but these two portions of the target do not occur in the same order. These portions may not occur at the same distance as the portions of the query.

BLAST: Altschul *et al.* (1990) developed this program. BLAST stands for Basic Local Alignment Search Tool. This algorithm uses a word-based heuristic to execute an approximate version of the Smith-Waterman algorithm known as the (maximal segment pairs) algorithm. The heuristics algorithms mean that use of approximation techniques. The first step in BLAST is to look for words of a certain fixed word length W that score higher than a certain threshold score (T). The value of W is 3 and 11 for protein and nucleic acid sequences, respectively. This is because since there are only four nucleotides, the amount of background noise is too large with similar word size. BLAST tries to find all double word hits along the same diagonal within some specified distance. BLAST extends individual word matches until the total score of the alignment falls from its maximum value by a certain amount producing high scoring segment pairs. BLAST is most frequently used over the Internet on the BLAST server (http://www.ncbi.nlm.nih.gov/BLAST/).

FAST: originally known as FAST. It is because that most popular version used is Fast A, hence named as FASTA. FASTA program achieves a high level of sensitivity fro similarity searching at high speed. In the first step words are made of specific K-tuple size. In K-tuple two sequences are aligned very quickly, first by searching for identical short stretches of sequences. The word means 3-6 nucleotides or 1 or 2 amino acids. These words are then joined into an alignment by the dynamic programming method. By default this is two for peptides. It then identifies all exact word matches between the query sequence and the database members. Sequences in FASTA formatted files are preceded by a line starting with a ">" symbol. The first word on this line is the name of the sequence. The remaining words of the line are description of the sequence.

Applications

The bioinformatics is coming as Mathematical Modelling of Life Processes. The commercial application of bioinformatics is concentrating in pharmaceutical and agricultural fields. However, it is also applied in variety of areas including environmental field. The aims and the possible results to be achieved by the application of bioinformatics is to provide faster drug research based on Computers and Biotechnology, genome-based therapy. It is also applied to understand epidemiology. It will play role in evolutionary biology structural prediction of macromolecules.

Bioinformatics in India

In India bioinformatics is at neonatal stage as compared to US and some European countries. The centers engaged in bioinformatics research are AstraZeneca Research Center, Banglore, GVK Biosciences, Hyderabad, DSQ Biotech, Banglore and reputed research institutes like the Center for Cellular and Molecular Biology (CCMB), Hyderabad and the Center for Development of Advanced Computing (C-CDAC), Pune.

Private companies such as Biocon (Banglore), Dr.Reddy's Lab (Hyderabad), and Ranbaxy (Delhi), Satyam Computers, Tata Consultancy Services (TCS), are giving good research. IBM's India Research Lab (1RL) biochip technology made it possible to simultaneously study expressions of thousands of genes or proteins in a single experiment in the laboratory. Bioinformatics Centre (BIC) at IMTECH (Chandigarh) has developed several computer programmes like FPMAP (Fingerprint Mapping) to assist biologists in classification

and identification of microorganisms. It is one of the nine centers established in 1987 under the programme Biotechnology Inforamtion Systems (BTIS) by Department of Biotechnology (DBT), Govt. of India. IMTECH has installed large number of these software programmes and are for users. BIC have number of public domain databases, which are free for academic users.

SUMMARY

Bioinformatics is a subject in which the computational technology is used to manage/ solve biological information/problems. The basic effort of Bioinformatics is to combine data from various sources. Bioinformatics companies and scientists work for (i) collecting and storing data, (ii) searching databases, (iii) classifying data, (iv) analyzing and interpreting the data and (v) distributing biological information derived from sequencing and functional analysis projects. Genomics is the study of all the genes of an organism. A worldwide initiative aimed at determining a large number of protein structures in a high throughput mode, is defined as Structural Genomcis. Functional genomics is the process of determining the function of genes and an understanding of how they affect normal physiology. A *Comparative Genomics* is morphological and physiological studies. *Pharmcogenomics* is an intersection of the fields of pharmacology and genetics. *Proteome* is the expressed protein complement of a genome. *Proteomics* is functional genomics at the protein level, and involves the separation, identification and characterization of proteins in order to understand their functions. The genetic and physical maps provide the likely order of items along a chromosome. In plants, significant discrepancies between genetic and physical distances have been observed. Any stable and inherited variation that can be measured or detected by a suitable method, and can be used subsequently to detect the presence of a specific genotype or phenotype other than itself, which otherwise is nonmeasurable or very difficult to detect is known as genetic marker. *Molecular markers* are used as versatile tool in plant genome analysis. An array is an orderly arrangement of samples. *The DNA microarray* is a small, solid supports onto which the sequences from thousands of different genes are immobilized, or attached, at fixed locations. The artificial or synthetic chromosomes were first created in yeast. It enables a scientist to understand the elements of a natural chromosome. A computer programme for finding genes will contain at least the following elements: Algorithms, Data Tables, Taxonomic differences, and Analysis rule.

GLOSSARY

Action spectra A graph of the intensity of some phenomenon, like rate of photosynthesis produced by monochromatic light as function of the wavelength of light.

Activation energy ($\Delta G^{\ddagger}$) The amount of energy required to convert all the molecules in 1 mole of a reacting substance from the ground state to the transition state.

Active site (catalytic site) region of enzyme surface, that binds substrate molecule at the time of catalytic activity.

Active Transport Movement of substance across the membrane against a thermodynamically unfavourable gradient with the coupled input of energy is known as active transport. It is endergonic. *Primary Active Transport* When solute accumulation is coupled directly to an exergonic reaction, like hydrolysis of ATP. *Secondary Active Transport* one solute moves uphill (endergonic) across the membrane is coupled to the downhill (exergonic) flow of a different solute.

Accessory pigments the photosynthetic pigment (carotenoids, xanthophylls, and phycobilins), which are complement to chlorophyll in trapping light.

Adenylate cyclase an enzyme catalyzing formation of cyclic AMP (cAMP), in which 5′ and 3′ positions of the sugar ring are connected by a phosphate group.

Algorithm is logical sequence of steps to perform a task. It is a complete and precise specification of a method for solving a problem. There is a set of rules for calculating or solving a problem which normally is carried out by a computer program. Input and output to the algorithm must be specified. Algorithm must be very effective and operation must be basic.

Amphipathic Lipids The lipid with both hydrophobic and hydrophilic regions.

Amphoteric molecule is a molecule that acts both as an acid or a base.

Ammonification is conversion of organic nitrogen to $NH4^{+}$ by soil microbes.

Amber codon is one of the nonsense codon, with sequence UAG.

Anomers and Anomeric carbon atom anomers are isomers of sugar which have difference only in the configuration about the carbonyl carbon atom. Two cyclic compounds have different optical rotations, but they will not be same because the compounds as a whole are not mirror images of each other. Compounds related in this way are called *anomers* and carbon 1 after cyclization becomes asymmetric is called now *anomeric carbon atom.*

Annealing renaturation of a duplex structure from single strands formed as a result of denaturation of double stranded DNA.

Antiport is movement of molecules in a counter-exchange mechanism. Same carrier is involved in movement of two molecules in opposite direction. One molecule flows down in its concentration gradient, thus exchanging one gradient for another.

Antiporters Tansporters involved in a system, where solutes and protons are exchanged in different directions.

Aquaporins are made up of tetramer subunits and each subunit has six transmembrane helices. The monomers form a water channel. These channels consist of a very narrow pore. They highly selective for water transport. Aquaporins are present on plasma membrane as well as on vacuolar membrane.

Asymmetric carbon atom four different groups are attached to a carbon atom by covalent bond and thus may exist in two different tetrahedral configurations.

ATP synthase is an enzyme catalyzing synthesis of ATP during phosphorylation.

ATPase is an enzyme hydrolyzing ATP.

β-oxidation is oxidative degradation of fatty acids into acetyl CoA. In this process, successive oxidation of fatty acids at takes place β-carbon atom takes place.

CAAT box it is one of the sequence that enhances binding of RNA polymerase. This is a part of conserved sequence of nucleotide is found in eukaryotic cells and is present approximately 75 base pair upstream from transcription start site. It is recognized by a large group of transcription factors.

Calmodulin is a soluble protein with molecular mass 15-17 kDa. It is the primary calcium receptor in both plant and animal cells. Each molecule of calmodulin contains four binding sites for calcium. The characterized calmodulin-binding proteins in plants are myosin V, kinesin, NAD^+kinase, glutamate dexarboxylase, protein kinases, and Ca^{2+}-ATPase. Plant calmodulin is also important in formation of actin filaments and cytoplasmic streaming, polarized growth, and the cell division cycle.

Callus An undifferentiated cluster of plant cells that is a first step in regeneration of plants from tissue culture.

Capping a modified guanine base it added at 5′ end of RNA in three enzymatic actions. The bond is unusual 5′→5′ involving three phosphatase.

Carbohydrates are defined as polyhydric alcohols of carbon possessing active aldehydic or ketone groups. It functions primarily as stores of chemical energy and as durable building materials for biological construction.

Central dogma In 1958 Francis Crick proposed clearly the genetic information is transferred to cytoplasm through mRNA. The latter is being translated in the language of protein. The information flows is unidirectional.

cDNA double-stranded DNA complement of a messenger RNA sequence synthesized in vitro by reverse transcriptase and DNA polymerase. This is used in DNA cloning.

Chemiosmotic theory in respiratory chain oxidation of components is coupled to the translocation of hydrogen ions (protons, H^+) from the inside to the outside of the inner mitochondrial membrane. It produces an electrochemical potential difference of protons across the coupling membrane. This potential was subsequently used to drive the dehydration of ADP and inorganic phosphate.

***cis*-dominant site** the properties only of its own molecule of DNA. *cis*-dominance indicates that a site does not code for a diffusible product.

Clone A group of genetically identical cells or organisms produced asexually from a common ancestor.

Cointegrate an intermediate in the migration of certain transposons in which the donor DNA and target DNA are covalently attached.

Comparative Genomics is the comparison of proteomes *i.e.,* the complete protein set of two or more organisms.

Composite transposable element have a central region flaked on each side by insertion sequences, either or both of which may enable the entire element to transpose.

Concentration gradient is a change in the concentration of a molecule or ion from one point to another. The gradient might be gradual.

Conformation is three dimensional arrangements of the atoms within a molecule. It gives biological properties to proteins and other molecules in living cell.

Core enzyme is RNA polymerase enzyme with a complex of subunits.

Cryptochrome pigment absorb violet and blue light, it has sequence similarity to DNA photo-lyases, a rare class of flavoproteins that catalyze blue light-dependent electron transfer reactions in bacteria. Cryptochromes may also modulate kinase activity.

C-value is the total amount of DNA in the genome (per haploid set of chromosomes).

C-value paradox is the lack of relationship between the DNA content of an organism and its coding potential.

Cyclic AMP (cAMP) AMP with the phosphate group joined to 3′ and 5′ positions of ribose sugar. It was discovered by Sutherland in 1965. A second messenger present in a cell. In plants cAMP is involved in some specific processes, like signals that initiate stomatal closure may be transduced through adenyl cyclase. The activity of guard cell channels is modified by cAMP-dependent phsophorylation. cAMP may also regulate pollen tube growth. Its main function is to transmit primary hormonal message from the cell surface to the cell interior.

Cytochromes are made up of tetrapyrrole ring, the porphyrin. The prosthetic group is heme They have Fe nucleus. It carries a single electron and oscillates between the ferrous

(Fe^{2+}) and ferric (Fe^{3+}) states. Five types of cytochromes are present in mitochondria. They are named as Cyt *b*, Cyt*c*, Cyt c_1, Cyt *a* and Cyta_3.

DNA gyrase or topiomerases catalyze the formation of negative supercoils in DNA.

DNA helicases enzyme which separates DNA strands at the time of replication. It moves along ssDNA in a defined direction and separate the two strands of DNA, and form single stranded DNA at the beginning of the replication fork.

D-series and L-series the orientation of the H and OH groups around the carbon atom first adjacent to the terminal primary alcohol carbon.

Deamination is loss of amino groups from amino acids. It is of two types *nonoxidative deamination* (Some amino acids are not oxidized while deamination) and *oxidative deamination* (Enzyme amino acid oxidase deaminase catalyzes the deamination of amino acid).

Denitrification The process in which, nitrates are reduced.

Density gradient used to separate macromolecules on the basis of differences in their density in a medium containing heavy soluble compound such as CsCl.

Diffusion is movement of substances from a region of high concentration to a region of low concentration. *Simple Diffusion* : when the compartments of two solutions of different concentration are separated by a permeable divider, the solute moves from the region of higher concentration to the region of lower concentration till the two compartments have equal solute concentration. *Facilitated Diffusion* impermeable substances bind to a membrane-spanning protein the facilitative transporter. The substance moves down a concentration gradient from a higher to lower concentration.

Domain a distinct structural unit of a polypeptide with separate functions and may fold as independent compact units.

Electrochemical gradient is overall difference in electrical charge and in solute concentration that determines the ability of an electrolyte to diffuse between two compartments.

Electroporation is a technique in which cell is electrically treated to induce transient pores through which DNA can pass into the cell.

Emerson effect (E) is the ratio of the rate of oxygen evolution (ΔO_2) in the far-red light in the presence of supplementary beam, and the same rate in the absence of it.

Endergonic Transport the uphill movement flow of substance across the membrane.

Energy State of a pigment A pigment reaches a number of states after absorbing a quantum. Molecule shows two types of excitation state. In *singlet state* electrons orbit move in opposite spins (antiparallel). While if the electrons start moving in the same spin, it is the *triplet state*.

Enantiomers are stereoisomers showing nonsuperimposable mirror images of each other.

Enthalpy (H) and Entropy (S) the heat content of a system is enthalpy while the extent of randomnbess or disorder in a system.

Epimers and Epimerism When two sugars differ only in the configuration around one specific carbon atom are called *epimers* of each other, and the phenomenon is called *epimerism*.

Exergonic Transport the downhill movement flow of substance across the membrane.

Exons and Introns The nucleotide sequences of mRNA that code for amino acids are known as exons, while noncoding sequences are called introns. The latter are removed by splicing.

Extra chromosomal genome is self replicating set of genes, not present in chromosome of bacteria.

Extra nuclear gene genes present in organelles other than nucleus (like mitochondria).

F-type ATPase takes part in energy conserving reactions in mitochondria and chloroplast. It performs the same function in bacteria also. The name F-type is due to identification as energy-coupling *f* factors. These trasportors catalyze the reversible transmembrane passage of protons, driven by ATP hydrolysis.

Fats and Fatty acid fats consists of a glycerol molecule linked by ester bonds to three fatty acids, the composite molecule is termed a triacylglycerol. Fatty acids are long unbranched hydrocarbon chains with a single carboxyl group at one end. Fatty acids that lack double bonds are described as saturated, and those possessing double bonds are unsaturated.

Flavoprotein enzymes have FMN (flavin mononucleotide) and FAD (flavin adenine dinucleo-5-phosphate) coenzyme. These groups are tightly bound to the protein and hence act as prosthetic group. Isoalloxazine ring of the riboflavin moiety accepts reducing equivalents.

Florigen (La *flora* flower; Gr *genno* to begtet) a hormone-like substance produced in the leaves, it may be also a complex of substances that is conducted to the shoot meristems in order to stimulate them to pass from vegetative growth to flower formation. It is not species-specific and also it not is physiologically specific.

Forster energy transfer mechanism or resonance Theoder Föster (1948) analyzed the mechanism of resonance transfer of energy between unlike molecules without overlapping absorption bands. In this mechanism, emission and re-absorption of photon is not required. Resonant energy transfer is directed towards pigments absorbing at longer wavelength.

Four Quanta theory van Neil formulated oxidation-reduction theory of photosynthesis and postulated that there is a single photochemical reaction i.e., decomposition of water. The reduction of CO_2 to carbohydrate requires the transfer of four hydrogen atoms **Eight Quanta theory** after a long controversy the quantum yield of photosynthesis, it was proved to be 1/8 i.e., that at least eight quanta must be absorbed to reduce one molecule of carbon dioxide.

Free energy is the energy that can do some work at constant temperature and pressure. The *standard free-energy difference* ($\Delta G^{o\prime}$) describes the difference in free energy when one mole of each reactant is converted to one mole of each product under standard conditions.

Functional genomics is the process of determining the function of genes and an understanding of how they affect normal physiology.

Functional Groups a specific atom or group of atoms that often behave as a unit and give biomolecules their physical properties, chemical reactivity, and solubility in aqueous solutions.

G-Protein (or **G-binding proteins**) bind with GTP or GDP and can alternate between two conformational states. GTPase are converted from an empty state to an active state by association with GTP. Hydrolysis of GTP to GDP inactivates the protein, and subsequent dissociation of GDP returns the protein to the empty state. These heteromeric proteins interact with receptor proteins located in the plasma membrane.

G0, G1 G2 states of cell cycle of eukaryotic cell. In G0 state cell ceases to divide. In G1 state is between the last mitosis and the start of DNA replication. In G2 state separates the replication of a cell's chromosome (S phase) from the following mitosis (M phase).

GenBank is a large database (GenBank), sequences were distributed on floppy disk. The arrangement of data in each entry in the GenBank is self-explanatory. GenBank entry consists of a number of keywords, relevant associated subkeywords, and an optional Feature Table. GenBank contains over 1 million sequences from approximately 18,000 organisms.

Gene library (Gene Bank, clone bank, library) A population of organisms, each of which carries a DNA molecule that was introduced into a cloning vector. It is sometimes also used for all of the vector molecules, each carrying a piece of the chromosomal DNA of an organism, prior to the insertion of these molecules into a population of host cells.

Genetic marker A DNA fragment representing an allele of interest coding for a known and well defined function. It is any stable and inherited variation that can be measured or detected by a suitable method. It can be used subsequently to detect the presence of a specific genotype or phenotype other than itself, which otherwise is non measurable or very difficult to detect.

Genomics is the study of all the genes of an organism. These studies include DNA sequences, structures, gene regulation, interaction, and products.

Glycolysis or glycolytic pathway (Gr. Glyco = sweet: lysis = splitting) the biochemical pathway of breakdown of hexose sugar into pyruvic acid.

Group Translocation movement certain substances which get chemically modified as they enter the cell. These substances are impermeable to membrane. During translocation, high concentration of the modified form is generated inside the cell and low concentration of unmodified substance outside the cell, and a metabolic function equivalent to active transport is accomplished by a trapping reaction.

Haworth Formula is commonly used to show the ring forms of monosaccharides. It suggests that the five and six member furanose and pyranose ring are planar. The pyranose ring exists in two conformations, the *chair* form and the *boat* conformations.

Hills reaction R. Hill (1937) observed that when isolated chloroplasts are supplied with ferric salts in the light, the ferric salts are reduced to the ferrous form and O_2 is evolved. Illuminated isolated chloroplasts were able to produce O_2.

Hydrogen bond when a hydrogen atom is covalently bonded to an electronegative atom, particularly an oxygen or nitrogen atom, the single pair of shared electrons is greatly displaced toward the nucleus of the electronegative atom, leaving the hydrogen atom with a partial positive charge.

Influx is the movement of the substance into the cell.

in vivo (in life) reaction or a process taking place in living cell.

in vitro (in glass) reaction or a process taking place in an isolated cell free extract i.e., in a test tube.

Iron sulphur proteins (FeS) enzymes have equal amount of iron and acid labile sulphur. Ferredoxin of mitochondria is an iron sulphur protein.

Kornberg enzyme In 1957, Arthur kornberg and his coworkers isolated an enzyme DNA polymerase from *E.coli*.

Kranz anatomy the leaves of C4 plants have two different types of chloroplast-containing cells: mesophyll cells and bundle sheath cells. Biochemical fractionation analyses have shown that enzymes unique to the Calvin cycle are located only in the bundle sheath chloroplasts, whereas oxalaoacetate is generated form HCO3- and phosphoenolpyruvate (PEP) in the cytosol of mesophyll cells.

Lagging and Leading strand during DNA replication the DNA strand synthesized in opposite direction in which replication fork moves is lagging strand, while strand synthesized in the same direction in which replication fork moves is known as leading strand.

Leader sequence a sequence of nucleotide present at 5′ end of mRNA, which is not translated.

Mechalis-Menton equation it is an equation describing the hyperbolic dependence of the initial reaction velocity on substrate concentration in many enzymatic reactions. The substrate concentration at which an enzyme-catalyzed reaction proceeds at one-half its maximal velocity is known as *Mechalis-Menton constant* (Km). A kinetic pattern in which the initial rate of an enzyme-catalyzed reaction exhibits a hyperbolic dependence on substrate concentration is called *Mechalis-Menton Kinetics*.

Membrane Potential it is the difference in electrical potential between two aqueous media separated by a membrane. electrophysiologist encounter it by V_m and bioenergeticists by $\Delta\psi$.

Metabolism (categorized in anabolism and catabolism), is the sum of all the chemical transformations that occur in a cell or organism.

Mutarotation When an aldohexoses is first dissolved in water and the solution is put in optical path so that plane polarized light is passed the initial optical rotation shown by the sugar gradually changes until a constant fixed rotation characteristic of the sugar is reached.

Nitirfication In warm, moist soils with near natural pH ammonium is further oxidized into nitric acid (nitrates).

Nuclein F. Miescher (1869) named "nuclein" to the material isolated from nucleus of the cell.

Ochre codon is one of the nonsense codon with sequence UAA.

Okazaki fragments In 1968 Reiji Okazaki and his colleagues worked on *E. coli,* and explained the synthesis of complementary chain of 3′→5′ strand. The small fragments of DNA strands (1-2kd), each attached with an RNA primer, were named as Okazaki fragments. It represents transient, nascent DNA molecules.

Opal codon is one of the nonsense codon with sequence UGA.

Optical activity it is due to the presence of asymmetric carbon atom. When a beam of plane polarized light is passed through a solution exhibiting optical activity, it will be rotated to the right or left in accordance with the type of compound, i.e. the *"optical isomers"* or *"enantiomorphs"*; when it is rotated to right, the compound is called *"dextrorotatory"* (d or + sign), when rotated to left, the compound is called *"Laevorotatory"* (l or –sign).

Outflux the movement of the substance out of the cell.

Oxidation-reduction (redox) potential the separation of charge measured in voltage for any given pair of oxidizing-reducing agents relative to a standard.

P-type ATPase. The transporter is an integral membrane protein, with amino acid homology. This homology is especially observed near the Asp residue. This residue is the site of phosphorylation. (e.g. P-type H^+ATPase). This transporter pumps proton out of the cell, and for the transport of each proton one ATP is consumed.

Partition Coefficient the ratio of solubility of a substance in oil as opposed to water. It is a measure of the relative polarity of a biological substance.

Passive Transport the movement of substance moves across the membrane down its concentration gradient.

Pentose Phosphate Pathway or Hexose Mono Phosphate Pathway or Phosphogluconate Pathway it is an alternate pathway (other than glycolysis and citric acid cycle) by which plants were able to obtain energy by oxidation of sugars. In this pathway, the pentose sugars are degraded and do not synthesize other sugars. Some intermediates of this pathway are same as that of Calvin cycle.

Pharmcogenomics is an intersection of the fields of pharmacology and genetics.

Phenomics is the studies on anatomy and function for cell and whole organism.

Phosphatidic acid is a molecule made up of two fatty acyl chains without any additional substitutions beyond the phosphate.

Phosphorylation is synthesis of ATP from ADP and inorganic phosphate. It is of three types *Substrate level phosphorylation*, an intermediated transfers a high energy phosphate moiety to ADP. *Photophosphorylation* in photosynthesis phosphorylation takes place in presence of light. Oxidative phosphorylation in respiration, ATP is generated by the electron transport system where presence of oxygen, ATP formation is driven by energy that is released from electrons removed during substrate oxidation.

Photoperiod is the length of exposure to light. Photoperiod regulates flowering in a wide variety of plants.

Photoperiodism is the response to the timing of light and darkness.

Photorespiration is evolution of carbon di oxide from photosynthetic tissues in the presence of light. Jackson and Volk (1970) defined it to describe all respiratory activity in the light regardless of the pathways by which CO_2 is released or O_2 consumed.

Physiomics the role of proteome within and among cells cooperates to produce the biochemistry and physiology of individual cells and organisms.

Phytochrome is a photoreceptor, a pigment that plants use to detect light. It absorb light (red and far-red) and effective in causing photomorphogenesis. Phytochromes have two isoforms P_r and P_{fr}, which absorb light at 660nm and 730 nm, respectively.

Polar and Nonpolar molecules molecules that have asymmetric distribution of charge are *polar molecules*, e.g., water. Polar molecules of biological importance contain one or more electronegative atoms, usually O, N, S, or P. Molecules that lack polarized bonds, such as those that consist almost entirely of carbon and hydrogen atoms, are *nonpolar molecule.*

Polyadenylation is the process in which, the polymerase CTD tail is involved in recruiting the enzymes necessary for polyadenylation. When polymerase reaches at the end of a gene, these factors [CPSF (cleavage and polyadenylation specificity factor) and CstF (cleavage stimulation factor)] activate poly-A signals. Polyadenylation is mediated by enzyme poly-A polymerase. It adds about 200 adenines to the 3′ end of RNA.

Pribnow box (TATAAT) a sequence of promoters essential for initiation of transcription. It is present ten base pair preceding the start point of transcription. It represents the part of the promoter over which polymerase binds tightly and opens the DNA helix to start the transcription.

Primary transcript is the original unmodified RNA molecule. The nucleotide sequence corresponds to a transcription unit.

Processivity is the ability of an enzyme to perform multiple catalytic cycles (synthesizing biological polymer) with a single template instead of dissociating after each cycle (from substrate).

Protein kinase is a protein which transfers the terminal phosphate group from ATP onto another protein, often having the effect of regulating the activity of the other protein.

Proteome is the expressed protein complement of a genome.

Protein conformation refers to a structural state that can, without breaking any covalent bonds, interconvert with other structural states.

Protein structure can be studied at four levels. The *primary structure* refers to the sequence of peptide bonded amino acids and locations of disulfide bonds. The *secondary structure* refers to a three dimensional secondary structures by way of folding or coiling consisting of a helically-coiled, zigzag, linear or mixed form. The *tertiary structure* is three dimensional arrangements of all atoms in a protein. *Quaternary structure* means that the two or more peptide chains held together by non-covalent interactions or by covalent cross-links.

Pseudocyclic photophosphorylation Arnon and his coworkers (1954) observed that if illuminated chlorophyll produces ATP even in the absence of CO_2 and $NADP^+$ and in presence of FMN or vit K and oxygen. This phenomenon is described as oxygen dependent FMN catalysed photophosphorylation.

Putrifaction is the anaerobic decomposition of protein.

Pyridine linked dehydrogenase these enzymes contain nicotinamide rings resembling pyridine as coenzyme. In a cell there are two pyridine nucleotide, namely NAD (nicotinamide adenine dinucleotide originally called coenzyme I, cozymase or DPN^+) and NADP (nicotinamide adenine dinucleotide phosphate originally coenzyme II, or TPN). The reducing equivalent is accepted as hydride ion.

RNA splicing is a process of removing the intervening DNA sequences (introns) from a primary transcript. This process is carried out by two successive transesterification reactions, mediated by a large molecular 'machine' known as spliceosome. In transesterification reactions phosphodiester linkages within the pre-mRNA are broken and new ones are formed.

Red drop Emerson and Lewis (1943) performed experiments on quantum yield using monochromatic light, and observed a decline in quantum yield in the red zone (at wavelength greater than 680 nm).

Reducing sugar A sugar that will reduce certain inorganic ions in solution, e.g., copper ions of Fehling's solution or Benedict's reagent.

Reductive deamination when ammonium ion combines directly with α-keto carboxylic acids to form amino acids.

Replisome and Replicon multiprotein complex (the multiple interacting machines) that synthesizes DNA at the replication fork is known as replisome. These enzymes function in a coordinated interaction between them. A replicon is a unit of DNA in which undergoes replication. It is defined by its possession of the control of the control elements needed for replication.

Ri plasmid found in *Agrobacterium rhizogenes* carry genes causes hairy root disease.

Scalar is transport of substances in which enzyme catalyzed reactions defined entirely by magnitude.

Shine-Dalgarno sequence (Ribosomal binding site) nucleotide sequence near the 3• end of mRNA that make possible in binding to small unit of ribosome.

Signal transduction is a complex process, and is defined as a phenomenon in which information is relayed across the plasma membrane to the cell interior and often to the cell nucleus.

Sphingolipids They are derivatives of sphinosine, an amino alcohol that contain a long hydrocarbon chain.

Splice site the 5′ and 3′ edges of each intron.

Steroids are built around a characteristic four-ringed hydrocarbon skeleton. One of the most important steroids is cholesterol.

Structural Genomcis A worldwide initiative aimed at determining a large number of protein structures in a high throughput mode.

Supercoiling coiling of a cloned duplex DNA in space so that it crosses over in own axis.

Substrate level phosphorylation direct synthesis of ATP through the transfer of a phosphate group from a substrate to ADP.

Symport is the mandatory simultaneous transport of two substances in the same direction. Same carrier is involved in the transport of both substances. One molecule flows down its previously established gradient and the other flow with it.

Symporters Carriers that catalyze solute flux in the same direction as H^+ or Na^+ flux.

TATA box (Hogness box) in eukaryotic cells it is a region within promoter that is a conserved A-T rich septamer found about 25 bp before the startpoint. The sequence is homologous to the bacterial Pribnow Box.

TFIID has two subunits TBP (TATA binding protein) and TAF. It is a transcription factor that binds to the TATA sequence upstream of the startpoint of promoters for RNA polymerase.

Thermodynamics is the study of the change in energy that accompanies events in the universe. *The First Law of Thermodynamics* is the law of conservation of energy. It states that energy can neither be created nor destroyed. Energy can, however, be converted (transduced) from one form to another. *The Second Law of Thermodynamics* expresses the concept that events in the universe have direction; they always proceed "downhill" from a state of higher energy to a state of lower energy.

Ti-plasmid (tumor inducing) this plasmid is present in *A. tumifaciens* which causes crown gall disease in dicot plants only. It contains a specific gene known as T-DNA (transferred DNA), it integrates into the plant cell genome and the proteins synthesized by it forms tumor. Ti plasmid containing bacterium is able to recognize wound phenolic compound, hence they are attracted strongly towards the wound.

Totiopotency is capacity of a higher organism cell to differentiate into an entire organism.

Transfection is a process that introduces DNA into cultured cells.

Transgene A cloned gene introduced in plant or animal. A gene from one source is stably incorporated into the genome of another organism, and is passed on to successive generations.

Transgenic plants carry transgene, i.e., DNA introduced into them by other means and not through gametes.

Transmination is a reversible reaction in which alpha amino group of one amino acid is transferred to alpha carbon atom of another keto acid resulting in the formation of new amino acid.

Transposition is movement of a transposon and joining to its target site. It is encoded by a protein named as Transposase.

Transposon (Transpoable elements) certain DNA sequences which can move from one place in a chromosome to an entirely different site (without having any sequence relationship), and effects on phenotypic expression.

Ubiquinone consists of a substituted 1, 4-benzoquinone nucleus plus a long polyisoprenoid side chain at C-2. The polyisoprenoid chain may contain 6-12 isoprene units, which depends upon the source of ubiquinone. It has isoprenoid side chain in their coenzyme.

Uniport is movement of a molecule across the membrane independent to other molecule. Flow of ions is driven directly by an electrostatic gradient.

Upstream and Downstream The base pair sequences prior to the start-point are described as upstream while those after the start-point are known as downstream.

Van der Waals force is a weak attractive force due to transient asymmetries of charge within adjacent atoms or molecules.

V-type ATPase proton pump is found within the vacuole, hence the name V-type (V=vacuole). This type of ATPase is responsible for acidifying intracellular compartments.

Vectoral events transport of substances in which enzyme catalyzed reactions defined by a magnitude and a direction in space.

***vir* region** has 25 genes with 8 operons, out of them six (excluding *virA* and *virG*) encode for various proteins involved in T-DNA transfer. Genetic analysis of the various *vir* genes suggested that *virA* and *virG* comprised a two-component system necessary for the expression of the *vir* regions. *virA* protein was postulated to be a sensor of the pheolic-inducing molecules. VirG protein is a transcriptional activator of the *vir* genes.

Warburg Effect Warburg (1920) observed that in algae photosynthesis is inhibited in the presence of O_2.

X-ray diffraction (x-ray crystallography) a technique that bombards protein crystals with a thin beam of a x-rays of a single (monochromatic) wavelength. The radiation that is scattered (diffracted) by the electrons of the protein atoms strides a photographic plate behind the crystal. The diffraction pattern produced b the crystal is determined by the structure within the proteins.

Yeast artificial chromosome (YAC) a synthetic DNA molecule (cloning elements that are artificial versions of a normal yeast chromosome) that contains an origin for replication, a centromere and telomere. It is used as a mean to propagate whatever genes it carries in yeast cells.

Z-DNA Alexander Rich examined Z-DNA also known as d(c-G)3. In Z-DNA regions, the glycoside bond connects the base to the 1′ position of 2′-deoxyribose. The alternating anti-syn conformation gives the backbone of left-handed DNA a zigzag look. In the left handed helix, the fundamental repeating unit usually is a purine-pyrimidine dinucleotide, with the glycosidic bond in the anti conformation at pyrimidine residues and in the syn conformation at purine residues. Z-DNA contains about 12 base/turns.

Zwitter ion in crystalline form and in neutral aqueous solutions amino acids occur as dipolar ions or zwitterions and shows both acid and base properties.

REFERENCES

Allen, G.J. and Sanders, D. (1997) Vacuolar ion channels of higher plants. *Adv. Bot. Res.* **25**: 217-252.

Allen, J.F. (2003) Cyclic pseudocyclic and non cyclic photophosphorylation: new links in the chain. *Trends Plant Sci* **8**: 15-19.

Arnon, D.I. (1960) The role of light in photosynthesis. *Scientific Americana* **203** : 104.

Bajaj, Y.P.S (1987) Cryopreservativon of plant cell cultures and its prospects in agriculture and forest biotechnology. In S. Natesh *et al.* (Eds). Biotechnology in Agriculture. Oxford & IBH Publishing Co. Pvt. Ltd. New Delhi.

Bassham, J.A. (1962) The path of carbon in photosynthesis : *Scientific Americana* **206** : 88.

Bassham, J.A. (1979) The reductive pentose phosphate cycle and its regulation. In Photosynthesis II-Photosynthetic Carbon Metabolsim and Related Processes. M. Gibbs and E.Latzko, eds. Springer-Verlag, Berlin.

Bhojwani, S.S. and Razdan, M.K. (1996) Plant Tissue Culture Theory and Practice. A Revised Edition. Elsevier Science Publishers, Amsterdam.

Bick, J.A. and Leustek, T. (1998) Plant sulfur metabolism-the reduction of sulfate of sulfite. *Curr. Opin. Plant Biol.* **1**: 240-244.

Blatt. M.R. (1997) Plant potassium channels double up. *Trends Plant Sci*. **2**: 224-246.

Boylan, M.T. and Quail, P.H. (1996) Are phytochromes protein kinases? *Protoplasma* **195**: 12-17.

Buchnan B. Bob, Gruissem, W. and Jones Russell, L. (2000) Biochemistry & Molecular Biology of Plants, I.K. International Pvt. Ltd.

Casal, J.J., Sanchez, R.A. and Yanovsky, M.J. (1997) The function of Phytochrome A. *Plant Cell Environ.* **20**: 813-819.

Campbell, W.H. (1999) Nitrate reductase, structure, function and regulation: bridging the gap between biochemistry and physiology. *Ann. Rev. Plant Physiol. Mol. Biol.* **50**: 277-303.

Charlotte J. Avers (1978) Basic cell Biology D.Van Nostrand Company.

Chen, E.F., Lapko, V.N., Lewis, J.W., Song, P.S. and Kliger, D.S. (1996) Mechanism of native oat phytochrome photoreversion - a time-resolved absorption investigation. *Biochemistry* **28**: 843-850.

Cheng, S.H., Willmann, M.R., Chen, H.C. and Sheen, J. (2002) Calcium signaling through protein kinases. The Arabidopsiscalcium-dependent protein kinase gene family. *Plant Physiol.* **129**: 469-485.

Cohen, J.D. and Bandurski (1982) Chemistry and physiology of the bound auxins. *Annual Review of Plant Physiology.* **33** : 403-454.

Coomber, Shirley A. and Feldmons, Kenneth A. (1993) Gene tagging in Transgenic plants. In Transgenic Plants. Vol I. Engineering and Utilization, Academic Press, London.

Crawfond, N.M. (1995) Nitrate nutrient and signal for plant growth. *Plant Cell* **7**: 859-868.

Crick, F.H.C. (1966) Codon-anticodon pairing: the wobble hypothesis. *J. Mol. Biol*. **19**: 548-555.

Dickerson, R.E. and Geis, I. (1969) The structure and action of proteins Menlo Park Calif: W.A. Benjamin.

Emerson, R. (1958) The quantum yield of photosynthesis. *Ann. Rev. Plant Physio* **1 9**: 1-24.

Eugene, Rabinowitch (1945) Photosynthesis and Related Processes, Vol I Wiley Easter Private Limited, New Delhi.

Eugene, Rabinowitch (1951) Photosynthesis and Related Processes, Vol II Wiley Easter Private Limited, New Delhi.

Eugene, Rabinowitch and Govindjee (1973) Photosynthesis. Wiley Easter Private Limited, New Delhi.

Fogg, G.E. (1976) Photosynthesis. The English Language Book Society and Hodder and Stoughton.

Fosket, D.A. (1994) Plant Growth and Development, A Molecular Approach. Academic Press, San Diego, New York.

Fox, C.F. (1972) The structure of cell membrane. *Scientific Americana* **226** : 30.

Furbank, R.T. and Taylor, W.C. (1995) Regulation of photosynthesis in C3 and C4 plants. *Plant Cell* **7**: 797-807.

Furuya, M. (1993) Phytochromes: their molecular species, gene families, and functions. *Adv. Biophys.* **25**:133-167.

Gelvin, Stanton B. (1993) Molecular genetics of T-DNA Transfer from Agrobacterium to plants. In Transgenic Plants. Vol I Engineering and Utilization, Academic Press, London.

Giese, Aurthur C. (1968) Cell Physiology. W.B. Saunders Company, Japan.

Givan, C.V. and Kleczkowski, L.A. (1992) The enzymic reduction of glyoxylate and hydroxypyruvate in leaves of higher plants. *Physiol* **100**: 552-556.

Glick, Bernard R. and Pasternak, Jack, J. (2002) Molecular Biotechnology: Principles and Applications of Recombinant DNA. Panima Publishing Corporation.

Govindjee and Govindjee, R. (1974) The absorption of light in photosynthesis : *Scientific Americana* **231** : 68.

Guillaume, T., Asai, T., Chiu, W.L. and Sheen, J. (2001) Plant nitrogen-activated protein kinase signaling cascades 4. *Curr. Opin Plant Biol*: **4** 329-400.

Gunning, B.E.S. and Steer, M.W. (1996) Plant Cell Biology, Structure and Function. Jones and Barlett, Sudbury, MA.

Halldal, P. (ed) (1970) Photobiology of Microorganisms. John Wiley & Sons Ltd. London, New York, Sydney. Toronto.

Harper, H.F., Hong, B., Hwang, I., Guo, H.Q., Stoddard, R., Huang, J.F., Palmgren,

M.G. and Sze, H. (1998) A novel calmodulin-regulated Ca2+-ATPase (ACA2) from Arabidopsis with an N-termina autoinhibitory domain. *J. Biol. Chem*. **273**: 1099-1106.

Hatch, M.D. and Slack, C.R. (1966) Photosynthesis by sugarcane leaves: A new carboxylation reaction and the pathway of sugar formation. *Biochemical Journal* **101**: 103.

Heldt Hans-Walter Heldt (2005) Plant Biochemistry. Elsevier Academic Press.

Hill, R. (1937) Oxygen evolved by isolated chloroplasts. *Nature* **139**: 881.

Hilman, W.S. (1967) The physiology of phytochrome *Ann. Rev. Plant Physiol.* **18**: 301-324.

Jacobs, William P. Plant Hormone and Plant Development. Cambridge University Press.

Jain, S.K. (2000) Textbook of Biotechnology. CBS Publishers & Distributers.

Jones, C.W. (1976) Out line studies in Biology. Biological Energy Conservation. Chapman and Hall, London.

Karp, G. (1996) Cell and Molecular Biology Concepts and Experiments. John Wiley & Sons.

Kende, H. and Zeevart, J.A.D. (1997) The five "classical" plant hormones. *Plant Cell* **9**: 1197-1210.

Kjellbom, P., Larsson, C., Johansson, I., Karlsson, M. and Johanson, U. (1999) Aquaporins and water homeostasis in plants. *Trends Plant Sci* **4**, 308-314.

Kornberg, A. (1974) DNA synthesis: San Francisco : W.H. Freeman.

Koshland, D.E., Jr. (1973) Protein shape and biological control. *Scientific American* **229** : 52.

Lea, P.J. (1993) Plants Biochemistry and Molecular Biology. John Wiley & Sons, New York.

Lehninger, A.L. (1971) Bioenergetics, MenloPark, Calif: W.A. Benjamin.

Lewin, Benjamin (2004) Genes VIII Pearson Education International.

Lin, C. and Shalitin, D. (2003) Cryptochrome structure and signal transduction. *Annu Rev. Plant Biol* **54**: 469-496.

Ma, H. (1994) GTP-binding proteins in plants: new members of an old family. *Plant Mol. Biol.* **26**: 1611-1636.

MacMillan, J. ed. (1980) Hormonal regulation of Development. I. Molecular aspects of Plant Hormones. Encyclopedia of Plant Physiology. New Series, Vol. 9, Springer-Verlag, Berlin.

Meselson, M. and Stahl, F.W. (1958) The replication of DNA is E.coli. *Proceeding of the National Academy of Science* **71**: 135.

Moore, T. Jr. ed. (1993) Lipid Metabolism in Plants. CRC Press. Boca Raton, FL.

Nerenberg, M.W. (1963) The Genetic code II. *Scientific Americana* **208**: 80.

Nomita Chowdhary Sen (2005) Photoassimilation. *Indian Hydrobiology* **8 (1)** :85-87.

Pauling, L. Corey, R.B. and Hayward R. (1974) The structure of protein molecules. *Scientific American* **191**: 51.

Ramakrishnan, V. (2002) Ribosome structure and the mechanism of translation. 108: 557-572.

Ramawat, K.G. (2002) Plant Biotechnology. S. Chand & Company Ltd. New Delhi.

Ranjan, S.S. and Balaji, R. (2003) Introduction to Bioinformatics. Himalaya Publishing House, Mumbai.

Roberts, M.R. (2000) Regulatory 14-3-3 protein-protein interactions in plant cells. *Cur Opin Plant Biol* **3**: 400-405.

Ranjan, R. (2002) Transgenic Plants. Agrobios (India).

Razdan, M.K. (2003) Introduction to Plant Tissue Culture. Oxford & IBH Publishing Co. Pvt. Ltd. New Delhi, India.

Reumann, S. (2000) The structural properties of plant peroxisomes and their metabolic significance. *Biol Chem* **381**: 639-648.

Robinson, D.G., (1985) Plant membranes. John Wiley and Sons, New York.

Roller, A. (1974) Discovering the Basic of life; An Introduction to Molecular Biology, McGraw-Hill, New York.

Russell, P.J. (1992) Genetics 3rd edn. Harper-Collins Publisher, New York.

Saedler, H. and Gieri, A. (1996) Transposable Elements. Current Topics in Microbiology and Immunology 204 Springer-Verlag, Heidelberg.

Salisbury Frank B. and Ross Cleon W. (2002) Plant Physiology. CBS Publishers & Distributors. New Delhi. India.

Scheel, D. and Wasternack, C. (eds.) (2002) Plant signal transduction. Oxford University Press. Oxford, UK.

Senger, H. and Schmidt, W. (1994) Diversity of photoreceptors In "Photomorphogenesis in plants". 2nd. Ed. Kendrick, R.E. and Kornberg, G.H.M. (Eds) Kluwer Dordrecht.

Singer, S.J. and Nicolson, G.L. (1972) The fluid mosaic model of the structure of cell membrane. *Science* **175** : 720.

Singhal, G.S., Renger, G. Sopory, S.K., Irrgang, K.D. and Govindjee (1999) Concepts in photobiology Photosynthesis and Photomorphogenesis Kluwer Academic Publishers.

Smith, Alan (1976) Protein Biosynthesis. Chapman and Hall, London.

Sprang, S.R. (1997) G protein mechanisms: insight from structural analysis, *Ann. Rev. Biochem*. **66**: 639-678.

Stitt, M. (1999) Nitrate regulation of metabolism and growth. *Curr Opin Plant Biol* **2**: 178-186.

Stone, J.M. and Walker, J.C. (1995) Plant protein kinase families and signal transduction. *Plant Physiol.* **108**: 451-457.

Strader, D. (1994) Studies and function of G-protein-coupled receptors. *Ann. Rev. Biochem.* **63**: 101-132.

Stryer, L. (1975) Biochemistry : San Francisco : W.H. Freeman.

Tanner, W. and Carpari, T. (1996) Membrane transport carriers. *Ann. Rev Plant Physiol Plant Mol Biol* **47**: 595-626.

Tinland, B. (1996) The integration of T-DNA into plant genomes. *Trends Plant Sci*. **1**: 178-184.

Tobin, A.K., ed. (1992) Plant Orgenelles. Cambridge University Press, Cambridge, UK.

Watson, J.D. (1970) Molecular Biology of the Gene 3rd edn. Reading Mass: W.A. Benjamin.

Watson, J.D. (1976) Molecular Biology of Gene 3rd edn. MenloPark, Calif: W.A. Benjamin.

Watson, J.D. and Crick, F.H.C. (1953) Molecular structure of nucleic acids : A structure for deoxyribonucleic acid. *Nature* 171: 737.

White, Frank F. (1993) Vectors and Gene Transfer in Higher plants: In Transgenic Plants. Vol I Engineering and Utilization, Academic Press, London.

Wray, J.L. (1993) Molecular biology, genetics and regulation of nitrite reduction in higher plants. *Physiol. Plant* **89**: 607-612.

Zelitch, Israel (1971) Photosynthesis, Photorespiration and Plant productivity. Academic Press, London.

Zhang, L. and Lu, Y.T. (2003) Calomodulin-binding protein kinase in plants. *Trends Plant Sci* **8**: 123-127.

Zielinski, R.E. (1998) Calmodulin and calmodulin-binding proteins in plants. *Annu. Rev. Plant Physiol. Plant Mol. Biol.* **49**: 697-727.

SUBJECT INDEX